Modern Metal Joining Techniques

621.791

LIBRARY
No. B 5110

25 MAR 1970

R.P.E. WESTCOTT

Modern Metal Joining Techniques

MEL M. SCHWARTZ
Martin Marietta Corporation

WILEY-INTERSCIENCE a division of
John Wiley & Sons   New York • London • Sydney • Toronto

Copyright © 1969, by John Wiley & Sons, Inc.

All rights reserved. No part of this book may be reproduced by any means, nor transmitted, nor translated into a machine language without the written permission of the publisher.

10 9 8 7 6 5 4 3 2 1

Library of Congress Catalogue Card Number: 71-82976

SBN 471 76615 1

Printed in the United States of America

To Carolyn and Anne-Marie

PREFACE

The accelerating pace of civilian and military developments in industry has resulted in an ever-increasing demand for a higher level of manufacturing capability. New technologies and process complexities will require greater educational and technical specialization for manufacturing the advanced product concepts that industry foresees in the next ten years.

The time cycle from conception to serious production is becoming shorter. Simultaneously, methods and equipment are becoming more complex and require extensive development to keep pace with advanced design concepts. These trends indicate that increasingly complex tools, methods, and equipment must be developed in shorter time spans to be ready for the production phase.

The materials used during the next ten years will present a variety of problems to manufacturers. Some materials will be extremely difficult to join. Many of them will require high fabricating temperatures and thus will need protective coatings and/or atmospheric control to prevent contamination. Within five years top service temperatures will be as high as 4000°F for brief exposures. Within ten years they will rise to 5000°F. Low temperature properties will also be important for both structural and fabrication applications.

The use of new materials will also point up the need for new facilities. New and more complex processes will require better and more expensive epuipment. In some cases automation and numerical control have to be adopted on a wide scale. Present handling and testing equipment will not be sensitive enough for future needs. New standards of quality control and data evaluation will also be essential.

It is the purpose of this book to provide the most up-to-date information on several joining processes. In undertaking this task I have drawn freely on the discoveries, formulations, and researches of the many scientists, engineers, and technicians who have studied, worked, and produced these "new" welding processes. Although original formulations of joining various materials have been made from personal experience and from observation of other scientists and engineers, I readily acknowledge a heavy debt to those who have made contributions to this important field.

I feel strongly that many of the "new" welding processes are based on re-examination and extended development of processes or applied principles that have been with us for many years. Others are truly new; that is, they are based on tried and proved principles newly applied to welding or at least now applied with practicability. Each has been, or is being, developed for specific application in terms of materials, design, or end-service requirements.

Each chapter contains most of the advantages and limitations of each process that is characteristic of its application. In referring to the entire group of processes the following general statement can be made. First, often one process provides the only feasible method of doing the job, even though equipment and operating costs are high. Second, some processes offer low operating costs as one of their major advantages. Here, too, the initial equipment cost may be high. Finally, some processes offer stronger welds with less heat-affected zone than competitive methods.

I have analyzed the need for new processes and found that several factors in modern technology may be said to be primarily responsible for new or renewed interest in these processes. All, however, can be traced to the requirements or the developments of our present atomic and space age. Thus, these two factors, requirements and developments, work together to accomplish continuing gains in welding technology.

None of the processes described here has reached its full potential. Welding equipment suppliers and many large companies that depend on welding to fabricate their products have investigated these processes, but only a few are being widely used in production today. The laser, electron beam, diffusion, vacuum—these and other exotic new welding processes have been much discussed and dissected. But do they contribute to manufacturing? Are they at work now in the production plants or are they mostly confined to the textbooks or the research laboratory for use in some unknown operation at some distant date?

In a recent survey I found that most applications of the so-called "new" processes lie in a middle ground between practical present-day application and experimental work that promises a good potential for the future. "New" is defined not in terms of years in existence, for some processes like numerical control have been available for a long period, but are new in terms of breadth of application and particularly in their use in production plants. The limits on sizes and thicknesses are determined only by the equipment available, not by the process. Larger and more powerful equipment can be, and will be, built when it is required. Thus no one can predict with any certainty the full potential of these new processes.

First, I want it to be made clear that there have been, and will continue to be, developments in the standard welding processes; nor have the standard processes been extended to their limits in terms of materials, designs, and applications. These standard processes and their further development will continue to be the foundation of the welding world.

As for "new" processes, there will certainly be more. In any welding operation the problem is simply one of achieving intimate contact that will promote interaction between atoms of the surfaces to be joined.

Many of the newer processes merely use variations of, or other sources of energy, for these physical phenomena. Others achieve the necessary contact by different methods of obtaining heat or pressure or by new and careful means of controlling the heat and pressure. In some processes, particularly the solid-state joining techniques, there is considerable overlapping of basic principles or methods.

I wish to express my appreciation to Mr. Elliott Miller, Manager of Public Relations, Martin Company, who encouraged me to undertake this task; his wise counsel has assisted me materially in the formulation of the book. I also wish to acknowledge the able assistance of Miss Augusta Scheerer, who assumed responsibility for the preparation of the manuscript.

Mel M. Schwartz

Baltimore, Maryland
March 7, 1967

CONTENTS

1 Electron-Beam Welding 1
2 Plasma-Arc Welding 143
3 Electric Blanket Brazing 168
4 Laser Welding 189
5 Fusion Spot Welding 223
6 Radiant Heat Brazing 250
7 Exothermic Brazing and Exo-flux Bonding 259
8 Vacuum Brazing 278
9 Diffusion Welding 370
   Index 473

Modern Metal Joining Techniques

Chapter 1

ELECTRON-BEAM WELDING

For a hundred years science has been concerned with studies of the interaction between energetic free electrons and matter. Perhaps the bulk of contemporary atomic theory has been influenced in one way or another by these investigations. From the materialistic point of view devices based on those principles that were an outgrowth of these efforts have had the greater impact on our technology. The x-ray and cathode-ray tubes are pertinent examples. There are factors that have often been considered deleterious to the performance of these and other electron-beam devices. Below several hundred keV energy electron beams are efficient means for converting kinetic energy to thermal energy. Although this has been well known for many years, practical utilization of electron beams as thermal sources has only recently been seriously considered.

This consideration has come about in the subtly shifting character of the whole welding industry. Until the 1940's welding was almost an art. It depended more on the talents of the man who did the welding than on the welding unit itself. Since the war, however, welding has become a science and the industry now is probing more deeply into amalgamation of molecules and the reason which makes two metals fuse and remain together. The result is that welding, with its broader horizon of acceptance, is surpassing the use of rivets, bolts, solders, and adhesives.

To the outsider the welding industry presents a baffling picture. There are almost fifty different welding processes, with each distinctly different in approach and applications. Almost the only thing most of them have in common is the generation of intense heat, either by gas or electricity, for softening up the surface of two metals so they can fuse together at the joint.

The invention and development of electron optical systems that permit the construction of versatile machine tools, such as electron-beam equipment, have added emphasis to the application of electron-beam processes in the fabricating industries. It is convenient to discuss electron-beam applications by considering the processes that are characteristic of the various states of matter. In the solid state, at temperatures from ambient to the melting point, the phenomena of diffusion and chemical reaction are observed. In diffusion interstitial and substitutional reactions involving both parent and foreign atoms or molecules can be expected. Chemical reactions taking place at these temperatures may be disassociations or recombinations that are temperature dependent or influenced by externally induced concentration gradients of the reaction components. At elevated temperatures in the solid state recrystallization occurs by the process of solid-state diffusion.

At the melting point, or with matter in the fluid state, the principal process is fusion. The more important fabrication processes that accrue as a result of bringing matter to these temperature ranges are fusion welding, welding, alloying, purification, and crystallization. Finally, in the vapor state, evaporation from the liquid or sublimation from the solid is anticipated.

Precisely controlled electron-beam machines such as the electron-beam welder are ideally suited for adding energy to material things in order to attain these thermal states. Moreover, the precision and controllability in time and volume with which these states can be achieved by a high voltage electron beam permit the fabrication of devices and systems of superior performance, reliability, and miniaturization.

Every day marks a new milestone for electron-beam welding when a new combination of metals is joined or another "impossible" weld is completed. As often happens to a promising new invention it becomes the subject of many enthusiastic and exaggerated claims, and electron-beam welding is no exception. The barrage of reported achievements makes it appear as though the invention of the electron-beam welder placed a magic wand in the hands of industry to solve its most difficult joining problems. Those who have experience with this tool know, however, that there are metals it cannot join and that many jobs are better accomplished by some other joining process. Yet electron-beam welding is assuming stature among the tools of industry and many have already, whereas others wait to reap its technological fallout.

Electron-beam metalworking, as a new technology, is still evolving at a rapid rate. Techniques that only a year or two ago represented the latest state-of-the-art have been superseded by innovations that enhance greatly the value and applicability of this technology.

This chapter therefore provides the latest up-to-date information on electron-beam welding including partial and nonvacuum welds and cold cathode or plasma electron-beam welding which are discussed as separate sections.

PRINCIPLES AND THEORY OF PROCESS

The birth of the utilization of the electron as a controlled source of heating can be traced back to the latter part of the nineteenth century. In 1869, Hittorf discovered electron radiation in Göttingen. In the 1870's Crookes described the principles of heating by electron bombardment. Nicoli Tesla refined these principles and applied them to heating of graphite in an evacuated glass bulb to produce light, as described in his patent application in the 1890's. Subsequently Edison discovered the glow emission of electrons. Further experimentation in electron bombardment of various materials yielded the development of artificially produced x-rays. The year 1900 marked the advent of the first electron-beam apparatus, the Roentgen tube, which projected a beam current sufficiently intense to melt the material of the anticathode and to vaporize it. Hence it sometimes occurred accidentally that the anticathode material, tungsten, was welded by the electron beam to the copper support. An x-ray pioneer, Dr. Marcello Von Pirani, converted the x-ray tube for use as a controlled heating process. In 1907 he was granted a patent on an electron-beam furnace that he used to melt refractory metals on a laboratory basis. This furnace could not be used for welding because of shallow penetration produced by the large area of uncontrolled electron bombardment.

In the first half of the twentieth century great strides were made in electronics and, in parallel, the technology in the control of electron emission and of the control of the electron beam. The first applications of the controlled electron beam were in the fields of high voltage x-ray tubes, cyclotrons, and electron-beam microscopy. In 1926 Busch discovered the optical properties of electrical and magnetic fields, which subsequently led to electron optics. In 1930 two independent groups in Berlin undertook basic research of electron microscopy. Zworykin, Hillier, and Gabor followed the lead of others. Extremely accurate control of the beam was maintained through two basic control elements, the electrostatic and the electromagnetic lenses. These are basic elements in the science of electron optics and comprise the focal components of the electron-beam welding gun. In 1939 M. v. Ardenne disclosed a device for the boring of diaphragms with corpuscular beams. The apparatus produced small diaphragm aperatures and the unit functioned on the principle of cathodic evaporation.

During the same period Ruehle constructed in Stuttgart the first equipment for the melting of metals and for their evaporation with the aid of electron beams, which at this early date anticipated all the properties of modern equipment. The equipment showed the separation between the beam-generating space and the melting space by means of a two-stage device and a diaphragm.

In the period following World War II increasing commercial use of the refractory metals introduced a requirement for a contaminant-free welding process to prevent reaction with oxygen, nitrogen, and hydrogen in both the welding and the postweld cycles. Because these metals react rapidly with these elements, a new method of welding was required, preferably with no atmospheric contaminants.

It was known that a vacuum of 10^{-4} torr would produce an atmosphere of less than 10 parts per million of atmospheric contaminants. Therefore a welding process, made to operate in this atmosphere, would be ideally suited to welding the reactive metals. In 1948 Mr. K. H. Steigerwald of the Carl Zeiss Foundation used an electron beam to cut and weld metal. Additional studies on electron-beam welding started in the French Atomic Energy Commission at Saclay in 1954. The first apparatus used was composed simply of a glass vessel, a small vacuum pumping unit, an electron gun of x-ray tube type, and a motor clock for moving the piece to be welded. The first application was for welding nuclear fuel elements composed of a uranium-molybdenum alloy tube canned with aluminum. At the Technical Symposium on Fuel Elements in Paris in 1957, Dr. J. A. Stohr, Chief of Technical Services of the C.E.A., announced the development of the first electron-beam welder. His original development utilized a work-accelerated electron gun operated at 25,000 V. Because the electron optics of this system could not focus an electron beam to very high specific energy, the weld exhibited a width-to-depth ratio of approximate unity. In addition, the electron-beam current was dependent on the position of the gun with respect to the work, and the close proximity of the filament to the work caused arcing whenever attempts were made to weld material that vaporized easily. Subsequently, the electron microscope was adapted for welding. Operating at about 100 kV, it produced a very high specific beam energy and demonstrated the possibility of narrow deep welds and reduced distortion.

In 1958 a drilling machine, incorporating a self-accelerated electron gun and rated at 100 kV and 1 kW, was brought to the United States and adapted to welding. A description of this unit was published by Burton and Matchett in 1959 [3]. Later in the same year Burton and Frankhouser

Principles and Theory of Process

reported on exploratory welding studies using this equipment in joining zirconium alloy components for nuclear applications [4].

Another early researcher in electron-beam welding was W. L. Wyman. His work, first reported in 1958, involved a work-accelerating gun similar to that used by Stohr. Wyman's electron-beam gun, which operated at approximately 15 kV, was the beginning of electron-beam welding equipment technology. The variety of equipment currently commercially available is far more complicated but more readily adaptable to various welding applications.

The basic types of electron-beam welding guns which have been developed are the Steigerwald and Pierce type guns. Mr. K. W. Steigerwald of the Carl Zeiss Foundation in West Germany patented the long focus electron optic lens. It is a triode-type gun with a cathode (the filament), the electron-beam-forming bias cup (which is a biased grid), and an angle (ground). It was first used in the electron-beam microscopes. With some modification this type gun was developed in an electron-beam welding tool. The Steigerwald gun is operated in a voltage range of 60 to 150 kV and a current range from less than 1 to 10 μA; it is classified as a high voltage electron-beam welder. Moreover, it is possible to control independently both current and voltage. This adds greatly to the extent of parameter selection, increasing flexibility and versatility of the process, and permits welding over the full range of capability without gun modification.

Low-voltage equipment was originally based on Stohr's work, but limitations with work-accelerated guns led to the substitution of a self-accelerated design incorporating an extremely efficient cathode based on research conducted by Dr. J. R. Pierce. He has published books on classical theory of electron optics [3] and his name has been associated with the low-voltage diode-type electron-beam gun. This gun has a filament electron emitter and geometrically associated cathode and anode. The Pierce-type guns currently used in industry are rated up to 60 kV and up to 1000 μA and 30 kW.

Basics of an Electron-Beam Setup

Electron-beam apparatus consists of a vacuum chamber in which electrons are produced and directed toward the workpiece that is also within the vacuum. Since electrons are easily scattered by gas molecules, the environment around the electron beam must be reasonably free of these molecules so that the beam can be controlled. Vacua specified for electron-beam processes vary widely, but an approximate upper limit is 10^{-4} torr (where 1 torr = 1 mm Hg = 10 μ).

The electron gun itself is basically a triode whose components are an electron-emitting cathode, a focusing electrode (sometimes omitted in the simplest systems), and an accelerating electrode. The focusing electrode runs at nearly cathode potential, with only a bias voltage between it and the cathode for use as a focusing control. The accelerating electrode, on the other hand, is maintained at a large positive potential difference from the cathode. When the workpiece itself serves as the positive electrode (top of Figure 1.1), the system is termed a work-accelerated gun. More commonly the accelerating electrode is separate and has a centered hole that allows the electrons to pass through it. This arrangement, shown at the bottom of Figure 1.1, is the self-accelerated gun.

The electron-beam guns can be arranged to provide one of two basic beam geometries—annular or columnar. In the annular emission system the emitter completely surrounds the work somewhat like an induction coil. It heats the work with a broad beam or a narrow beam depending on whether a focusing electrode is included. Annular systems are useful in applications like zone refining of metals as well as in some melting furnaces in which the work to be heated is in the shape of a bar that moves slowly through the heating annulus. These systems are simple and easy to scale up. However, since the workpiece is usually quite close to the emitter, sudden discharges of gas impurities or even metal vapor can easily raise the pressure near the gun. This scatters the beam and interrupts the process. Moreover, electron emission can be gradually destroyed by metal vapors depositing on the cathode.

To avoid these problems guns producing columnar beams are often employed because they can operate a safe distance away from the work. This, however, requires an auxiliary focusing lens to keep the beam converging at the right place since without it the beam comes to a natural focus and then diverges. When emission interruption resulting from outgassing remains serious (for example, when the work contains considerable quantities of dissolved gases), the vacuum around the gun in a columnar setup has to be kept at 10^{-4} torr or lower by special techniques. In these situations the columnar setup has the disadvantage of complexity.

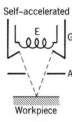

E—Emitter
G—Grid
A—Accelerating electrode

Figure 1.1 Sketch shows differences in work-accelerated and self-accelerated guns.

Theory

Thermal energy is transferred by electron bombardment through conversion of electron kinetic energy into heat. This high velocity beam of electrons

Principles and Theory of Process

traveling up to 60% the speed of light strikes the metal to be welded. (Some of the energy is converted into radiation, which may be an accompanying hazard.) A high vacuum is the only efficient environment for maintaining the focused stream of electrons and preventing arcing between the high voltage elements. The slightest pressure of air, 10 μ or more, diffuses the electron beam too much. At the high vacuum required (less than 1 μ), there is no chance for entrainment of active or inert gases in the weld zone. Furthermore, any pressure rises above the critical level breaks down the beam; thus oxidation or contamination of the work is virtually impossible.

After the welding chamber has been evacuated by a mechanical vacuum pump and a diffusion pump, a tungsten filament is heated to a temperature above 4000°F and thermionic emission of electrons takes place. By carefully shaping the surfaces of the cathode electrode and the anode, an electric field is established and causes the electrons to pass through a hole and focus slightly beyond the center of curvature of the spherical anode. Once past the anode the electrons travel at a constant velocity and the beam diverges due to mutual electrostatic repulsion of the electrons. A magnetic field is used to refocus the beam on the workpiece. Spot diameters for welding generally range from $\frac{1}{32}$ to $\frac{1}{8}$ in.

This electromagnetic focusing system allows an increase in the distance between the welding point and the gun as well as further focusing the beam to a spot as small as 0.010-in. diameter. This system also permits the insertion of a shield or modification of the gun column (discussed under the section on equipment) to prevent gases and metallic vapors from reaching the high voltage regions in sufficient concentration to cause arcing.

Whatever the particular design, the thermal energy is highly concentrated at the metal surface and is sufficient to overcome the capacity of the adjacent body to absorb it through conduction.

To realize effectively the impact of electron beams a very important principle was established which pointed out that suitably designed electrostatic and electromagnetic fields influence electrons in the same manner as glasss lenses or prisms influence light. The same fundamental mathematical laws that predict the performance of light optics therefore apply to electron control techniques and result in the common reference of "electron optics." There is, however, an important difference between electron radiation and light radiation. Light is an electromagnetic wave radiation and its energy content is dependent on the temperature of the light source. Once the light ray has been produced its energy cannot be increased. Electron emission is somewhat different. The initial energy of the negatively charged particles (electrons) is also dependent on the temperature of their source, but it is relatively easy, through the use of charged fields, to accelerate the particles and thus increase their energy. This possibility of accelerating and focusing a stream of electrons provides a source of heat with a high order of energy density, precision, and mobility.

If one wanted to explain electron-beam welding in simpler terms, the following might be sufficient. In many ways an electron-beam welder is similar to a television set. Electrons in a picture tube are emitted by a heated tungsten filament, concentrated by an electron optics system to a small diameter beam and moved so rapidly by a deflection system that a picture is produced on a fluorescent screen. Although a properly designed electron-beam welder has several thousand times the beam intensity of a picture tube, it has very similar operating features and is almost as simple to operate.

To change television stations, volume, brightness and contrast, you merely adjust knobs while examining the picture. You do not disassemble the set. With usable electron beam-welding equipment, you can change weld settings and all other necessary variables by simple knob adjustment while looking directly at the weld joint.

Principles of the Electron Beam

In spite of their insignificant mass electrons acquire a large kinetic energy by passing through a high potential difference. This is caused by their high velocity which, at 100,000 V is approximately 100,000 mps (half the velocity of light). When these electrons impinge on a material, they are stopped and, in the process, give up their kinetic energy [6]. From classical considerations of collision processes this transfer cannot be direct to the lattice atoms of mass M, but they are first to the lattice electrons of mass m, since the ratio K of transferred energy and initial energy before collision is equal to [7]:

$$K = \frac{2mM}{(M+m)^2}.$$

The lattice electrons then partially transmit the vibrational energy to the total lattice. The amplitude of the lattice vibrations is therefore increased, which means that the material can reach a very high temperature. In fact, the temperature becomes so high that the material melts and even evaporates. This evaporation in turn permits deep penetrations. It is well known that electrons of 100,000 V, for example, can penetrate in iron only about one-thousandth of an inch unless the material density differs considerably from that of the solid or liquid state. As a result of the previous discussion of electron beam-theory, the schematics of Figure 1.2 attempt to describe, in a simple step fashion, the physics of the electron-beam process.

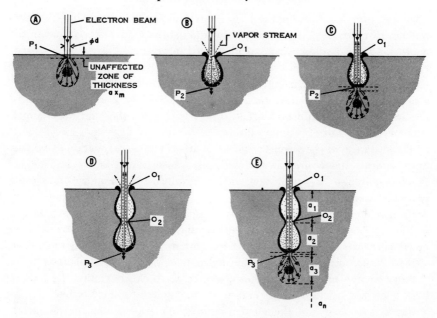

Figure 1.2 Schematics of electron–beam weld mechanism [6].

In Figure 1.2A an electron beam of diameter d and accelerating voltage V penetrates the surface of a material to a depth of ax_m ($a < 1$; x_m is the depth of penetration) with little effect on this thin layer that is practically transparent to electrons of higher velocities. However, in the deeper regions the electrons are scattered and stopped, thus heating a pear-shaped volume (cross-hatched area). The unaffected thin surface then ruptures as illustrated in Figure 1.2B. With this rupture a channel, O_1, is opened, releasing the high internal pressure developed plus a rapid stream of evaporated material. The escaping material serves to keep the channel, O_1, open.

As a result of the rupture and as illustrated in Figure 1.2B, a liquid, circular dike is formed and kept from flowing back into the channel by the escaping vapor stream. The action and reaction forces resulting from this vapor stream create a jet-propulsion effect that contributes to the deep penetration. Because these forces are of higher magnitude than the pressure exerted by the electrons impinging on the material [6], electron pressure can be neglected.

As the vapor density decreases, electron scattering becomes less and the formation of vapor ions contributes to a refocusing of the electron beam on the bottom, P_2, of the first cavity. The entire process is now repeated, starting from P_2. (See Figure 1.2C.) The electron beam again penetrates a short distance, ax_m, into the partly liquid and solid material, heats the next pear-shaped volume, and, as illustrated in Figure 1.2D, partially evaporates this volume through the ruptured opening, O_2, and so on. The electron scattering and gas focusing effects probably account for the resulting wave shape. At the end of the operation, when the beam is shut off, part of the liquid dike mentioned earlier draws back into the narrow channel, probably as a result of capillary forces.

The distance, a_n, between the nodal points is mainly a function of voltage and material and is greater with increased acceleration voltages. The distance, a_n, is also dependent on the uniformity of the material and quite irregular shapes are sometimes observed. These irregularities cannot be explained solely on the basis of the gas-focusing pinch effect. However, irregularities of the material results in irregular scattering. The explanation given here can only be considered as working hypothesis since the whole process is, because of its nonequilibrium nature, far more complicated [8].

Based on the previous hypothesis, researchers have concluded the following: once the electron beam enters the metal its geometry is fixed by its entrance geometry and the elements and their proportions in the metal. The penetration is determined by the power density, the total power and the elements, and their proportions in the metal. It is believed that once the high power density electron beam enters the metal a plasma is produced from the gassifying and ionizing of elements in the metal whose positive ions and electrons act as a focusing device to retain the high electron density in the beam. A single electron with 150 kV of energy penetrates not more than a few atomic layers of the metal. An analysis of the potential energy in an electron beam as discussed readily reveals the power that produces penetration. At an arbitrary set of parameters of 150 kV and 10 μA the electron beam delivers 6.28×10^{16} electrons/sec.

CALCULATING THE HEAT OUTPUT FROM AN ELECTRON BEAM

    1 electron volt = 1.6×10^{-12} erg,
    1 erg = 2.39×10^{-8} cal,
    1 electron volt = $1.6 \times 10^{-12} \times 2.39 \times 10^{-8} = 3.7 \times 10^{-20}$ cal.

At 150,000 V, each electron carries

    $3.7 \times 10^{-20} \times 1.5 \times 10^5 = 5.56 \times 10^{-15}$ cal,
    1 A = 6.28×10^{18} electrons/sec,
    0.010 A = 6.28×10^{16} electrons/sec,
    beam energy = $6.28 \times 10^{16} \times 5.56 \times 10^{-15}$ = 349 cal/sec.

An electron beam at 150,000 V and 0.010 A current delivers 349 cal/sec of energy at the work [9].

Therefore on impact with the workpiece 349 cal are released as heat energy. At a beam focal diameter of 0.100-in. this energy is capable of heating ¼-in. thick tungsten at 16,950°C/sec. A beam diameter of 0.010-in. can theoretically yield a one-hundredfold increase in the rate of heating. Although conduction, vaporization of material, and radiation losses reduce the actual efficiency of energy transmittal, the indicated power is sufficiently large to explain the weld-penetration-to-weld-width ratio obtained with electron beams [9].

High Voltage Versus Low Voltage

Before concluding this section the following should be noted. In the early period of electron-beam welding (1957–1960), an opinion was widely maintained that the ability to produce electron-beam welds that had a high depth-to-width ratio was a characteristic of, and only possible with, equipment operating in the "high voltage" region. That conclusion resulted in large measure from the work produced by some early investigators in the "low voltage" field who used work-accelerated electron guns. Meanwhile other efforts using self-accelerated electron guns soon demonstrated the ability to make narrow, deep penetrating welds using "low voltage."

In effect the high power density electron beam drills a hole by vaporizing the metal and producing a molten lining of the hole. When the electron beam moves relative to the workpiece, possibly a combination of two things happens: (a) the gas on the trailing side of the hole condenses to metal and the metal on the leading side vaporizes and (b) the molten metal on the leading side of the hole flows to the trailing side of the hole. In low energy density electron-beam welding that produces depth-to-width ratios of approximately one to one the electrons are stopped near the surface of the metal and their kinetic energy is converted to thermal energy.

For any electromagnetically or electrostatically focused electron gun, changing of the electron-beam current requires changing the focusing in order to maintain the same beam diameter at the workpiece surface. Changing the kilovoltage also requires similar changes to maintain the same beam diameter.

The Steigerwald electron gun, by the use of its grid, maintains a constant current with change in kilovoltage. Inherent in the early design of the Pierce low voltage gun design was changing current with changing kilovoltage [10]. Today the Pierce low voltage gun is capable of maintaining constant current with changing kilovoltage.

It is now recognized that high or low voltage is in itself not a sufficient or necessary condition for creating electron-beam welds having the high depth-to-width ratios. Rather, this characteristic result depends on many factors including the total power and the specific power-delivering capabilities of the welding equipment (of the electron beam) and the effect of these two parameters in combination with welding speed. Thus for a given material and weld depth characteristically narrow welds result if sufficient total beam power is available to permit using rapid welding speeds and if the beam power density is great enough to develop and maintain a hole equivalent to the depth of the weld as the welding progresses.

MECHANISMS AND KEY VARIABLES

Numerous factors can affect electron beam welds. Among them are the electron-beam gun and its design for high and low voltage machines, the fixed and movable gun, focusing, accelerating voltage, beam current, power density, welding speed, and vacuum environment.

Electron Gun

DESIGN

It is interesting and important to view some of the factors that control the design of the electron-beam gun itself. Of course, these depend to a large degree on the particular job to be done: the size of the workpiece, the volume of the vacuum chamber, the gas evolution expected (if any). These in turn determine beam power needed, spot size at the work surface, and distance from the gun to the work.

An electron gun is a device which generates, accelerates, and, to some extent, focuses a beam of electrons. The constituents of an electron gun are logically divided in two categories: (a) the elements necessary for the generation of free electrons or cathode elements and (b) the field-shaping elements necessary for the production of a useful beam. Although the major design effort is directed toward generation of the electrode configuration, and hence beam shaping, careful attention has been paid to the cathode elements from the beginning to make sure that usefulness of the final design is not hampered. In conventional guns the cathode elements may be subdivided into groups by function. The most important element is the cathode proper which serves as the source of electrons. With this knowledge two problems emerge for consideration: the cathode material and its shape and the voltage that will be used to accelerate the electrons to the target.

A cathode must meet five criteria. It must be more or less self-supporting; it must not be poisoned by the gases released from the metal or by the metal vapor itself; it must have a high electron emission, which implies a high operating temperature—above 2000°C for the less-efficient, pure metal emitters. This useful emission current must be adequate over a period of time. Next it is desirable to have a small power input (implying a low work function; see formula below), good thermal efficiency, a small cathode, and a cathode which is simple to construct. Finally, the cathode must operate in the gun environment. It is, however, usually impossible to fulfill all these requirements completely.

When the beam is used as a heat source, the materials that meet these conditions best are the refractory metals molybdenum, tantalum, and tungsten, particularly the last two. The more common emitter materials used in low-power devices—nickel coated with barium oxide, porous tungsten impregnated with barium aluminate, and thoriated tungsten—all are poisoned too easily by the small amounts of oxygen, nitrogen, metal vapor, and oil vapor likely to be released in the electron-beam welding process.

Tantalum has the higher emission efficiency, but lower resistance to sagging, which places an upper limit on the operating temperature. Thus tungsten is used for delicate filament-type emitters where emission is generated by current passing through the filament. Tantalum emitters use a mechanically stronger curved-disk geometry, heated by electron bombardment from behind the emitting surface, and thus requires an additional power supply. A Pierce gun with a space-charge limited, bombardment-type cathode has a very high efficiency, whereby 99% of the emitted cathode current can be focused through the anode aperture and produces a uniform current density over the beam cross section. In addition, the minimum diameter will lie well beyond the anode and thereafter the beam will diverge uniformly.

The Steigerwald or telefocus gun in Figure 1.3 is designed to produce a primary gun focus at a relatively large distance from the anode. The long-focus effect is because of the hollow shape and negative bias of the grid electrode. Near the cathode the electric field is diverging; consequently, emerging electrons are given an outward radial velocity. Between the grid electrode and the anode the equipotentials first become flat and then converge toward the anode. The beam acquires a net radial velocity inward, of smaller magnitude than the initial outward velocity because of the higher energy of the electrons. Consequently, the beam converges quite slowly and has a long focal length. An increase in the bias increases the focal length by increasing the curvature in the cathode region and starting the beam with more divergence. This increase has also the effect of reducing the beam current. In addition, space-charge forces are present over the relatively greater length and the spot size is thereby limited by beam current.

These guns are all focused to their final spot size by electromagnetic coils and can have their beams oscillated, interrupted, or deflected by suit-

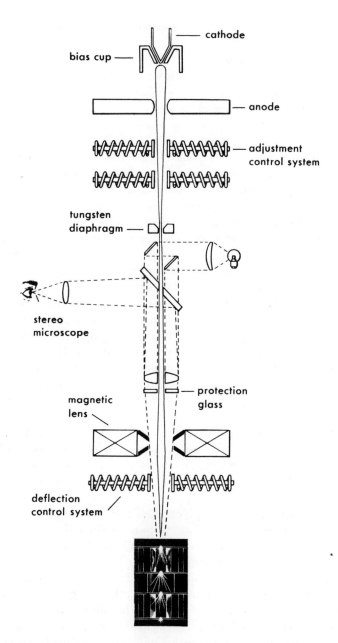

Figure 1.3 Schematic of high voltage electron–beam system.

able power-supply controls. Using a beam deflection and pulsing system can generate circular, sine wave, saw tooth, pulsing, and other special patterns. Oscillation and deflection have been particularly useful in welding. With an oscillating beam (perpendicular to linear weld joint on workpiece surface), the final weld is less sensitive to tightness of the initial joint-fit-up, and seam tracking is less critical, particularly at higher speeds. Deflection likewise aids in seam tracing since nonlinear seams can be effectively traced by electromagnetic control of beam position; otherwise, the more difficult approach of precisely translating the workpiece or the gun through the nonlinear joint geometry would be necessary with a fixed beam. Oscillations $\pm\frac{1}{4}$ in. from a neutral axis and deflections for $\frac{1}{4}$ in. on either side of the initial seam position are easily obtained.

The mechanism by which a metal is able to give up electrons is explained most accurately by quantum dynamics. At 0°K the energy levels of electrons in any material are well-defined bands of finite width. When the temperature of the material is increased the width of these bands increases. In particular, for a metal, the upper limit on the conduction band becomes "fuzzy," and some of the conduction electrons acquire enough energy to overcome the potential barrier at the surface of the metal (the value of the energy required to overcome the potential barrier is termed the work function of the metal). These electrons may then be drawn off by the application of a suitable field. If the field is of sufficient strength to draw all the available electrons from a cathode of work function ψ, the saturation current density obtained at temperature T is*

$$J_s = A_0 T^2 e - \frac{e\psi}{kT}$$

$$J_s = A_0 T^2 e - \frac{11{,}610\psi}{T}$$

The term A_0 is a constant determined by the material used and usually has a theoretical value of 120 A/(cm^{2}deg^2) although recent research indicates 60 for tungsten, 55 for molybdenum, and 37 for tantalum [11].

In practical gun design it is usual to draw less than saturation current. This operation, termed space-charge-limited, has the advantage that a smaller virtual cathode is formed slightly in front of the cathode which has a stable charge density, essentially independent of cathode temperature. The current that flows between parallel electrodes is given by the following Childs-Langmiur equation which shows the relationship between the current flow and the voltage between two electrodes.

$$I = KV^{3/2}$$

* Richardson equation as modified by Dushman.

in which I is the beam current density, V the accelerating voltage, and K a factor that depends on the spacing between the electrodes and the ratio e/m (the charge and mass of an electron).

Accelerating Voltage

Accelerating voltage is a major gun parameter. Its importance lies in the fact that it is one good way to control the spot size, that is, the diameter of the beam at the impact point. The beam is not, of course, a single file of electrons—one following the other to the target area. Rather, it is a controlled bundle of electrons. When the accelerating voltage is increased, the beam current needed for a given power setting decreases in proportion. Thus, with fewer electrons in the beam to repel each other, they can form a narrower beam, as shown by

$$d \sim \left(\frac{I}{V}\right)^{3/8},$$

where d = spot diameter,
 I = beam current, and
 V = beam-accelerating voltage.

The electron optics of the systems, too, have a great influence on controllability of spot size. Consequently, the great variation in construction of electron-beam guns makes it difficult to generalize about voltage. However, for guns operating in the 60 kV range, spot sizes down to 0.002 in. diameters are currently possible. Moving to 150 kV, spot size drops to the order of 0.001 in. or less. Practical limitations on voltage, of course, are the problems of voltage breakdown of insulation, X-ray hazards, and power costs.

Beam Current and Focusing

In the Pierce-type electron-beam gun the function of the focusing coil is to establish the point in space at which the smallest possible electron-beam cross section occurs. The greater the current through the focusing coil, the closer the beam focus is to the face of the gun. For a long focal length of 15 in. the focusing current is at a minimum value.

Thus total beam power is the product of maximum beam current and maximum accelerating voltage expressed in kilowatts. When varying beam power by programming accelerating voltage, the focus coil current must also be programmed to maintain the beam in focus. This is true because the focal length of the beam is related to accelerating voltage and focus coil current by the following equation [12]:

$$F = \frac{K'V}{(N_1)^2}.$$

Thus
$$I = K''(V)^{1/2},$$
where F = focal length,
N = number of turns in the focus coil,
I = focus coil current,
K' and K'' = constant of proportionality,
V = accelerating voltage.

It can be seen therefore that the current in the focus coil is a function of the square root of the accelerating voltage. In a recently completed electron beam-welding program [13] this knowledge and relationship permitted the use of an analog programmer for the focus coil current as well as for the accelerating voltage.

In the Steigerwald gun, beam current is secondary compared with accelerated voltage. This can be seen in the following series of equations:

$$\text{beam power density} = D = \frac{IV}{A}, \qquad (1)$$

where I = beam current,
V = electron accelerating voltage,
A = area of beam impingement on workpiece.

The basic formula for the electron-beam current in an electron optical system has been derived [14] from Langmuir's formula and reads

$$i = C_0 V^{8/3} \qquad (2)$$

where C_0 depends only on the electron optics and filament heating [6] and $i^{3/8}/V = d$ = diameter of beam impingement on workpiece.

Solving this equation for d^2 and combining with $A = \pi d^2/4$, the beam impingement area is found to be

$$A = \frac{1}{S_0} \frac{(i)^{3/4}}{(V)}, \qquad (3)$$

where
$$S_0 = \frac{4 C_0^{3/4}}{\pi}. \qquad (4)$$

When Equations (1) and (3) are combined, the power density becomes

$$D = S_0 i^{1/4} V^{7/4}.$$

This equation clearly shows that the beam power density is much more dependent on accelerating voltage than on beam current.

Fixed and Movable Electron-Beam Guns

Welding chambers that employ the fixed gun design generally have the electron beam gun mounted on the top center of the welding chamber with the electron beam projecting vertically downward. The gun, mounted externally, can be valved to isolate it while the work chamber is open, thus reducing contamination of the gun and increasing filament life. Furthermore, the valve permits changing the filament and cleaning the gun without requiring the chamber to be vented to atmospheric pressure.

With the fixed gun concept it is possible to incorporate visual optics that permit viewing of the entire weld joint before, during, and after welding, regardless of joint configuration. Also the gun is out of the work chamber and is protected from metal vapors or other contaminants.

Work motion in two axes of control is provided in the table at the bottom of the chamber. Vertical control is provided by the gun focus system which produces a focal range variable from $\frac{1}{2}$ to 25 in. below the base of the gun.

Designs have been developed and are in use that permit moving the gun with a plate and seal system with the gun mounted on the plate. Generally this technique is used for complete travel across the chamber. Mounting of the fixed position gun is not limited to the vertical position with the beam projecting downward. The gun may be mounted in any position. Provision can be made for mounting the gun in several radial and axial positions with tooling for angular mounting. In a two-gun system in use, one gun projects a beam vertically upward and the other gun projects a beam vertically downward.

Welding chambers that employ the movable gun design usually have two axes of control on the gun carriage. They are generally front-to-back and vertical with the right-to-left motion being with a carriage. It is just as feasible to use any two of the motions for the gun carriage with equal facility. Gun carriage systems that have all three motions are feasible but are difficult to design so that adequate high voltage clearances and positioning accuracies can be maintained.

Vacuum Environment

Controlled electron beams can be produced in either a high vacuum or a low pressure gas. A vacuum of 1×10^{-4} torr or less is required for electron beam control for all welding with electron beams produced by thermionic emission. For electron beams produced by a gas cathode a pressure of approximately 0.1 to 0.5 torr is used, depending on the gas used for the electron source. The gas cathode is discussed later in the chapter.

Research has been carried on and production equipment is available for electron-beam welding at atmospheric pressure by passing the beam from a thermionic electron source electron gun through a differentially pumped labrynth. The beam then passes through a thin inert gas shield and into the weld. The versatility of an electron gun that operates without a large high vacuum chamber and that produces welds at atmospheric pressure is discussed later in the chapter.

As a result of the variation in some of the mentioned factors, weld penetration has increased, weld shape has changed, and distortion has been reduced and weld depth-to-width ratios have exhibited a significant increase dependent on the material and thickness.

Penetration–Weld Shape–Depth-to-Width–Distortion

Because of the deep electron penetration, the shape of the fusion zone produced during electron-beam welding is quite different from that produced by normal surface-heating techniques. With conventional surface heating, penetration is dependent on thermal conductivity. If one compares the typical fusion zone produced by conventional heating with that produced during electron beam welding it can be shown that, to penetrate a given thickness of material, about $\frac{1}{25}$ as much material must be melted during electron-beam welding. Thus only approximately $\frac{1}{25}$ as much energy is absorbed by the workpiece, which results in significantly less distortion and material property change.

The effects of electron accelerating voltage, beam current, and welding speed have been investigated for several materials. Figure 1.4 shows the effect of both voltage and current on penetration in AISI Type 302 (18-8) austenitic stainless steel welded at a constant speed. It can readily be seen that penetration increases with both current and voltage and it can be shown that penetration is approximately proportional to the product of current and voltage, or power. As expected, welding speed affects penetration. Penetration decreases with increased welding speed; it should be noted, however, that excellent penetration can be obtained at relatively high speeds. For example, stainless steel can be welded at 30 ipm in thicknesses up to 0.400 in.

Specific weld characteristics at any given welding condition are, of course, dependent to some extent on the physical properties of the workpiece. Research work and practical experience indicate that there is a strong dependency of penetration on the thermal conductivity of the material

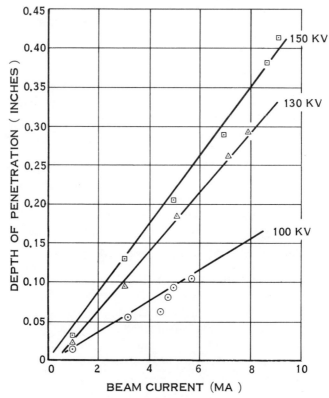

Figure 1.4 Depth of penetration versus beam current. Type 302 stainless steel; welding speed = 27 in./min [17].

being welded. Other data, however, clearly indicate that thermal conductivity is not the only controlling factor. For example, the penetration in AISI 4340 low-alloy steel is roughly double that observed in pure columbium under the same operating conditions and yet the thermal conductivities of the two materials are about equal. It is quite possible that other physical properties, such as melting point and vapor pressure at the melting point, have a significant effect on weld-zone characteristics.

A brief comparison should be made comparing the energy input from electron-beam welding with that required for the other fusion welding processes. Electron-beam welds are compared with inert gas shielded arc welds in Figures 1.5a and b.

Thus it is seen that about ten to fifteen times as much energy was required to MIG or GTA weld as compared with electron-beam welding for these

Mechanisms and Key Variables 21

two examples. Examination of the figures shows that the MIG and GTA welds are very much wider than are the electron beam welds as would be expected.

Thermal energy transferred to the unmelted base metal during the welding process is quite generally detrimental. This thermal energy can cause buckling and warping, undesirable grain growth, metallurgical transformations, physical changes reducing strength or ductility or both, and residual stress. It can be therefore concluded that the lower the quantity of energy that must be transferred to the work in order to produce a fusion weld, the greater the merit of the welding process.

Depth-to-width ratios are meaningless unless they are related to a specific example. For example, in welding microcircuitry the depth-to-width ratio may be as low as one to two, whereas in welding thick metals, it may be 20 to 1.

In view of the extreme importance of weld distortion and heat-affected zones because of welding, a brief discussion will serve to explain their effects. A combination of factors contribute to weld distortion, among which

	E.B.	M.I.G.
Voltage, volts	30×10^3	28.7
Current, amperes	225×10^{-3}	360
Welding speed, ipm	37	6.5
Power, kilowatts	6.75	10.3
Energy, kilojoules/in.	9.1	95.5

Figure 1.5a Electron-beam weld superimposed on metal inert gas weld. Material: 0.500 in. thick, type 304 stainless steel [13].

 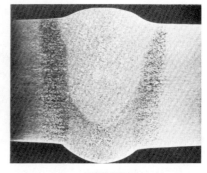

	EB Weld	1st Pass*	TIG Weld 2nd Pass†	Total
Voltage, volts	30×10^3	11.7	13.0	—
Current, amperes	200×10^{-3}	270	270	—
Welding speed, ipm	95	6.5	7.5	—
Power, kilowatts	6	3.2	3.5	6.7
Energy, kilojoules/in.	3.8	29.6	28.0	57.6

* w/0 wire
† $\frac{1}{16}$ in., type 2319 wire used

Figure 1.5b Comparison of electron beam and tungsten inert gas welds in $\frac{1}{2}$ in. 2219 aluminum [13].

are the thermal effects acting in the heat-affected zone, the molten-pool shrinkage, the relaxation of residual stresses, the introduction of new stresses, and, in some materials, phase transformations. There are undoubtedly other reasons for distortion such as poor fitup, but these are the principal contributing factors. Distortion is not peculiar to any individual welding process, be it arc, gas, electric-resistance, pressure-bonding, or electron-beam welding. If heat is used to enact a weld, one or more of these distortion factors will occur.

The heated area surrounding the molten pool is known as the heat-affected area. It is here the forces causing much of the distortion occur. Any radial line originating in the molten pool and contained in the workpiece exhibits a thermal gradient ranging from the melting temperature at the pool's edge to the base-metal temperature some distance away. The gradient is a function of the thermal properties of the metal, such as the thermal conductivity, specific heat, density, phase transformations, heat of fusion, emissivity, and of the rate of heat-energy input. The higher the heat-energy input, the steeper the thermal gradient, all other factors remain constant. The thermal gradient also varies as a function of its direction with respect to the movement of the weld puddle, the gradients in front being much steeper than those at the rear.

Why all this concern about thermal gradients? Because metals have several temperature-dependent properties that act together to cause distortion. One of these, the thermal expansion property, although not always linear, usually shows an increase in size with an increase in temperature. The volume changes generally occur simultaneously with the thermal changes; it is therefore important to consider the dynamic thermal behavior of a weld.

The volume changes associated with the heat of welding and acting alone would probably not contribute to distortion if this were the only thermal-dependent property of the metal. Unfortunately, yield strength is another metal property that changes with temperature. Note that as the temperature increases, the yield strength decreases.

From the previous discussion it was shown that the greatest volume change occurs nearest the weld puddle and decreases radially from it until the area of no change is reached. If the workpiece is large enough, it will resist the tendency for the heat-affected area to expand. This resistance to natural movement creates compressive stresses in the area around the weld puddle. As long as the yield strength is not exceeded the metal will only experience elastic strain and return to its original shape on cooling. But in the area nearest the weld, where the expansion tendency is the highest and the yield strength the lowest, the compressive forces will usually exceed the yield strength, thus causing plastic deformation.

It should be noted that when reduced to standard conditions, no real volume change takes place, only a geometric change. The hot metal is squeezed smaller in the plane of the workpiece and on cooling occupies less area than before heating. Since during cooling the metal also regains its strength, it will not plastically flow back to its original shape, but will be forced into elastic tensile strain. These tensile forces cause the distortion common to welding.

If the workpiece is not bulky enough to restrain the expanding heat-affected area completely, the expansion forces may cause additional distortion in the form of buckling such as often occurs when welding thin-gage materials. When rigid fixturing is used to effect a cure, the fixture may provide the necessary bulk to resist expansion effectively, thus allowing the tensile-stress-type distortion to be greater than if no fixturing were provided. Good fixturing can be designed to permit expansion but prevent buckling so that the tensile distortion forces can be kept to a minimum.

The molten puddle contributes to distortion in several ways. It offers no resistance to the expanding heat-affected metal and therefore provides a virtual void into which the heat-affected metal can flow plastically. The molten puddle readily conforms to these changes in its perimeter and on cooling solidifies to a shape different from what it occupied before melting. At the time of solidification the cast metal is essentially without strain. On cooling it both increases in strength and decreases in size and although some plastic deformation may take place at the higher temperatures, the cast structure will usually be left with a high level of tensile stress by the time it cools to room temperature. These tensile stresses also contribute to the weld distortion.

Some metals experience one or more phase changes on being heated to their melting points, which are usually accompanied by volume changes. Sometimes these phase changes can add to the plastic flow by increasing the volume when the metal is weak enough to yield. If, however, the phase change produces a smaller volume, less distortion is likely.

Residual stresses contained in the workpiece before welding, either because of forming or straightening operations or from forcing poorly prepared components into proper fitup, can be the cause of some distortion. Dimensional equilibrium is maintained by a balance of stresses in a structure, but if the heat of welding modifies the residual stress pattern, change in shape occurs.

When welds become more complicated than a simple butt weld, the same basic distortion factors still apply; for example, when the metal thickness increases to require two passes (one from each side), the total amount of distortion will most likely increase even though some of the first-pass-induced residual stresses causing distortion are relieved by the heat of the second pass. When welding with filler additions, the heat-affected area may be smaller than when welding the same size part without the filler additions because usually less parent metal is melted, with much of the available weld energy going into the melting of the metal addition. But when heavy multipass welds are made, the geometry disturbing factors are at work during each weld pass required. On such welds it is possible to keep some forms of distortion to a minimum by careful planning of the weld-pass schedule, but distortion does occur.

The total distortion can be seen to be a function of both puddle size and the heat-affected zone size. The larger the molten pool, the greater the volume to shrink to a new shape; the larger the heat-affected area, the greater the volume subjected to plastic flow. Similarly, the larger the heat-affected area, the greater the change through relief of residual stress. Thus it can be reasoned that minimum distortion can be achieved through techniques that tend to reduce all these factors. It is not quite fair to make a direct comparison in puddle size since the arc-weld puddle is not all molten at one time. Still the additive affect of multiple-pass shrinkage would be much greater than for the single electron-beam weld. The heat-affected areas are similarly much different in size. Each pass slowly dissipates heat to the surrounding area. A corner butt weld made in a zirconium alloy shrank 0.021 to 0.041 in. when welded by the tungsten inert-gas process, but only shrank 0.004 to 0.006 in. when welded by the electron-beam process.

Until now very little consideration was given to the shape of a weld and how it might influence distortion, but an arc butt weld is much wider at the top than at the bottom. Therefore the top surface tends to shrink more than the bottom and causes a buckling of the plate in addition to straight or overall shrinkage. An electron-beam weld through this same thickness of material would produce little or no bending moment and only slight straight shrinkage.

TYPES OF EQUIPMENT AND TOOLING

Electron-beam welding manufacturers were quite numerous in the early 1960's. Numerous pieces of equipment were commercially available. However, as the industry grew so did the equipment and eventually resulted in two major manufacturers, one specializing in high voltage and the other in low voltage. Today both manufacturers have overlapped each other in regard to high and low voltage equipment.

High Voltage Equipment

The evolution of high voltage electron-beam (usually considered to be above 50 to 60 kV), vacuum-welding equipment proceeded from a 1-kW maximum power output machine to the present production 25-kW maximum output machine. A prototype of this machine is shown in Figure 1.6. In addition to the obvious advantage of deeper weld penetrations, other improvements have been made to the electron optical column, chamber size, and the welding controls and power supply.

ELECTRON OPTICAL COLUMN–CHAMBER–DEFLECTION

Arcing has been a serious and general problem with the electron-beam welding of aluminum and its alloys, particularly in heavy sections. This problem has not been confined to one type or design of equipment but, as recorded in the literature, has been encountered with all basic types. For reasons which are not yet clearly understood, aluminum is unique in this respect. Apparently, the aluminum vapor formed during the welding process contaminates the electron gun and causes breakdown of its insulating characteristics. This results in arcing, which in turn causes loss of gun parameter control. Recently, a modification of an existing electron-beam optical

Figure 1.6 25-kW test-bench unit.

column was developed; this completely eliminates arcing due to metal vapor contamination.

The basic principle behind this modification is quite simple. When metal vapor is formed at the point of impingement of the electron beam on the material being welded, the molecules of vapor travel in a straight line until they impinge on a solid object. Furthermore, most of these vapor molecules do not carry an electrical charge. Thus, if the electron optical system contains a bend that completely shadows the gun region from the workpiece, the vapor particles will not reach this region where damage can occur. It is possible to incorporate such a bend in the electron optics. This is because the electrons themselves do carry an electric charge and therefore the electron beam can be turned around this bend by using a suitable magnetic field without disturbing the straight-line path of the vapor particles.

All of the gun region is completely protected from the straight-line paths of the metal vapor, and the electron beam is bent easily with a relatively simple magnetic coil. The bend is placed above the optical viewing system and thus does not affect this important feature of the equipment.

In order to make the bent column simple to operate, it is necessary to compensate automatically for different voltages normally used in welding. Since the electrons are moving faster at higher accelerating voltages, a stronger magnetic field is required to produce the same bend in the beam at higher voltages as compared with lower voltages. To accomplish this the current in the bending coil is programmed automatically with changes in accelerating voltages so that the bend angle remains constant. Figure 1.6 shows the prototype bent column installed on a 25-kW welding unit.

Another annoying problem is the burnout of tungsten filaments and the need for a continued available supply. Techniques have been developed by an aerospace company to fabricate their own cathodes and thus insure continuous operation of the electron-beam equipment. Another additional improvement has been the development of the gun column to permit changing the filament while the chamber is evacuated.

Inherent in the design of the high voltage gun is its length of useful beam from the gun orifice. The maximum distance of maximum energy density from the gun orifice was 15 in.; this, however, has been increased to over 20 in. and by adjusting the focusing current the same power density may be maintained over this distance. This permits welding down deep holes or welding at varying distances from the electron gun without changing the workpiece to gun distance.

A deflection coil located below the magnetic lens is used to deflect the focused beam about its centerline on either the "X" or "Y" axis by ac (oscillating deflection) or dc (steady deflection) means at variable amplitudes for several inches below the column. This deflection system includes provisions for producing circular or elliptical welds in the horizontal plane, and special signal inputs at the deflection coil will allow for any regular or irregular deflection shape desired. A beam-pulsing control is also available to interrupt or pulse the electron beam. Pulsing control allows for selection of a number of pulse frequencies and pulse widths. After the parameters have been preset, beam pulsing is automatic. The beam is turned on and off as a square-wave function. In addition, electronic controllers have been developed that can generate circles, squares, rectangles, triangles, hexagons, and optional patterns.

Numerous types of control devices for automatic welding and tracing the electron-beam weld have been developed and applied to various machines. A unique tracer system permits the operator to program the machine to automatically "steer" the desired weld path under the beam. The programming is accomplished by steering the work under the beam location with the beam turned off. The beam is turned on and off automatically as programmed to make the desired weld.

Tracing (recording of the joint path) can be performed at speeds of $\frac{1}{2}$ to 8 ipm; welding is done at 4 to 50 ipm in the programmed mode and 4 to 120 ipm in the manual mode. The speed at which the program was recorded has no effect on welding speed in the automatic mode. The program as traced over the joint is then played back into the system to reproduce the previous recorded motion of the table, causing the weld to be made exactly on the joint.

Another traversing device is a numerically controlled three-axis contour control unit which provides full programmed control of X, Y, and rotary motion. In addition, eight auxiliary functions are provided in the system, which controls both the speed and direction of the workpiece motion for either continuous or intermittent welds over any irregular weld path. This automatic system is being used for repetitive welding of complex parts. The costs involved in preparing the tape are quickly repaid by the savings resulting from making complex weld joints accurately and repeatedly.

Chambers today range up to 40 ft long with a television viewing system as seen in Figure 1.7, whereas still others have been modified to make electron-beam welding portable. One aerospace manufacturer developed a "walking seal." The unit took the vacuum environment to the work—a small vacuum that could be moved along the joint as the weld progressed.

Figure 1.7 A 40-ft chamber with its own television viewing system [19].

That is what the "walking seal" does; it maintains the necessary vacuum in a local area of a large weldment while traversing long weld joints. The device employs a system of flexible chambers terminating in seals that literally "walk" along the joint. In operation the gun chamber is evacuated after the unit is positioned on the work to be welded. At the same time the outer chamber is also evacuated with its seal tight against the work. Meanwhile the inner chamber is also under vacuum, of course, but it is capable of a sliding movement over the joint. The movement is limited to 1-in. increments.

When the inner chamber has made this "step," argon gas at 80 psi is injected into the outer chamber to break its seal with the work. This makes it capable of making its own sliding movement while the inner chamber maintains the vacuum environment for welding. After the outer chamber has moved 1 in. along the weld path, the process is repeated by again evacuating the outer chamber and freeing the inner one for movement. Travel speed along the weld path depends on the cycle rate for evacuating the outer chamber and then injecting argon into it. At present this is 40 cycles per minute, which gives a maximum welding speed of 20 ipm.

An added factor in welding with this equipment is that a backup piece must be held firmly against the root of the joint, and the power of the electron beam must be regulated so that it penetrates into, but not through, the backup piece. Otherwise, the molten metal would be forced into the welding chamber by the outside pressure.

Another type of portable machine was developed for use on the B-70 airplane. In this machine, the part itself acted as the bottom of the vacuum chamber. In other words, the chamber was really like a giant suction cup. The concept in this instance was to keep both the part and the chamber stationary for any one weld length and mount the gun on a sliding lid on top of the chamber. In order for a lid to slide the length of a chamber and still maintain a vacuum, the lid must be twice the length of the chamber. Therefore production welds were limited to 5-ft lengths in order to still maintain some portability. A system of specially designed seals maintained a vacuum between the chamber and the weld and the chamber and the sliding lid. The lid was moved at speeds up to 60 in./min without affecting the vacuum. Figure 1.8 illustrates the equipment in operation.

The latest portable machine is trailer mounted and was designed for electron-beam welding wing beam caps, windshield frames, inlet frames, and longerons for the SST. This machine is relatively new and still under test and rated up to 60 kW. An exterior adapter allows buttwelds in members of any length. In addition, a number of special seal configurations have been devised to match part geometry. The gun, viewing systems, pumps, and beam controls are in one assembly that is hung from a crane boom and lowered onto smaller chambers built around the parts to be welded. This approach could probably be applied to a variety of other situations.

Finally, we discuss the hand-held electron-beam welder which can be carried to the weld site. One potential use is the fabrication and repair of space stations. (See Figure 1.9.) This conceptual electron-beam space welding system can make use either of the on-board spacecraft power supply or of a separate power pack of its own which can be recharged from the on-board power supply. For most in-space welding missions, this system, it appears, could be made fully portable when it is operated from its own power pack. In this mode the system consists of an electron-beam gun column, a power supply and controls package, and a connecting cable. The prototype electron-beam gun column for this system, shown in Figure 1.9, is about 9 in. in diameter and 18 in. long. For its welding mode an extended pulse operation was selected in experimental testing, which can also be called "on-off" welding. In this mode the pulse duration was on the order of minutes instead of milliseconds. On-off welding has been shown to combine effectively the metallurgical advantages of continuous-mode welding with the weight advantages of a small high-rating transformer. In addition, it requires no complex pulsing network, so that minimum welding parameters can be used, which in turn improves the inherent reliability of both the welder and the weld beads it produces.

The power system has provisions for operation in any of three modes: from spacecraft 28 vdc, from spacecraft 208 vac, or from the system's own power pack which consists of batteries and an inverter. The battery unit as well as the whole power pack can be removed from the system; in power pack the pack is replaced by a low voltage cable that can be

Figure 1.8 Production portable vacuum chamber—swarf position [20].

Types of Equipment and Tooling

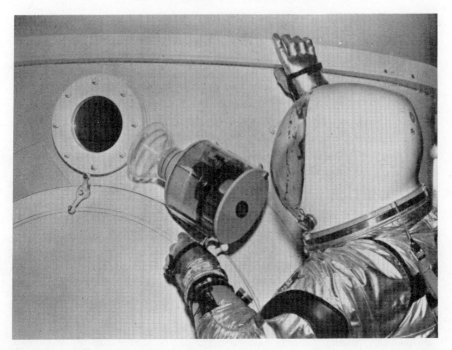

Figure 1.9 Self-contained electron-beam welder uses "free" vacuum of outer space. Astronaut is using mockup of gun [19].

connected directly to the spacecraft power supply. The materials which have been welded and test results are discussed later in the chapter.

Low-Voltage Equipment

The evolution of low-voltage, electron-beam machines proceeded from a 15-kW output machine to the present 60-kW powered machine. In addition to denoting the operating voltage region, the term "low voltage" also identifies the equipment in which, by virtue of its lower operating voltage, the electron gun is mounted inside the welding chamber and is mobile along two axes.

A typical low voltage electron-beam welding machine is shown in Figure 1.10 together with its motor control cabinet and its operator's console. The vacuum chamber is made in the shape of a rectangular box because this form generally provides much more working space for a given volume than does a cylindrical shape of equal length. Within the chamber at the top is the electron gun mounted on its carriage and below, mounted on

Figure 1.10 Chamber extensions are attached to the electron-beam welder. They are made of aluminum tubing and are in sections so that length can be tailored to requirement.

rails on the floor of the chamber, is the work fixture carriage. The gun travel parallels the width ("Y"-axis) and the height ("Z"-axis) of the chamber, and the work fixture carriage travels the length ("X"-axis) of the chamber.

CHAMBER-CONTROL DEVICES

Several methods of future expansion should be kept in mind after an initial choice of chamber size. An actual application is shown in Figure 1.10 where lengthy and small tapered tubes of dual wall thickness are welded.

A unique application of electron-beam welding was programming the power of the electron beam in welding varying thickness material and a vacuum chamber that housed only the aligning fixture, the electron guns, and a small area of the Y-ring in the vicinity of the joint to be welded. The actual sealed chamber before welding is shown in Figure 1.11.

The modes of travel longitudinal (X-axis), transverse (Y-axis), up or down (Z-axis), and circular around Z-axis are available for a movable electron-beam gun. Recently a movable electron-beam gun rated at 60 kW and 500 µA utilizing a triode electron optical system similar to the high voltage electron-beam welder was tested. This unit, which also contains a deflection coil to provide total welding capabilities, joins the numerous electron-beam welding systems currently in production plants.

FILLER METAL ADDITIONS

Unlike more conventional fusion welding processes electron-beam welding does not require the addition of filler metal in the weld seam. There are exceptions to this statement, particularly when dissimilar metals are joined together. In this instance the filler metal can be preplaced or a filler wire feeding system can be used to supply filler metal to the joint. Filler wire feeders, such as those supplied for automatic GTA welding systems, are totally satisfactory and require only minor installation innovations. Filler wire addition is necessary when there is incomplete filling of the joint and when it is necessary to add oxidizers, scavengers, or absorbants to prevent the formation of brittle intermetallics when welding dissimilar metals.

Conventional filler wire feeding mechanisms are both practical and feasible. Some installations place the wire feed motor drive within the vacuum

Figure 1.11 Y-ring electron-beam welder showing Y-ring sealed into vacuum chamber [13].

chamber, which eliminates complex port openings and adjustment devices extending through the chamber. Extensive use of the internally mounted wire drive mechanism has proven simplicity of operation and minimum setup time, for the technician simply jogs filler wire to the center of the joint with a light vertical wire tension. The very fine stream of electrons permits very close proximity of the wire guide tube and relative to the molten weld puddle; this eliminates wire spiraling and/or "pig tailing" tendencies common to adding filler wires that require the wire guide tube to remain some distance from the weld puddle.

Safety

As with many "new" processes, electron beams present an occupational hazard that must be dealt with to assure the safety of the user. This hazard involves exposure of personnel both to the accelerated electrons or to secondary radiation in the form of X-rays or neutrons, with X-rays by far the more common problem. Basically, an electron-beam facility may be thought of as a giant X-ray tube with the electron gun acting as the filament and the workpiece as the target. Instead of producing X-rays for useful purposes as with the X-ray tube, the primary goal of the electron-beam facility is the use of the electron beam itself and X-rays become a byproduct.

The early work involving the use of electron beams in the metallurgical and related fields usually included the use of accelerating potentials under 20 kV. At these potentials the wall thicknesses of chambers necessary for high-vacuum welding generally provided adequate X-ray shielding. As voltages increased, however, lead shielding was added to the chambers interior. When a viewing window was provided, it was always a possible source of X-ray leakage in steel casing units. A thickness of lead glass was added over an ordinary thickness of plate glass to provide the necessary shielding in this area. It is recommended that the following suggestions are observed.

1. In compliance with existing regulations and good general practice concerning equipment capable of producing X-rays, a radiation survey should be made annually commencing one year from completion of the machine installation.
2. Any time that a major disassembly or rework of the equipment is performed, it is mandatory that a radiation survey be made. This is applicable to any work performed (or changes made) on equipment, at which time lead shielding is either disturbed or modified, regardless of location; for example, whenever the column shroud is removed or the column is relocated to its alternate position, a survey is required.

Tooling Considerations

Tooling and fixturing rank second only to design in importance to electron-beam welding. Proper weld joint design and good tooling control the success or failure of any welding process. The electron-beam welding process employs weld joints with contacting edges which have been machined to a relatively close tolerance and generally to a 125-microin. machine finish. This is particularly true of butt-type joint configurations. Although distortion occurs far less than that encountered by most other fusion welding processes, it is still a factor and must be compensated for. It is indeed foolhardy to expend the required capital outlay for a highly sophisticated fusion welding system and expect it to perform miracles in metal joining if tooling requirements and consideration have not been included in the initial planning stages when procuring an electron-beam welding system.

The tooling required by the electron-beam process is similar, but does not have to be as strong as tooling requirements for automatic GTA and MIG fusion welding processes. Since joint configurations are, except for minor innovations, identical, the tool required to maintain joint alignment with minimum residual restraint is applicable to all automatic fusion welding processes. Of particular interest to electron-beam welding fabricators is the concept of loading more than one part in an electron-beam welding chamber at one time. Figure 1.12 is a multispindle rotary fixture used

Figure 1.12 Multispindle welding fixture [21].

to produce vacuum tubes. When tooling is designed, fabricated, and used, which permits multiple-part welding in one chamber pump down cycle, the process then actually produces higher quality welded configurations and at possibly less cost than those produced by other fusion welding processes.

The ingenuity of the tool designer has become more important in electron-beam welding than in any other welding process. All work is done remotely and in a vacuum, the weld details must be held in practically intimate contact, the motors for movement of the parts must be capable of operation in vacuum, and all types of assist tools, subfixtures, and work-handling systems must be accurate and function with practically zero error tolerances. Shown in Figure 1.13 is a component for a thermionic generator with the subassist tools used to fabricate the complete generator.

As requirements for electron-beam welding arise, imaginative fixturing must evolve. A recent need to produce TZM molybdenum alloy honeycomb

Figure 1.13 Fixturing for rotating components for electron-beam welding.

Types of Equipment and Tooling   37

Figure 1.14 Electron–beam welding fixture for honeycomb core fabrication. The spring-loaded upper fixture exerts pressure on the strips of foil in the lower tool to insure consistent welds.

resulted in the fixture shown in Figure 1.14. Could D-43 columbium alloy sine wave corrugations be electron-beam welded and have ductility and strength? Figure 1.15 illustrates the sine wave tooling fixture and its accompanying traversing mechanism including a nitrogen-cooled motor which is operable in the electron-beam chamber.

Tooling can be complex as well as simple. Consider, for example, the welding of pressure vessels. Titanium spheres, 14 in. in diameter, have been electron-beam welded with the tools shown in Figure 1.16. The small plug in the center of Figure 1.16 is tack welded lightly to the sphere, which in turn is threaded for attachment to the rotary fixture within the electron-beam chamber. The ring with the slots throughout is tightened to hold the two halves of the sphere butted together. As the sphere is rotated within the chamber, a light pass of the beam (nonpenetrating) touches each open slot in the ring and fuses the two halves similar to a tackweld in fusion welding. On completion the ring is removed and then a full penetration weld is applied to the sphere. As a result there is no distortion or tooling.

Another prime example of simplicity is the electron-beam welding of C-120AV (6AL-4V) titanium elevon tracks.

Figure 1.17 illustrates the fixturing required for gas-tungsten-arc welding of these parts. The fixture is custom-manufactured for welding of the tracks

Figure 1.15 The tooling operating within the electron-beam welding chamber.

and includes expensive contour clamps, many locating pins, a contoured base, and other costly manufactured parts. In contrast, the electron-beam welding fixture, also illustrated, consists of a simple U-channel and readily available C-clamps and threaded parts. Not only is the electron-beam welding fixture much less expensive than the gas-tungsten-arc welding fixture but its simplicity decreases significantly setup time.

Another important application of the electron beam is microwelding of

Figure 1.16 Simplified tools for welding spherical motor cases.

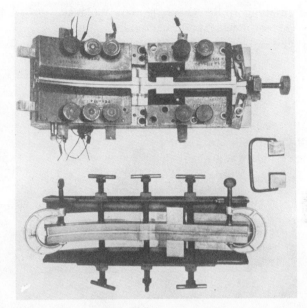

Figure 1.17 Fixturing of elevon tracks for gas-tungsten-arc welding (top) and for electron-beam welding (bottom) [23].

electronic components. Most fixtures used are different; one of these is illustrated in Figure 1.18, in which the components are readied for welding on a metallized wafer. The miniature fingers that retain the microcomponents on the wafer may consist of flat knife edges, U-shaped prongs, or may contain a 0.015-in. diameter hole drilled at the very end of each finger. The beam, which has a diameter varying from 0.003 to 0.005 in., can be easily programmed through the openings at the end of these microfingers. A miniature wheel device can also be used to weld multiple components for semiautomated or fully automated production. The component leads are moved under the wheel, which assures contact in the close proximity of the weld area. The end of the component lead is then welded to the desired surface. A vacuum chamber can be fitted with a micromanipulator arm that can be operated automatically or manually from the exterior of the chamber. The functioning of this manipulator is similar in principle to that shown for the copper fingers in Figure 1.18. The manipulator also contains a universal-type head with several types of probes that can be selected as required.

A caternary-type fixture can be used for welding fine wires or ribbons to the edge of a metal substrate. By preloading the ribbon to the nickel surface, the resulting spring action permits the metal ribbon to flow in toward the center of the nugget during welding.

Another fixturing concept used for welding modules is an X-type work table [24]. The wafers are installed and supported at four points. The wires are kept in contact with the edge of the wafers by taping down their ends and using a tungsten microfinger to hold down each wire to be welded. Each weld is then rapidly made by using the micromanipulator previously discussed. Rapid positioning of the assembly for proper location of electron-beam impingement can also be accomplished semi-automatically by using a set of relays and a step motor that programs the table mechanism. Finally, in Figure 1.19 is the work dolly table with tooling used in fabricating different core components for commercial reactors.

Whereas a few years ago most time was spent by electron-beam manufacturers in improving beam generation and control, today the emphasis has

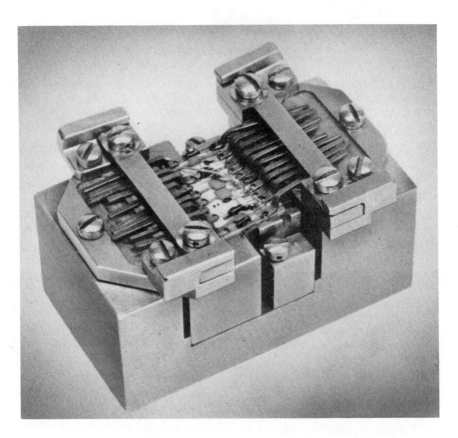

Figure 1.18 Wafer fixture for welding microcomponents [24].

Figure 1.19 One of the earliest, largest, and most complex welding fixtures ever designed for electron–beam welding.

shifted to work-handling systems, tooling, and special beam control devices to turn welders into true production tools.

ADVANTAGES–DISADVANTAGES–ECONOMICS

Advantages

The use of electron-beam welding equipment for metal-working purposes is growing every year. The electron-beam machine is particularly well suited for welding operations. Its primary advantages stem from the characteristic high depth-to-width ratio of the fusion zone and the ability of the process to operate in a vacuum.

The deep penetration feature of electron-beam welds is desirable from the standpoint of minimizing distortion and limiting any aging type metallurgical reaction that could occur during a weld cycle. In an electron-beam weld there exist a fusion zone and a heat-affected zone, which are also in an arc weld. The difference is that these zones are narrower in an electron-beam weld because of the highly concentrated but lower energy input rate in comparison with an arc weld. (See Figures 1.5a and b.) Metallurgical reactions that could occur in an arc weld may not have sufficient time to occur in an electron-beam weld. When welding hardened steels, that portion of the heat-affected zone that may be overtempered and softened is very narrow and it is strengthened by the adjacent harder material. In addition, the time in this temperature range is very short so that less overtempering occurs than in an arc weld. The higher temperature regions of the heat-affected zone and fusion zone will become hardened by quenching as would occur in an arc weld, but the smaller amount of shrinkage usually prevents cracking. These hardened regions will normally require a tempering treatment after welding. Other materials that are subject to an embrittling reaction either by precipitation hardening or by formation of a weak grain boundary phase can be joined more successfully by the electron-beam process. Here again the rapid heating and cooling reduce the time for the reaction to occur and the resulting fine-grained structure is affected less by a grain boundary phase. When handling heavily cold-worked materials, electron-beam welding will preserve the cold-worked properties.

The deep penetration feature of electron-beam welding is also advantageous from an economic point of view. Fewer passes are required than in conventional arc-welding processes. Besides requiring less weld time, in work where each pass must be radiographed, the amount of radiography is reduced.

Thus these special advantages come partly from the beam itself. Several additional ones include (a) precise heat control; when welding unfamiliar metals or one-of-a-kind assemblies, the operator can gradually increase beam energy to obtain proper welding conditions, (b) reproducibility; once the welding conditions are established for one assembly, the remaining assemblies can be welded with a high degree of uniformity; (c) design advantages; because of the close control of heat input possible, the chance of burn-through is greatly reduced. Section thickness in the joint area need not be increased to provide the margin of safety needed with torches, and (d) speed of welding; welding rates are variable between 0 and 120 ipm. The other advantages come from the vacuum; atmosphere contamination levels corresponding to the normal vacuum pressure used for electron-beam welding are nearly three orders of magnitude less than the contamination level of welding grade inert gases supplied commercially. This higher atmosphere purity enhances the properties of welds by reducing atmospheric reaction, particularly when reactive metals such as titanium are welded.

In addition to these, there are (a) weld quality; weld zone is completely free of inclusions and porosity. The combined effect of heat and vacuum drives off any gases and breaks down and removes most impurities present from earlier processing. Since the environment is free of gases, voids cannot form in the weld, (b) corrosion resistance; the nitrides of certain metals, such as zirconium, titanium, and hafnium, catalyze corrosion. Absence of these inclusions prevents this possibility. Also, welds in 300 and 400 series stainless steels are more corrosion resistant by the elimination of carbide inclusions.

Further advantages include the process itself. The process is adaptable to many applications not feasible with conventional processes, the most important of which are welding foil material, thick-to-thin materials, and dissimilar metals. It is also important in complex structures where little distortion can be tolerated; structures where welding is required in deep holes, deep grooves, and other relatively inaccessible areas. Another, more subtle advantage to the process is that the amount of heating can be varied smoothly by varying the voltage; this means preheating and postheating are easily accomplished and spatter is seldom a problem. Finally, a minimum of operator skills is required in order to weld successfully. The training period for operators is usually 40 hr of classroom instruction and practical training with the equipment.

Within the last five years considerable progress has been made in the development of welding techniques for the fabrication of microbonds. This area was completely dominated by the resistance-welding technique of joining circuit assemblies; however, conversion to electron-beam microwelding is now taking place in the electronics industry.

The principal advantage of electron-beam welding when compared to resistance welding is the elimination of electrode upset—a principal parameter in resistance welding. The upset pressure required for fusion welding with the electron beam is provided by thermal expansion while the materials are at welding temperatures. Therefore electron-beam welding is made with a "feather touch" imparted by the momentum of the impinging beam of high velocity electrons.

Other advantages, inherent in electron-beam microwelding, are as follows:

1. It is less sensitive than resistance welding to surface irregularities and impurities.
2. Because electrodes are not required, electron-beam joining facilitates microminiaturization.
3. Electron-beam welding is less sensitive and therefore requires a minimum of machine parameter rescheduling for various material combinations.

4. Vacuum environment minimizes joint contamination.
5. It is not necessary to avoid closed loops and particular polarities of components because currents flowing during the welding process are exceedingly small.
6. The inherent vacuum environment permits direct hermetic encapsulation of completed circuit assemblies without introducing an additional process step.

Limitations

The limitations of the electron-beam welding process are associated with the same factors that make it desirable, namely, the narrow fusion zone and the requirement to operate in vacuum. Another major limitation, which quickly becomes apparent to potential users of the process, is the high cost of the equipment. The lowest price of a production electron-beam welding facility is about $80,000. An automatic tungsten-arc welding setup, however, can run as high as $100,000. Whether or not an electron-beam welding machine costs too much depends on the type of welding intended and the required accuracy of the finished parts.

One of the disadvantages of a very narrow fused zone is that the short time that the metal is molten does not allow for adequate removal of interstitial impurities already present in the material. Narrow beams require very precise fitup of joints and alignment of the joint with the gun. When joint fitup is poor, a wider weld can be made by broadening the electron beam by defocusing or by oscillating it normal to the weld direction. The advantage of a high depth-to-width ratio, however, may be reduced.

Metals or alloys having a high gas content or metals with a high vapor pressure are difficult with the electron beam because the high metal temperatures results in enough pressure to be developed within the molten metal to cause it to be expelled from the joint as spatter. This usually results in a joint having some porosity and top and bottom undercutting from the loss of metal. In this type of situation the material may be welded at lower speeds and lower energy density to allow the gas to escape without expulsion of metal. Here again the penetration-to-width ratio may be reduced.

Although overall stresses and distortion are lower in electron-beam welds compared with arc welds, local thermal and transformation stresses may be sufficient to cause cracking when brittle phases are present. An example would be cracking in hardenable steels.

The requirement that the process operate in a vacuum places a physical limitation on the size of a piece that can be welded in a particular chamber. Attempts have been made to overcome this limitation by the development of the nonvacuum electron-beam process and by the use of local vacuum chambers which enclose the area to be welded. In the nonvacuum welding process, however, results thus far have shown that the penetration-to-width ratio is much smaller than that obtained by welding in vacuum. (This is discussed later.) Local vacuum chamber welds appear to have characteristics and quality similar to standard electron-beam welds, but the chambers are normally limited to one particular application.

In some instances the nature of the electron beam may present problems in welding. Since the electrons are moving charges, they are deflected from their main path by any transverse components of a magnetic field. The electron-beam current produces a uniform concentric magnetic field about itself provided that the magnetic permeability of the surroundings are uniform. If the magnetic permeability is not uniform because of fixturing or different materials in the joint, transverse deflection of the beam occur.

In electron-beam welding, X-rays are generated when the electrons strike the metal to be welded. Every electron beam welder contains the same components found in an X-ray tube, a cathode, an anode or a target and a source of high voltage. Characteristics and continuous X-ray spectra are emitted by the material being joined. The intensities depend on the material, the accelerating voltage, and the current. The peak X-ray intensity varies directly as the beam current and as the square of the accelerating voltage. Up to about 35,000 V the X-rays generated can be stopped effectively by approximately $\frac{1}{2}$ in. of steel in the vacuum chamber walls and by leaded glass. Above this voltage it becomes necessary to increase the wall thickness to reduce the radiation to a tolerable level. Usually it is more practical to line the chamber with lead than increase wall thickness. With nonvacuum electron-beam welding it is necessary to provide shielding to protect the operator from radiation generated at the weld area because there is no vacuum chamber to absorb the radiation. There is no electron-beam welding machine on the market today that poses any health hazard because of X-rays.

Economics

Although the initial cost and capital investment of electron-beam welders are high, their commercial value has proven their worth many times. While extensive cost figures concerning electron-beam welding are being compiled, isolated examples predict favorable comparison with typical automatic fusion welding techniques. Comparative floor-to-floor times, using both automatic GTA welding and electron-beam welding to fabricate a 4340, 5-in. diameter pressure vessel and to join a cobalt-base alloy turbine wheel to its shaft are shown in Table 1.1. Other examples include the following:

1. Two complex experimental titanium housings, completely machined to tolerances of ±.0001 in., were destined to be scrapped because

of the misplacement of a hole. These two housings were worth $13,000 and had a three-month raw material delivery lead time. The electron-beam machine was able to weld plugs in the subject holes without distortion of the rest of the piece or without affecting significantly the heat treat. The holes were then correctly machined, saving both the time and dollars.

2. Many of the 138 machined aluminum alloy castings worth $25,000 were subject to scrap when a machined slot had to be narrowed. The electron-beam machine was able to weld a small plate in the slot, which was then machined to the proper dimension, resulting in both large dollar savings and critical time saving.

3. Eighteen alloy castings valued at $9800 were salvaged when a mandatory engineering change required the addition of a wedge-shaped boss to the casting. This was done with the maximum distortion of the piece being under 0.00025 in.

4. A firm has estimated that it saves $1000 to $12,000 daily with EB welding. It formerly cost the firm $25 to copper braze a pressure cap onto the body of a high altitude transducer. Electron beam does the job for 25 cents.

5. Another company supplied a titanium shaft for the Surveyor that is now on the moon. Switching from solid to a two-piece electron beam-welded design halved the price, as shown by this breakdown:

	Solid	Welded
Material	$80	$26
Machining	$60	$20
EB weld		$20
Total	$140	$66

Table 1.1 Two Examples of Floor-to-Floor Times for Electron Beam and Gas Tungsten Arc Welds.

Pressure Vessel

	Joint Configuration	Filler Wire	Inert Gas	Number Passes	Floor Time
Electron Beam	Square Butt	None	None	1	9 minutes
Automatic Tig	Chamfer V or U Butt	Yes	Yes	3	15 minutes

Turbine Wheel to Shaft

	Joint Configuration	Filler Wire	Inert Gas	Number Passes	Floor Time
Electron Beam	Square Butt	None	None	1	8.5 minutes
Automatic Tig	Chamfer V or U Butt	Yes	Yes	2	13.5 minutes

Table 1.2 Estimated Cost to Weld 24-in. Diameter Cylinder to End Dome by the Electron Beam, MIG, and TIG Welding Process Maraging Steels [26].

Cost Item		1/4" Plate			1" Plate		
		E. B.	TIG	MIG	E. B.	TIG	MIG
Joint Preparation	(hours)	(0.75)	(1.00)	(1.00)	(2.00)	(3.50)	(3.50)
	cost	7.50	10.00	10.00	20.00	35.00	35.00
Weld and Setup	(hours)	(0.80)	(1.42)	(0.65)	(0.82)	(11.78)	(4.67)
	cost	11.75	12.25	5.60	12.05	101.40	40.30
Inspection	(hours)	(1.00)	(1.00)	(1.00)	(1.00)	(2.00)	(2.00)
	cost	10.00	10.00	10.00	10.00	20.00	20.00
Total Labor and Overhead		29.25	32.25	25.60	42.05	156.40	95.30
Tooling		5.00	5.00	5.00	7.50	7.50	7.50
Filler Wire		-	5.40	8.40	-	76.20	84.00
Inert Gas		-	2.80	0.53	-	26.35	4.96
Filament		0.14	-	-	0.19	-	-
Vacuum Pump Oil		0.34	-	-	0.34	-	-
Power		0.24	0.03	0.03	0.24	0.32	0.24
Total Cost		34.97	45.48	39.56	50.32	266.77	192.00

A government installation recently purchased an electron-beam unit for $175,000 and expects to pay for it in two years through savings in repairing aircraft parts that would otherwise have to be scrapped. For instance, it salvaged $1700 carburetor parts at a cost of $50 each; also 48 magnesium wheels were repaired at a net saving of $750 per wheel.

In conclusion, a recent survey was conducted to develop a detailed cost breakdown of the electron beam, tungsten inert gas, and metallic inert gas welding processes as applied to thick plate. From this a comparison was made of the relative economics of joining thick plate by each process. The cost breakdown for welding included all items that contributed to the cost of making a joint, starting with the cost of preparing the plate edges through inspection of the joint. Other factors considered were tooling, setup time, welding time, materials, and equipment costs. For the purposes of this program it was assumed that eliptical ends were to be welded onto 24-in. diameter cylinders, 36 in. long. The analysis was performed for two plate thicknesses, ¼ in. and 1 in. to show the effect of thickness on comparative costs.

For this analysis the cost of the electron-beam welder was estimated to be $120,000 and was to be written off over a ten-year period at a rate of $1000 per month or $6.20 per hour. The arc-welding equipment for either GTA or MIG was estimated to cost $12,000 and to be written off over a ten-year period at a rate of $100 per month or 62 cents per hour. The equipment contribution to the cost of welding by either method was based on the total time the equipment was required for setup, making the weld, and cleaning after welding. In addition, it was assumed that the cylinders were fabricated from plate by rolling and welding plate thicker than the required thickness. These cylinders were then bored and turned to give the correct wall thickness. The end domes were assumed to be formed with sufficient thickness to permit machining for the proper wall thickness and joint configuration.

The summary of costs to make the cylinder to dome weld for both thicknesses of material and by both electron-beam or arc-welding methods is given in Table 1.2 for maraging steel.

MATERIALS

The initial use of electron-beam welding was for joining metals whose mechanical or chemical properties were seriously impaired by even a minute amount of atmospheric contaminants (principally oxygen and nitrogen). These were refractory or highly active materials such as molybdenum, tantalum, tungsten, beryllium, columbium, and zirconium.

Today, as Table 1.3 indicates, the electron-beam welding process is applied not only to the refractory metals but also to a wide range of other metals. Significantly, the major emphasis has now passed from the refractory metals to those that might be best described as the structural metals, particularly in view of their extensive use in various types of structures. These applications take advantage of the higher weld joint efficiencies and reduced distortion and shrinkage, compared with other types of fusion welds, that result from the electron-beam weld's narrower weld and heat-affected zones.

Aluminum and Its Alloys

As seen in Table 1.3 practically all the aluminums have been welded. However, certain aluminum alloys, particularly the 7000 series, are very difficult to join. Moreover, even with the readily weldable alloys, such as 6061, electron-beam welding is sometimes unsatisfactory for a particular application because of unacceptable distortion, excessive heat input, or related problems. In tests of the tensile strength of this alloy welded in the solution-treated condition and aged to the T-6 condition after welding approximately 85% of expected phase-metal strength was obtained.

Visual and radiographic inspection of aluminum welds indicates no incidence of cracking or porosity. The high purity of the vacuum-welding environment undoubtedly contributed to this favorable condition. The tensile-test and bend-test results for 7075 aluminum are listed in Table 1.4. For comparison, typical properties of the base metal and recently reported data for conventional welding are also included.

The electron-beam weld results are far superior to the corresponding

Table 1.3 Materials Joined by Electron–Beam Welding.

STEELS	ALUMINUM ALLOYS	
300 Series Stainless	1100	5456
400 Series Stainless	2014	6061
Maraging	2017	7039
17-4 PH	2024	7075
17-7 PH	2219	7079
PH 15-7 Mo	3003	7178
14-8 Mo	5005	Cast 355
A286	5052	Cast 356
AM 350 and 355	5083	Cast 357
Carbon Steels (Including, but not limited to 1010, 1025, 1035, 1065, 1095 and Uddeholm UHB-15)	5086	Cast AMS 4291
	5254	Cast RED-X-20
Low Alloy Steels (Including, but not limited to 4130, 4140, 4340, 52100, 8640)	RARE AND PRECIOUS METALS	
Tool Steels (Including but not limited to H-11, M-2, and W-2)	Gold and Alloys	
HY-80	Silver and Alloys	
HY-150	Platinum and Alloys	
D6AC	Palladium	
300M	Iridium	
Rocoloy	Gallium	
Simonds #73		
Vascomax	FERROUS ALLOYS	
	300 Series to 400 Series Stainless Steel	
SUPER ALLOYS	300 Series to Austenitic Precipitation Hardenable Stainless	
Cosmoloy	300 Series Martensitic Precipitation Hardenable Stainless	
Hastelloy C	300 Series Stainless to Mild Steel	
Hastelloy N	300 Series Stainless Steel to Beryllium Copper	
Hastelloy W	400 Series to Austenitic Precipitation Hardenable Stainless	
Hastelloy X		
Haynes Stellite 25		
Haynes Stellite 31	400 Series to Martensitic Precipitation Hardenable Stainless	
Haynes Stellite 36		
Inconel 600	400 Series Stainless to Mild Steel	
Inconel X-750	Austenitic Precipitation Hardenable to Martensitic Precipitation Hardenable Stainless	
Inconel 718		
Inconel 722		
L-605	4130 to 17-22 (V)	
N-155	4130 to 4140	
Rene 41	4130 to Trancor T	
Udimet 500	6150 to M-2 Tool Steel	
Waspaloy	Kovar to Steel	

tungsten-inert-gas (GTA) results. This is not surprising since the GTA welds were made with nonheat-treatable 4043 filler. The comparison is valid, however, since the 7075 alloy has not been successfully welded without filler, and the 7075/4043 welds are among the highest-strength welds produced to date. Finally, the tensile-elongation and bend-test results indicate good ductility for all welds.

In electron-beam welding of heavier thicknesses (1.5 in.) of 7075 alumi-

Table 1.3 (Continued)

SUPER ALLOYS	TITANIUM AND TITANIUM ALLOYS
A286 to Inconel Alloy 713	Commercially Pure
A286 to Inco 100	5Al - 2.5SN
A286 to Udimet 500	6Al - 4V
A286 to Waspaloy	13V - 11Cr - 3 Al
Elgiloy to Beryllium Copper	7Al - 4 Mo
Hastelloy X to Inconel Alloy 713	3Al - 1 Mo - 1V
Haynes Stellite #6 to Steel	8Al - 1 Mo - 1V
Haynes Stellite #21 to Steel	6Al - 6V - 2 Sn
Haynes Stellite #21 to Nitralloy	
Haynes Stellite #31 to AISI 8640	COPPER AND COPPER ALLOYS
Incoloy Alloy 901 to Inconel Alloy 713	
Inconel Alloy 713 to Mild Steel	OFHC
Inconel Alloy 713 to Udimet 500	Electrolytic Tough Pitch
Inconel Alloy 713 to Waspaloy	Beryllium Copper
Inconel Alloy X-750 to Mallory 1000	Aluminum Bronze
Inconel Alloy X-750 to Mild Steel	Cupro-Nickel
Inconel Alloy X-750 to Molybdenum	Constantan
Inconel Alloy X-50 to Simonds #73	
Inconel Alloy X-750 to Tungsten	
Inconel Alloy X-750 to Waspaloy	NON-METALS
N-155 to 17-4 PH	
N-155 to 347 Stainless Steel	Alumina
Udimet 500 to Low Alloy Steel	Beryllia
Udimet 500 to Inco 713	Vycor
Waspaloy to Low Alloy Steel	Magnesia
Waspaloy to Inco 100	Thoria
	Pyrex
REFRACTORY METALS	MISCELLANEOUS METALS AND ALLOYS
Beryllium	
Columbium and Alloys	Kovar
Molybdenum and Alloys	Invar
Rhenium	Rodar
Tantalum and Alloys	Karmawire
Tungsten and Alloys	Mu-Metal
Vanadium	Nickel
Zirconium and Alloys	Vicalloy
	Cadmium
	Uranium
MAGNESIUM ALLOYS	Mallory 1000
	Ni Span "C"
AZ 31B	Nickel Silver
AZ 91A & C	Monel (Nickel-Copper)
ZK 60A	Havar
Cast AMS 4442 (EZ 33A)	
HK 31A	
ZRE-1	

num, several precautions must be taken. A slower welding speed must be established in order to allow more time for gas or vapors to escape from the molten puddle. The speed is usually of the order of 8 to 10 in./min, and to reduce the level of oxides the abutting edges and adjacent surfaces of the weld joint are brushed immediately before welding.

The results of welding 7075 aluminum in the -T651 condition indicate that porosity-free welds can be produced with the electron beam-welding

Materials

process. The transverse tensile strength of the weldment is low in comparison to parent metal tensile strength. It is apparent from tensile test results that low transverse joint strength results when thick sections of 7075 aluminum are joined by the electron beam-weld process. Low transverse joint strength must be allowed for in the design of components requiring the joining of thick sections.

As indicated previously, electron-beam welding of the 5000 series of aluminum alloys has been successfully accomplished. Shown in Figure 1.20 is a heavy section weld in 5083 aluminum alloy. This weld was accomplished with a single pass, and when one considers the number of passes and the time required to produce such a weld using conventional techniques, this

Table 1.3 (Continued)

REFRACTORY METALS AND ALLOYS

Molybdenum to Platinum
Molybdenum to Titanium
Molybdenum to Tungsten
Molybdenum to Kovar
Molybdenum to Tantalum
Columbium to Tantalum
Tungsten to Beryllium Copper
Tungsten to Copper
Tungsten to Stainless Steel
Tungsten to Steel
Tungsten to Titanium
Vanadium to 347 Stainless Steel
Vanadium to AISI 4340
Vanadium to Titanium

RARE AND PRECIOUS METALS

Gold to Germanium
Gold to Nickel Plate
Gold to Silicon
Iridium to Nickel
Palladium to 347 Stainless Steel
Platinum to Nickel
Platinum to Platinum-Rhodium

Chromel to 347 Stainless Steel
Constantan to Copper
Copper to Kovar
Copper to Mild Steel
Copper to Nickel Plate
Copper to Silver
Copper to Silver Alloys
Cupro-Nickel to SAE 1010 Steel
Dumet to Nickel
Elkonite to Simonds #73
Kovar to Nickel Plate
Mallory 1000 to Simonds #73
Nickel to Copper Plate
Nickel to Nichrome Plate
Nickel to Silicon
Nickel to Steel
Nickel Silver to SAE 1010 Steel
Nickel Silver to Silver

MISCELLANEOUS MATERIALS

Alumel to Chromel
Alumel to 347 Stainless Steel
Aluminum to Beryllium
Aluminum to Copper (Hermetic Seal Only)
Aluminum to Stainless Steel (Hermetic Seal Only)
Aluminum Bronze to Mild Steel
Aluminum Bronze to Stainless Steel
Austenitic Stainless Steel to Tungsten Carbide
Beryllium to Copper
Beryllium to 52100 Steel
Beryllium to Uranium
Brass to 94% Pb - 4.5% Sn Foil

is quite remarkable. The application of a 5000 series aluminum alloy was recently demonstrated in the fabrication of a 5254 aluminum alloy torpedo component.

The aluminum alloy 2219, a newly developed high-strength alloy, is heat treatable and shows a definite response to cold work. Hence the properties of 2219 are a result of thermal treatment and/or cold work. Strengths in the -0, -T42, and -T62 tempers depend on thermal treatment alone, whereas strengths in the -T31, -T37, -T81, and -T87 tempers depend on a combination of thermal treatment and cold work. Obviously, properties can vary; however, recent work on 2219 aluminum alloy has shown that significantly higher strength and joint efficiency can be developed in the plate gages welded by the electron-beam process than by either gas-tungsten-arc or gas-metal-arc welding, that is, tensile strength efficiency 70 to 80% for electron-beam welds and 50 to 65% for gas-tungsten-arc and gas-metal-arc.

Table 1.4   Tensile and Bend Results for Electron-beam Welded 7075 Aluminum.

Specimen Condition	Ultimate Strength (psi)	Yield Strength (psi)	Elongation (% in 1 in.)	Elongation (% in $\frac{1}{4}$ in.)	Bend Angle at Failure (degrees)
EB welded in T6, Tested as Welded	50,500	43,300	2.0	8.0	—
	51,000	43,200	2.0	8.0	—
	—	—	—	—	41
	—	—	—	—	46
EB welded in W, Aged to T6	49,400	44,000	4.0	12.0	—
	50,700	44,000	2.0	8.0	—
	—	—	—	—	37
	—	—	—	—	26
Welded in O; Heat-treated to T6	70,500	61,000	6.0	8.0	—
	69,700	59,700	6.0	8.0	—
	—	—	—	—	25
	—	—	—	—	27
O Base Metal	32,000	14,000	17	—	90
T6 Base Metal	76,000	67,000	11	—	58
TIG Welded in T6 4043 Filler Tested as Welded	37,000	29,000	2	—	—
TIG Welded in T6 Reheat-Treated to T6	44,800	29,100	4	—	—

Figure 1.20 Electron-beam butt weld in 5-in. thick 5083 aluminum alloy [18].

A comparison of welding conditions and strength for satisfactory gas-tungsten-arc and electron-beam welds in $\frac{3}{4}$-in. thick 2219 plate is shown in Figure 1.21. It is evident that, for the gas-tungsten-arc process, the greater heat input lowers the tensile strength somewhat but drastically lowers the yield strength.

One of the most interesting indications with electron-beam welding 2219 is that higher yield strengths are obtained with heavier gages than with the thinner material.

Sintered aluminum powders with the dispersion strengthening effect of submicroscopic particles of aluminum oxide (Al_2O_3) have been successfully electron-beam welded. The application of these sintered aluminum products (SAP or Frittoxal in Europe, APM in the United States) [29] offer interesting possibilities for use as structural and canning materials in organic-cooled or organic-cooled and moderated reactor systems.

Beryllium and Its Alloys

Beryllium, with its low density, high modulus of elasticity, and high specific heat, is being used to an increasing extent in numerous applications.

Use of beryllium for structural applications, however, has been impeded partly by the lack of satisfactory welding techniques for the production of sound and reliable joints. Conventional fusion-welding techniques have been hindered by contamination, cracking, hot-shortness, and related problems, which have all led to low room-temperature strength and poor ductility. Some success has been realized with brazing and braze-welding techniques; brazed joints, however, have limited applications because of the difference in the melting point of the braze metal and the beryllium.

In view of the difficulties encountered in welding and brazing beryllium by conventional techniques, electron-beam welding has produced high-strength welds. Contrary to almost all other metals, fine grain size and random grain orientations in beryllium electron-beam welds are not dependent on fast welding speeds. Instead, best results to date have been achieved with relatively low welding speeds (15 ipm). Controlled cooling, which is required to produce the optimum weld-zone microstructures, is accomplished by moderating the power density and decreasing the welding speed. Of course, the ideal weld zone would be small in overall size while exhibiting a fine-grained, random microstructure. The results illustrated in Figure 1.22 indicate that this weld structure is attainable by using moderately high voltages in combination with lower beam currents while maintaining the slow welding speeds.

Results obtained from tensile testing beryllium welds at room temperature and 1000°F show that the room-temperature ultimate strength of the electron-beam welds is consistently about 65% of the base-metal strength. However, the base-metal strength is considerably higher than that reported in other beryllium welding investigations [30,31]. Thus, even with the 35%

	Gas tung-sten-arc	Electron beam	
Kilojoules, in./in.	171.3	14.2	[a]
Travel, ipm	3.25	60	[b]
Tensile strength, ksi	41	43	49
Yield strength, ksi	21	33	39
Elongation, %	8	2	..
Specimen	½ in. flat	[c]	[d]

[a] Optimum settings.
[b] Expected.
[c] ⅛ in. slice.
[d] 1½ in. wide.

Figure 1.21 Noticeable differences exist between electron–beam and gas-tungsten-arc welds.

Figure 1.22 Weld cross section in 0.040-in. beryllium.

drop in ultimate strength, the electron-beam welds have strengths among the highest reported. At 1000°F, even though all the welded specimens failed in the welds, the base metal and weld ultimate and yield strengths are essentially the same. Similar trends have been reported for GTA welding.

Ductility appears to be the property that has suffered most from the welding process. It should be noted that in weld fractures, however, a very high proportion of the strain is absorbed within the very narrow weld zone. Thus a very high degree of local ductility could be masked by the overall measurement. In light of the continued development of electron-beam welding of beryllium, the relationship between composition and quality of beryllium sheet and properties of electron-beam welds has determined that the quality of welds in beryllium is linked directly to the oxide content of beryllium.

Copper and Its Alloys

Electron-beam welding of the beryllium-copper alloy 25 has been accomplished and a comparison of test results, although limited, shows the increase in strength exhibited over the conventional shielded gas processes. Copper has been electron-beam welded to itself and the melted zone extends completely through the two copper strips. Limited applications have restricted the use of electron beam in joining copper.

Columbium and Its Alloys

Numerous companies have evaluated the electron-beam welding characteristics of first and second generation columbium alloys. The results have shown that all the alloys are weldable, but primarily with the first generation alloys (FS82, D14, D31, F48) electron-beam welding has provided a significant improvement in ductility over gas-tungsten-arc welding. The improvement in ductility of the electron-beam welds in columbium alloys is attributed to the greater weld purity than is attainable with gas-tungsten-arc welding.

Electron-beam welds in the second generation columbium alloys (Cb752, FS85, C129Y, D43, B66) have shown excellent ductility, improved weld transition temperatures, and joint efficiency. Joint ductilities at 70 and 2200°F in D-43 are equivalent to those for the base metal, and joint efficiencies at these temperatures are within the range of 91 to 102%.

Magnesium and Its Alloys

In spite of the high vapor pressure at its melting point, the electron-beam welding process has been applied successfully to magnesium and its alloys. Most applications have been for the fabrication of small precision components and for the salvaging of mismachined parts. In these instances more conventional joining methods could not be employed because of the accompanying excessive weld distortion.

Molybdenum and Its Alloys

The main technical objections to the more extensive use of molybdenum have been its general brittleness, low ductility, and low impact strength. All these problems are aggravated by joining operations. Since the detrimental effects of high temperature heating are primarily dependent on the purity of the molybdenum, successful fusion welds can only be obtained on metal of the maximum purity together with the fastest possible welding techniques, the highest cooling rates, and shielding gases of high purity. For these reasons electron-beam welding, with its low total energy input providing fast heating and cooling and its vacuum atmosphere providing a high purity surrounding atmosphere, has been investigated for application

Figure 1.23 A container for nuclear fuel which shows the tube electron–beam welded to the cap and subsequently to the cylinder.

to molybdenum and its alloys which have been primarily Mo-0.5% titanium and TZM. Figure 1.23 illustrates a welded TZM container and lid for a nuclear fuel, which indicates that the material exhibits excellent electron-beam weldability. Weldments so produced in Mo-0.5% titanium alloy also exhibit relatively good mechanical properties.

Joint efficiencies as determined by room temperature tensile tests of electron-beam welds of 0.040-in. TZM have been found to be 71.1%, whereas GTA welds have exhibited 63.6%. Tensile test data show that the ultimate and yield strengths of the Mo-0.5% Ti electron-beam welds are substantially higher than for the gas-tungsten-arc drybox welds at 1600°F and above. The strength of the electron-beam weld is approximately 20% higher than for the tungsten-arc welds. This improvement in strength is attributed to the much finer grain size of the electron-beam welds as compared with the coarse structure of the tungsten-arc welds.

Superalloys–Nickel–Cobalt–Iron Base

In the family of iron-base superalloys, A-286 has been successfully electron-beam welded and tensile and bend data recorded in Table 1.5 show the optimum strength properties obtained in the electron-beam weldments which are welded in the solution-treated condition and postweld aged. The ultimate and yield strengths are 88 and 98%, respectively, of the base metal strength. These strength levels compare favorably with those obtained from a gas-tungsten-arc weld using the same heat treatments and Hastelloy-W filler wire. The ultimate strength and yield strength joint efficiencies in the GTA welds are 85 and 98%, respectively. The bend tests indicate somewhat better bend ductility for the electron-beam welds compared with the tungsten-arc welds as shown by the data in Table 1.5.

The electron-beam welding of several nickel-base alloys, Inconel 600, Waspalloy, and Hastelloy-X, Table 1.3, again indicates the relative versatility of the process to handle all types of metals.

René-41, a precipitation-hardening nickel-base austenitic alloy with good high temperature properties which, combined with its relatively good formability and its availability as a sheet material, makes it attractive as a structural material. Satisfactory fusion welds have been produced in René-41 using the electron-beam process. The alloy, 0.062 in. thick, which was welded in the as-received condition and tested as-welded, exhibits base metal properties as shown in Table 1.6. Successful welds have been made with one pass in 1 in. René-41 plate. Welds, made in solution-annealed plate and then postweld solution annealed and aged, have exhibited tensile and stress rupture properties approximately equal to those of parent metal.

Two pass welds in 1.5 and 1.75 in. René-41 have been found to be free of surface cracks; however, a slight amount of fine porosity has been observed. There is a tendency for cold shuts to occur at the tip of the fusion zone of the second weld pass. This tendency does not seem to be significant. The cold shuts in René-41 appear to occur at random and

Table 1.5 Tensile and Bend Test Results of Electron–Beam Welds in A286 Sheet [32].

Preweld condition	Post weld heat treat	Ultimate tensile strength, psi	Yield strength 0.2% offset, psi	Elongation in 2 in., %	Bend, deg[a] Face	Root
Solution treated	Aged	132,600	113,100	4.9	47	65
		133,700	111,700	6.2	76	36
Average		133,250	112,400	5.6		
ST and aged	Reaged	128,900	...	13.3	37	10
		132,000	...	5.5	31	20
Average		130,450	...	9.4		
ST and aged	None	100,300	99,500	0.9	16	16
		100,300	95,900	0.7	27	10
		100,300	97,700	0.8		
Tungsten-arc weld[b]						
Solution treated	Aged	129,500	111,500	9.2	15	
		129,500	114,100	9.0	18	
Average		129,500	112,800	9.1		
Base metal ST and aged		151,800	115,000	23.0	160	

[a] Loaded with face or root in tension over a 2T bend radius on 1 in. span.
[b] Welded with Hastelloy W filler metal; bend specimen size—0.100 x 3/8 x 2 1/2 in.

Materials

Table 1.6 Mechanical Properties of René 41 [33].

Condition	Tensile yield strength 0.2% offset, psi	Ultimate tensile strength, psi	Elongation in 2 in., %	Elongation in 3/8 in., %	Joint efficiency, % Yield strength	Joint efficiency, % Tensile strength
Annealed[a]	78,700	137,200	30	..	..	..
Soln. Treated[b]	115,450	172,875	31.3	..	..	..
Aged[c]	140,650	176,750	3	..	..	..
As-welded in ann. cond. (gas tungsten-arc weld)	70,260	136,400	30	39	89	100
As-welded in ann. cond. (electron beam weld)	80,800	139,000	28	38.7	100	100
Soln. treat after welding in ann. cond. (gas tungsten-arc weld)	110,333	163,833	19.3	26.7	98	95
Soln. treat after welding in ann. cond. (electron beam weld)	116,000	167,000	18.3	24	100	97
Welded in soln. treat cond. & aged after weld (gas tungsten-arc weld)	128,425	148,200	1.5	8	92	84
As-welded in soln. treat & aged cond. (gas tungsten-arc weld)	90,760	114,150	2.5	12.5	62.5	65

[a] Annealed at 1975°F for 30 min–water quenched.
[b] Solution treated at 1950°F for 4 hr–air cooled.
[c] Aged at 1400°F for 16 hr–air cooled.

not continuously; therefore in electron-beam welding of 1.5 in. or greater René-41, extra care should be taken due to the tendency for cold shuts to occur and the increased tendency toward strain age cracking.

Another nickel-base alloy, Inconel alloy 718, exhibits excellent electron-beam welding characteristics. Actually, the electron-beam welder is capable of penetrating all the way through from one side, but a higher quality weld can be obtained with the pass from each side when welding $1\frac{1}{4}$-in. thickness.

The relative ease of electron-beam welding cobalt-base alloys is illustrated in Figure 1.24, which shows the cap and inner web of a 6-ft spar member. The weld was a penetration weld through the cap, and the excellent fillet weld on the corrugation web is seen. The tracking of the weld was done with the equipment seen in Figure 1.15.

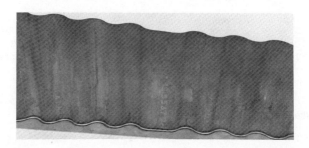

Figure 1.24 Underside of the cobalt-base welded component.

The growth and usage of the new series of thoria-dispersion hardened alloys have required that all joining media be evaluated to determine their joining qualities. Electron-beam welds have been made on 0.025- and 0.050-in. thick thoriated-nickel sheet. The welds obtained were of good quality although there is some thoria agglomeration. Bends in specimens were accomplished after welding, which indicates some ductility in the weld zone, although it did not show up in elongation measurements. The weld strength is about 80% of normal room temperature material strength.

Steels–Stainless–Precipitation Hardened–Low Alloy Ultra High Strength–Maraging

All steels that have been fusion welded by either one arc process or another can be electron-beam welded. Therefore since the varieties of steels and gages are so numerous, only a few examples of the more prominent families of steels are illustrated.

STAINLESS

Tensile tests have been conducted on butt-welded material in the as-welded condition with no surface preparation after welding. Figure 1.25 is a statistical plot of the tensile strength of welded 0.062-in. material. Since the data fall on a straight line, a normal distribution of strength is indicated, and since the line is very steep, the standard deviation or spread in strength is small. The random distribution of △'s and ○'s, all on the straight line, show conclusively that the weldments and the base metal have the same strength. The solid ▲'s and ●'s are further significance. These specimens were subjected, subsequent to welding, to the Strauss test, which is a corrosion test to determine if austenitic stainless steel has been sensitized and thus has lost its corrosion resistance. Since the strength was not affected by the corrosive environment as shown by the distribution of these solid points, it is obvious that no sensitization occurred. Similar results have been obtained with 0.275-in. material. The extremely small fusion zone and the high speed possible with electron-beam welding undoubtedly permitted the material to pass through the critical temperature range at a rate sufficient to prevent sensitization.

PRECIPITATION HARDENED

In electron-beam welding of the precipitation-hardened (PH) steel alloys, no unusual problems have been encountered. As expected, fractures occurred in the weldment at strengths slightly above those for annealed material. Data obtained for 17-7PH welded in condition A (annealed) and subsequently solution treated and aged have shown base-metal strength was obtained in the weldment. The tensile strength obtained with material welded in the solution-treated condition and aged to the TH1050 condition after welding has been recorded, and weldment strengths were only slightly below the base-metal properties. Foil thicknesses weld just as readily as thicker materials.

In thicknesses up to 1-in. of 15-7 Mo PH steel, crack-free electron-beam welds have been produced with excellent tensile properties.

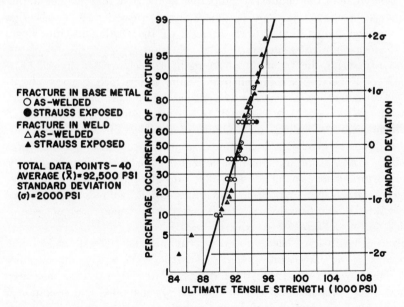

Figure 1.25 Ultimate tensile strength electron–beam butt-welded type 302 stainless steel (0.062 in. annealed sheet) [17].

A new PH steel, PH14-8Mo, which is currently finding its way into new products like pressure tanks, supersonic aircraft applications, and the Apollo moon vehicle, has exhibited weld tensile strengths of 205 ksi and yields of 203 ksi.

LOW ALLOY ULTRA HIGH STRENGTH

AISI 4340 and 4130, widely used deep-hardening low-alloy structural steels, respond very well to electron-beam welding. No significant problems are encountered in producing sound, crack-free welds without preheating or postheating. These welds display excellent mechanical properties. Furthermore, as with austenitic stainless steel, deep narrow welds can be produced.

Tensile and fatigue tests conducted on 4340 $\frac{1}{4}$-in. thick welded both before and after heat treatment (R_c 26–32) shows weldment strength equal to that of the base material. Similar weldments heat treated to R_c 47–50 exhibit the same base-metal strength.

Tensile tests have been conducted on welds made in heat-treated 4340. The results were quite remarkable in that, once again, base-metal strength was maintained in the weld zone. Results of the fatigue testing of 4340 weldments also have been outstanding. The endurance properties of weldments produced in annealed material and heat treated to R_c 26–32 after welding approach those of the smooth base material, and they are significantly better than the endurance properties of the base material containing a standard notch. Similar data for material heat treated to R_c 40–44 after welding have shown that fatigue strength of the weldments approaches that of the smooth base material.

The reason for the retention of strength in steels welded in the heat-treated condition can be deduced from Figure 1.26, which is a microhardness traverse through a typical arc weld and a comparable electron-beam weld. It can be seen that the small fusion zone and high welding speed associated with the voltage and density of the electron beam weld have limited the overtempered or heat-affected zone to a very thin sandwich-layer of softer, weaker material. This thin layer is supported by adjacent stronger material, producing triaxial stressing that prevents reduction in area. Thus the true, rather than apparent, tensile strength is realized. With normal arc welding a significantly wider and softer overtempered region is produced; thus tensile strength is reduced.

Electron beams have been applied to gear materials. Tensile and fatigue properties of the carburizing grade SAE 9310 (AMS 6265), the nitriding grade Nitralloy 135 (AMS 6470) and SAE 4340 (AMS 6415), have been evaluated for comparison. The materials were heat treated and conditioned as applicable to the fabrication of actual gears.

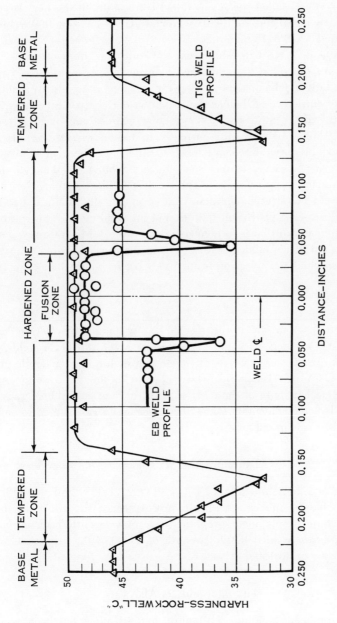

Figure 1.26 Hardness profile comparison: tungsten inert gas weld (TIG) and electron-beam weld. AISI 4340 (0.062-in. R$_c$ 40-44 sheet, stress relieved after welding at 650°F) [17].

The results of tensile and fatigue tests indicate that electron-beam welded joints develop properties quite adequate for fabricated gears, as shown in Table 1.7. The welded specimens failed consistently in the heat-affected zone for AMS 6265 and in the base metal removed from the weld for AMS 6470 and AMS 6415 materials.

Fatigue tests of AMS 6265 show that the endurance limit of the welded specimens with the respective unwelded base metal specimens have a joint efficiency in fatigue which is above 85% for all three materials. In addition, the service performance of the electron-beam welded joints has been verified for several specific gear assemblies by actual gearbox and engine tests.

The H-11 hot work die steel currently in use in pressure vessels and containers exhibits welding characteristics similar to those of 4340 steel discussed previously. Several other steels have exhibited the same excellent strength characteristics as noted in Table 1.8. The biaxial strength of electron-beam weld metal is higher than that of chemically identical wrought material and substantially higher than that of arc-weld metal. This comparison relates to material that is fully heat-treated after welding. Consequently, in nine cases out of ten failure of fully machined and heat-treated transverse tensile test specimens of electron-beam welds occurs more than $\frac{1}{4}$ in. from the weld.

MARAGING

In a recent comparison of the mechanical properties of electron beam welds in maraging steel with the properties of MIG and GTA welds, the electron-beam welds were capable of developing slightly higher tensile properties than GTA or MIG welds. Two pass welds are not recommended in maraging steel because of the possible occurrence of cold shuts, cracking, or porosity. It is recommended that sufficient beam energy be used to enable the joint to be made in one pass.

Tantalum and Its Alloys

Tantalum and tantalum alloys -T-111, T-222, T-333, and GE-473 are readily weldable by the electron-beam process. Since the alloys are new, intensive welding development programs have only recently been initiated. The electron-beam welded tensile strength of T-111 is the same as Cb-752 columbium alloy up to 1500°F.

Titanium and Its Alloys

The electron-beam welding of titanium and its alloys has been accomplished with relative ease. Since titanium, a very reactive metal, must be

Table 1.7 Average Tensile Test Results for Electron-Beam Welds in AMS 6265, AMS 6415, AMS 6470 Materials [44].

	Ultimate tensile strength, ksi	Yield strength (0.2% offset), ksi	Elongation in 1 in., %	Fracture location[a]	Joint efficiency, % of UTS[b]
AMS 6265 (9310)					
Base metal heat treated to R_c 36-40	179.2	152.0	13.7	...	...
Heat treated to R_c 36-40, EB welded and stress relieved 350° F, 2 hr	166.2	148.0	11.7	HAZ	92
AMS 6470 (Nitralloy 135)					
Base metal heat treated to R_c 33-37	152.1	131.3	15.7	...	...
Heat treated to R_c 33-37 EB welded and stress relieved 960° F, 2 hr	157.5	135.9	11.1	BM	100
AMS 6415 (4340)					
Base metal heat treated to R_c 32-36	164.2	155.4	11.7	...	...
EB welded, then heat treated to R_c 32-36	168.5	158.2	11.1	BM	100
Heat treated to R_c 32-36, EB welded and stress relieved 1000° F, 1 hr	161.7	151.6	11.1	BM	100

[a] HAZ—heat-affected zone; BM—base metal.
[b] UTS—ultimate tensile strength.

Table 1.8 Comparison of Uniaxial and Biaxial Properties [34].

Material	Welding process	Tempering temperature, °F	Yield strength (0.2% offset), 1000 psi	Uniaxial ultimate tensile strength, 1000 psi	Biaxial ultimate tensile strength, 1000 psi	Strength increase, %
H-11[a]	Gas tungsten-arc	400	223.0	287.0	306.0	8.5
H-11[a]	Electron beam	400	224.1	270.0	326.5	20.8
H-11[b]	Gas tungsten-arc (manual)	1000	180.0	245.0	...	...
H-11[b]	Electron beam refusion of gas tungsten-arc weld	1000	196.0	271.0	337.5	24.7
H-11[b]	Electron beam	1000	186.0	262.5	357.0	36.2
MX-2	Electron beam	400	171.0	277.0	340.0	22.3
4335	Electron beam	400	147.5	252.0	303.5	20.4
300M	Electron beam	600	...	286.0	343.5	21.4
300M	Gas tungsten-arc	600	237.4	277.2	300.8	8.5
300M	[c]	600	245.5	289.1	311.0	8.5
D6AC	Electron beam	600	222.0	260.0	292.0	12.3
D6AC	[c]	600	205.0	234.0	...	...
D6AC	Gas tungsten-arc	800	221.7	246.9	266.0	7.7
D6AC	[c]	800	242.5	258.5	287.6	11.3
X-200	Electron beam	600	205.0	274.5	339.0	23.5
X-200	[c]	700	252.3	289.1	317.9	10.0
X-200	Gas tungsten-arc	700	217.0	268.8	305.1	13.6
MBMC No. 1	Electron beam	600	229.5	278.3	332.0	19.4
MBMC No. 1	Gas tungsten-arc	600	200.3	232.2	273.7	17.9

[a] All values are an average of two or more tests.
[b] 1850° F, 1/8 hr austenitize.
[c] 1900° F, 1/4 hr austenitize.

welded with protective inert gases to produce satisfactory weld joints, the vacuum of the electron-beam chamber is a natural environment to insure clean, void-free welds.

Pure Titanium (A-70)–A 110AT (5AL-2.5 Sn)–C120AV (6AL–4V)

Welded tensile data and the tensile properties of these three alloys are presented in Table 1.9. It is feasible to join different titanium alloys together as well as different thicknesses.

Sound welds have been produced in one pass in 1.0- and 1.75-in. Ti-6AL–4V using the electron-beam welding process. The solution and aged parent metal exhibited ultimate and yield strengths of 159.1 and 148.5 ksi, respectively. Comparable transverse weldment strengths and longitudinal weld metal strengths from 1-in. thick weldments have exhibited 148.3 and 138.8 ksi average ultimate and yield strength, respectively.

A comparison of electron beam welds made in 1-in. annealed 6AL-4V titanium material with MIG welds made in 1-in. annealed material indicates that electron-beam welding offers porosity free welds with tensile and fracture toughness properties comparable to parent metal. Tensile and impact toughness properties comparable to parent metal have been obtained by the MIG welding process. However, electron beam offers a porosity-free weld, and the weld can be made in one pass without filler material and without extensive joint preparation other than grinding.

Table 1.9 Tensile Properties of Electron–Beam Welded Titanium [23].

Specimen	Alloy and thickness	Ultimate strength, psi	Yield strength, psi	Elongation, % in 1 in.	Fracture location
1	C120AV—0.050 in.	136,000	130,000	14	Base
2	C120AV—0.050 in.	140,000	132,400	17	Base
3	C120AV—0.050 in.	140,800	132,000	17	Base
4	C120AV—0.050 in.	138,000	131,200	17	Base
5	C120AV—0.050 in.	139,200	132,000	17	Base
6	C120AV—0.050 in.	135,600	131,200	6	Weld
7	C120AV—0.050 in.	137,200	132,000	14	Base
8	C120AV—0.050 in.	136,000	128,800	14	Base
9	C120AV—0.050 in.	138,000	132,000	16	Base
10	C120AV—0.050 in.	137,200	132,000	20	Base
11	C120AV—0.050 in.	136,400	128,400	18	Base
12	C120AV—0.050 in.	138,400	132,000	18	Base
13	C120AV—2.0 in.	134,300	120,500	8	Base
14	C120AV—2.0 in.	136,000	122,500	8	Base
15	C120AV—2.0 in.	135,900	124,500	8	Base
16	A110AT—2.0 in.	135,900	125,600	14	Weld
17	A110AT—2.0 in.	132,800	125,600	13	Weld
18	A70—2.0 in.	101,700	87,700	12	Weld
19	A70—2.0 in	100,800	87,700	26	Base

B120VCA(13V-11Cr-3AL)/7AL-2Cb-1Ta/6AL-6V-2Sn/8AL-1Mo-1V

The 13V–11Cr–3AL all beta titanium alloy as well as these three Ti alloys have been welded by the high voltage and low voltage electron-beam processes and also by the gas-tungsten-arc vacuum purged drybox technique in an argon atmosphere. Both the as-welded and aged electron-beam welds sustained base metal strength; tensile and bend test results of the as-welded are recorded in Table 1.10.

Table 1.10   Tensile and Bend Test Results of Welds in 13V–11Cr–3Al Titanium Alloy [32].

Welding process	Condition	Ultimate tensile strength, psi	Yield strength 0.2% offset, psi	Elongation in 2 in., %	Fracture location	Bend, deg[a] Face	Root
Hv E-beam	As-welded	143,400	143,400	13.9	PM	180	180
		138,200	138,200	17.0	PM		
Average		140,800	140,800	15.4			
Lv E-beam	As-welded	133,300	131,000	9.2	PM	132	170
		131,900	127,900	14.7	PM		
Average		132,600	129,450	11.9			
Tungsten-arc drybox	As-welded	135,300	132,000	1.9	W	42	9
		134,000	133,500	0.8	W		
Average		134,650	132,750	1.3			
Base metal	As-received	137,700	130,000	19.2	...	180	...
Hv E-beam	Aged	185,700	173,900	4.8	W	7	30
		186,200	173,500	5.0	W	6	33
Average		185,950	173,700	4.9			
Lv E-beam	Aged	182,800	177,300	1.6	W	17	7
		181,600	178,500	2.0	W	12	12
Average		182,200	177,900	1.8			
Tungsten-arc drybox	Aged	178,500	172,800	1.4	W	Nil	Nil
		170,600	170,600	1.3	W		
Average		174,500	171,700	1.3			
Base metal	Aged	186,000	170,000	5.7	...	26	...
		186,000	164,000	6.5	...	26	...
Average		186,000	167,000	6.1			

[a] Loaded with face or root in tension over a 2T radius on 1 in. span; bend specimen size— 0.100 x 3/8 x 2 1/2 in.

Tungsten and Its Alloys

Tungsten sheet can be welded successfully by the electron-beam process; the welded joints, however, are brittle at room temperature and must be handled very carefully to avoid fracture. Elevated temperature tests of

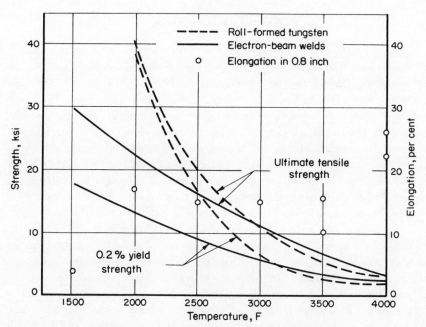

Figure 1.27 Comparison of elevated-temperature tensile properties of roll-formed tungsten and electron–beam welds in tungsten sheet [40].

0.040-in. thick tungsten sheet are shown in Figure 1.27. The welds have tensile strengths equivalent to the base metal strength at temperatures approaching the recrystallization temperature.

Zirconium and Its Alloys

Another very reactive metal is zirconium. It is rarely used as a pure metal, but is used as an alloy, Zircaloy. Zircaloy-2, which is zirconium-base alloy containing small amounts of tin, iron, and chromium, is well known in nuclear applications. Zircaloy-2 is highly reactive at fabrication temperatures and can dissolve both oxygen and nitrogen with deleterious effects on its mechanical, physical, and corrosion properties. Particular precautions must be taken in welding this alloy to prevent its reaction with the atmosphere. The fusion of Zircaloy-2, with its 1800°C melting point, by the tungsten-arc process requires the use of relatively high power input with a wide fusion zone in relation to the depth of penetration.

In contrast, electron-beam welding provides an inherent dynamic operating vacuum of 10^{-4} torr, which is an ideally pure atmosphere for joining Zircaloy-2 and has produced welds whose properties are equivalent to or better than the base metal.

NonMetallics–Ceramics

Ceramic materials are being used extensively by the electronics industry in the construction of power electron tubes, microwave windows, and similar components. The advantages of ceramics in these applications include high dielectric strength and superior elevated temperature properties. These advantages have not been fully realized, however, because of difficulties in joining separate ceramic sections into structurally sound, vacuum-tight components. Current metallized and brazed joints present problems of matching thermal expansion. Furthermore, brazed joints exhibit lowered arcover thresholds and reduced useful service temperatures. In addition, the joints are subject to electrode capacitance effects that limit electrical performance.

Electron-beam welding is one of the few methods available for direct fusion welding of ceramics. Actual welding of ceramics is attractive, for it can minimize many of the difficulties associated with brazing. Gas or plasma torches can fuse certain ceramics; however, these methods lack the precise controllability and cleanliness available with electron-beam techniques. Because of these considerations, electron-beam welding procedures have been developed for the joining of ceramic materials, including alumina, quartz, and magnesia, to themselves and to metals.

A requirement for preheating and controlled cooling were two immediate and apparent prerequisites. Without preheating welds invariably show evidence of cracking. Therefore two preheating techniques have been used, electron-beam heating and resistance heating. Resistance preheating techniques have been more successful because of more precise control. The heating element is a tungsten filament wound on a zirconia coil form. Multiple-heat shields of tantalum and molybdenum surround and cover the unit except for a small opening at the top to allow optical viewing and welding. Weld specimens are supported on a tungsten platform inside the coil form. Temperatures as high as 3300°F have been obtained and field effects on the electron beam have been reduced.

Welding speed is quite critical and best results are achieved at higher travel rates. Lower speeds (15 ipm) produce a glassy structure. Although no cracks are evident in the weld zone, flexural testing has indicated that glassy welds are relatively weak. The tendency toward formation of glassy welds increases in the less pure aluminas.

Figure 1.28 illustrates a weld made in 96% alumina, with an 1800°F preheat. The increased welding speed (30 ipm) has limited glass formation. The advantage of higher welding speed appears to result from the decreased time that the weld zone was above the 1800°F preheat temperature. This limited the time available for segregation of the glassy constituents and for grain growth.

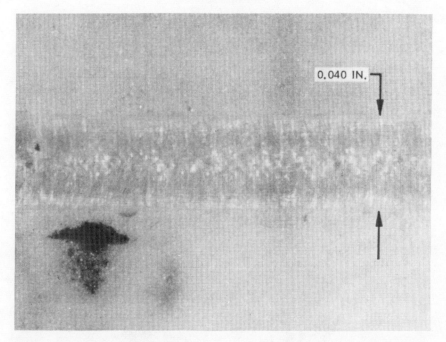

Figure 1.28 Al$_2$O$_3$ weld-bead surface (1800°F preheat, 90 kV, 2 mA, 30 ipm, 0.020 in. transverse oscillation) [42].

Flexural bend testing of welded ceramics has indicated that relatively good joint strengths can be obtained. Under simple beam bending the base ceramic achieved a surface stress of approximately 50,000 psi. Electron-beam welded joints withstood as much as 19,700 psi. Retention of 40% of the base strength is very good.

Limited tests have been conducted on the joining of alumina to metals. Using techniques similar to those used in pure alumina welding, crack-free joints have been produced between 96% alumina and tungsten, molybdenum, and columbium. Undoubtedly the successful joints resulted in part from the relatively closely matched coefficients of thermal expansion.

Various joint characteristics are primarily dependent on the beam location. Glassy-phase joints result from impingement of the beam directly on the ceramic member. Locating the beam to fuse the metallic member adjacent to the ceramic has produced interdiffusion of the two materials and yielded stronger joints.

Molybdenum feed-throughs welded into alumina electron-tube supports are illustrated in Figure 1.29. The welds were accomplished by circular

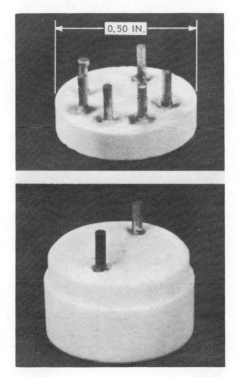

Figure 1.29 Electron tube molybdenum feed-throughs [42].

oscillation of the beam about the base of the 0.040-in. diameter pins. Joints of this type have been tested and were helium-leak tight.

Dissimilar Metals

In the fabrication of welded assemblies with the electron beam, it is not essential that welded members be of identical material. In many applications of electron beam technology dissimilar materials are joined as effectively as identical materials. In some instances problems arise as a result of differences in melting temperatures and mutual solubility, but generally these can be overcome through proper tooling or by the manner in which the electron gun is handled. For example, aluminum and stainless steel comprise a difficult combination because their melting temperatures are 1200 and 2600°F, respectively. Beams of sufficient density to melt stainless steel will reduce aluminum to a plasma, thus making any direct joining of the two impossible. This problem is resolved by concentrating the beam on the stainless steel at an angle at a short distance from the joint. This

Materials

enables the stainless steel to act as a high temperature soldering iron that melts the aluminum. The resulting joint is a brazed type of low efficiency, but it proves effective for many purposes.

Problems arising from mutual insolubility of two materials can also be solved by the manner in which the beam is positioned. One example is copper and titanium which have limited solubility with each other and form mechanical mixtures if melted together. Strict control of melting of these two materials by the electron beam is necessary because a high percentage of titanium in the mixture causes the joint to become excessively brittle and subject to transverse cracking. If the percentage of copper is high, however, a good weld can be produced. Additional difficulties that must be resolved arise from the fact that heat applied at the juncture of the two metals will be dissipated unevenly since copper has a higher thermal conductivity than titanium. As in the preceding example of stainless steel and aluminum, this problem can be overcome by directing beam impingement on the copper at an angle to the joint. This compensates for differences in rate of heat dissipation while providing a high percentage of copper in the joint. Virtually all such problems in the welding of dissimilar materials can be overcome by varying the angle of beam impingement.

Table 1.3 lists numerous combinations of dissimilar metal joints. Figure 1.30 illustrates a series of various dissimilar metal joints.

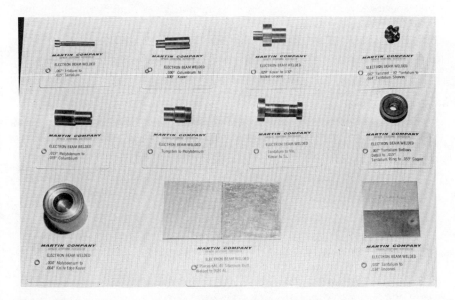

Figure 1.30 Numerous dissimilar metal combinations in various joint configurations.

The application of the electron beam for welding dissimilar metals has been particularly successful in the fabrication of turbine wheel and shaft assemblies. They use a cast wheel configuration of GMR 235, a high-nickel cast alloy welded to a shaft of SAE 8620 material. The shaft is joined to the turbine wheel in a butt-type joint. As a result the weld nugget properties are totally dependent on the realloying of the GMR 235 alloy with the SAE 8620 shaft.

Current usable applications include various refractory metals, iron and nickel-base superalloys, platinum and phosphor bronze to copper and beryllium copper to steel.

Thus unusual and useful combinations of bimetallic joints have been welded and used. The following paragraphs and concluding references [21, 43] should enable the reader to decide what he can and cannot join by electron-beam welding.

Preparation of useful bimetallic weld joints requires that the constituents of the joint do not form undesirable products while molten or during cooling. The usefulness of any weldment, however, is based on the properties required for the joint; therefore some normally unacceptable bimetallic combinations may be usable, depending on the joint requirements.

All bimetallic joints listed in the chart in Figure 1.31 are those made without cracks or porosity. Furthermore, all joints had sufficient strength and/or ductility to make them useful for a wide variety of applications.

In order to evaluate the possible usefulness of a bimetallic combination not specifically listed in the chart, a theoretical examination can be performed relying on binary equilibrium diagrams. These diagrams indicate the composition of the various phases that are formed when two elements are melted together (as in a weld joint). Many of these phases are ductile, whereas others are extremely brittle and will fracture under application of very small stresses. These diagrams do not indicate the strength of any of the phases; thus the usefulness of a joint, based on load-carrying capacity, cannot be predicted. However, those phases that have no ductility can be recognized from binary equilibrium diagrams. This criterion is the basis for the chart in Figure 1.31. Any combination forming one or more intermetallic compounds (which generally are devoid of ductility) are indicated as being undesirable bimetallic combinations.

In the accompanying chart the various binary combinations are characterized by the structures produced.

S   The combinations that form "solid solutions" generally have high ductility; the strength is generally equal to or greater than that of the weaker of the two constituents.

C   The combinations that form "complex structures" can have almost

Ag	—	SILVER
Al	—	ALUMINUM
Au	—	GOLD
Be	—	BERYLLIUM
Cd	—	CADMIUM
Co	—	COBALT
Cr	—	CHROMIUM
Cu	—	COPPER
Fe	—	IRON
Mg	—	MAGNESIUM
Mn	—	MANGANESE
Mo	—	MOLYBDENUM
Nb	—	NIOBIUM (COLUMBIUM)
Ni	—	NICKEL
Pb	—	LEAD
Pt	—	PLATINUM
Re	—	RHENIUM
Sn	—	TIN
Ta	—	TANTALUM
Ti	—	TITANIUM
V	—	VANADIUM
W	—	TUNGSTEN
Zr	—	ZIRCONIUM

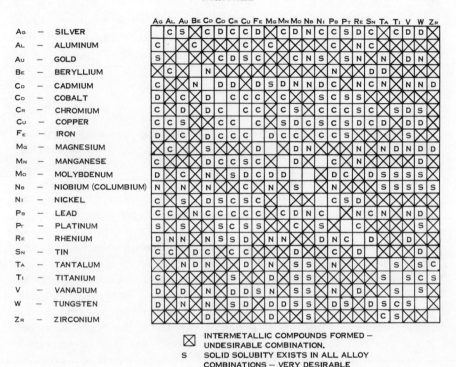

- ⊠ INTERMETALLIC COMPOUNDS FORMED — UNDESIRABLE COMBINATION.
- S SOLID SOLUBITY EXISTS IN ALL ALLOY COMBINATIONS — VERY DESIRABLE COMBINATION
- C COMPLEX STRUCTURES MAY EXIST — PROBABLY ACCEPTABLE COMBINATION
- D INSUFFICIENT DATA FOR PROPER EVALUATION — USE WITH CAUTION
- N NO DATA AVAILABLE — USE WITH EXTREME CAUTION

Figure 1.31 Chart of bimetallic weld joint combinations [43].

any ductility and may consist of discrete grains of each of the two constituents of any number of new phases. Use of these combinations generally would require an evaluation of the joint, but should be useful for many engineering applications.

D The combinations that have not yet been fully documented must be examined with care; none has been reported to form brittle intermetallic compounds. The lack of complete data for these systems makes them suspect and investigation of their properties is suggested.

N No data are available and these combinations will require careful study before use.

X The combinations that form brittle intermetallic compounds should be joined only when the weld is not subjected to stresses or when no mechanical strength is desired in the joint. These joints may crack during cooling from fusion temperatures as a result of cooling stresses [43].

Space-Gun Welded Materials

The electron-beam gun for welding in space shown in Figure 1.9 has made preliminary welding tests with the following typical penetrations: 0.270 in. in 304 stainless, 0.205 in. in 2219 aluminum, 0.072 in. in TZM molybdenum, and 0.255 in. in titanium of better than 99% purity. In addition to tests of maximum penetration, sample butt, lap, and T-welds have been made to evaluate the effect of welding parameters on the weld zone characteristics. The parameters considered included working distance, beam current, accelerating voltage and welding speed; the sample welds were inspected visually, radiographically, metallographically, and by dye penetrant.

In all the materials tested and simulated in-space electron-beam welds had greater depth-width ratios and smaller grain sizes than generally achieved in conventional fusion welding of comparable specimens. Wider fusion zones could have been produced by varying the beam focus, but it was felt that the important point for in-space welding was to achieve a reasonable depth-width ratio, for the fusion area is related directly to the energy and power requirements. Test results have confirmed that electron-beam welding offers maximum efficiency and minimum power requirements for in-space welding. The ultimate tensile and yield strengths of the stainless steel and aluminum weldments nearly equaled the strengths of the base metal.

JOINT DESIGN—TEST RESULTS

Joint Design

As with any welding process joint design for electron-beam welding must receive careful attention. Since the process is used to produce extremely narrow welds, it must be remembered that the beam is also very narrow. It is therefore mandatory that the preparation and fixturing of the workpiece are accurate.

The most useful characteristic of the electron beam is its power of penetration. Figure 1.32 shows some of the standard weld joints that can be made with an electron-beam welding system. Even when no joint exists on the accessible side of the material the beam penetrates the solid top piece to reach the hidden member.

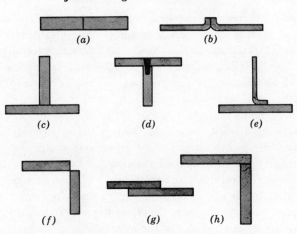

Figure 1.32 Weld joints that can be made with electron–beam welder: (a) butt, (b) standing edge, (c) straight T, (d) blind T, (e) modified T, (f) open corner, (g) lap, and (h) closed corner.

It is important to remember that the electron beam itself has infinite stiffness and can be aimed in any direction to make horizontal, vertical, overhead, circular, or flat welds. Because of its inherent stiffness, the beam produces a weld zone that assumes the same angle as the beam direction. This directional characteristic improves the joint efficiency when the part is loaded in tension because only a small amount of weld metal is involved for each increment of loading.

Studies of other welding processes show that most failures do not originate in the weld, but rather in the zone immediately adjacent to the weld, usually referred to as "the heat-affected zone." Because of its extremely small focal spot, electron-beam welding can dramatically reduce the heat-affected zone in weldments. The size of the spot can be varied, however, by simply changing the focusing current.

To make full use of the electron-beam process it should be kept in mind that welds of greater width can be obtained for joints that are to be loaded in tension. Deep, narrow welds are often more applicable to true butt joints or joints loaded in shear because the volume of cast material is reduced. However, the shear strength of metal is lower than its tensile strength, and the possibility of joining members in tension rather than in shear should be considered.

In welded joints stresses are introduced during welding by the unequal expansion and contraction of the molten weld zone, the heat-affected zone, and the unaffected base metal. Therefore the electron-beam welding process, which appreciably reduces the heat input to the material, also minimizes the origin of such stresses. Characteristically, the electron beam produces welds that are rectangular as compared with the triangular shape of conventional arc welds. The configuration of the welds minimize or, in many instances, eliminate angular distortion of the welded parts. With distortion held to reasonable limits, the parts may often be welded in the finish-machined condition.

In arc-welded T-joint or corner joint, welding stresses acting on the two members frequently present distortion and design problems. These stresses are primarily the result of the unequal shrinkage of the fillet face and root. To overcome this problem the joint is often designed with opposing fillets that balance, but lock in, the stresses. A better design would place the weld penetration more deeply in the material so that the penetration approaches the neutral axis of the joint.

It would be highly desirable to make a T-joint the same way that an electron-beam butt joint is made. When two very thin materials are being joined, this is sometimes possible, but, for heavy materials, the optimum joint could be made only by forging or machining the horizontal plate to provide a land for the vertical member to "butt" against and be welded. (See Figure 1.33a.) This solution, however, is not as practical as simply producing a deep penetration weld through the horizontal member into the neutral axis of the vertical member. (See Figure 1.33b.)

The two welds in Figure 1.33 are optimum for joint efficiency. Since they are 90 degrees displaced, an angular weld position that approximates either condition also approximates the ideal weld. Although this position is the same as for a conventional arc-fillet weld, the advantage of the

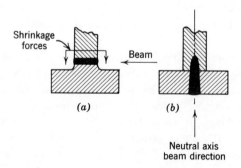

Figure 1.33 Ideal electron–beam welded T-joints. At (*a*), vertical member is butted to machined or cast land on horizontal member and standard electron beam butt weld is made. At (*b*), electron beam is directed through horizontal member along neutral axis of vertical member.

Joint Design—Test Results

Weld No.*	E_b, kV	I_b, mA	I_{fc}, A	Travel, ipm	Direction of Welding	Depth of Weld, in.
64911-1	27.0	260	9.8	15.4	Vert. up	1.3
64911-2	27.0	260	9.8	11.3	Vert. up	1.5+
64912[b]	27.0	255	9.8	15	Vert. up	1.3
64914[c]	28.0	250	9.8	15	Vert. up	1.3

* Corresponding welds appear below: Kroll's reagent: X1 (reduced 40% on reproduction):

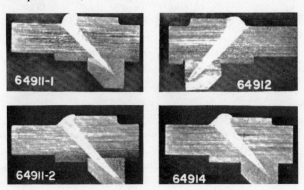

[b] Specimen with 20 mil gap.
[c] Specimen weld alternate station.

Figure 1.34 Electron–beam welds in simulated tube-to-head seam of navol tankweld penetration [27].

electron-beam weld is that by deep penetration the weld will have little or no fillet and will be lighter than a larger fillet weld having little penetration. This joint is illustrated by the 5254 aluminum welds in Figure 1.34.

The joint stresses discussed previously have been primarily transverse in nature. Electron-beam welding, however, can also help to solve some of the problems of longitudinal stresses. The amount of material melted is less and therefore the shrinkage in the longitudinal plane is less. Also, the beam may be used to prestress the joint. This is accomplished by defocusing the beam so that it does not melt the joint, but rather heats the area near the joint, expending the metal in a longitudinal direction. After welding, the cooling weld metal and the heated margins contract more uniformly.

In lap joints unfavorable stress conditions can also be relieved. To arc weld a lap joint, when both sides of the material are accessible, a double fillet is used. This same joint can be duplicated with an electron-beam weld with only one surface accessible by producing a double fillet effect with a penetration weld at each free edge.

The strength of a joint should be determined by its design. The addition of filler metal, unless a reduction in thickness is involved, adds nothing useful to the strength of the joint but increase its weight. In many applications electron-beam welding eliminates the need for filler metals, unless a situation occurs when a filler metal is used to create more favorable metallurgy in the weld. As an example, 713C, a vacuum-casting nickel alloy, when welded to itself, develops a condition that results in microcracking. However, the insertion of a strip of Udimet 500, a nickel-base super alloy, of proper thickness, into the joint produces an apparently crack-free weld. Usually elimination of filler metal reduces the effective shrinkage forces acting on the joint as does the placing of the welds as near as possible to the neutral axis of the assembly. The closer the weld is to the neutral axis, the less is the leverage exerted to pull the section out of alignment.

Joint Gap

A process that generates heat by electron bombardment creates no heat where there is nothing to bombard. That is, the electron beam must be intercepted by a physical body to create heat. Therefore gaps along the length of the joint result in reduced heating immediately at these discontinuities because a portion of the beam is not intercepted. To compensate for this loss the diameter of the beam must be increased to bridge the gap.

The size of gap that can be tolerated depends largely on the thickness of the material and on the fluidity and coalescence of the melted material. Joints in foils (less than 0.005 in. thick) may require a perfect match or even a slight overlap. Metals up to 0.500 in. thick should be aligned to less than 0.005 in. gap. Heavier materials may tolerate slightly larger gaps. These values are only approximate and do not provide any allowance for the use of special techniques. Table 1.11 shows the results of a study on gap size and the marked effect on strength.

Table 1.11   Effect of Gap on Strength of Electron-Beam Welds.

17-7 PH Annealed	0.020" Thick
Parent Metal Strength	125000 psi
Gap .....001"	123300 psi
.....002"	115800
.....003"	116100
.....004"	111000
.....005"	103200
.....006"	81850

Table 1.12 Tensile Test Results: Electron–Beam Welded Dissimilar Turbine Alloys.

ALLOY COMBINATION	ULTIMATE STRENGTH (psi)	YIELD STRENGTH (psi)	ELONGATION (% in 2")	REDUCTION IN AREA (%)	FAILURE ** LOCATION
U 500/713C	102,600*	97,300	1.5	5.5	713C BM
U 500/713C	103,000*	96,800	5.5	19.7	713C BM
U 500/713C	110,800	100,500	4.0	9.3	713C BM
U 500/713C	113,000	99,000	5.0	8.6	713C BM
A286/713C	87,600*	84,500	2.5	10.1	A286 HAZ
A286/713C	107,000*	86,600	11.0	17.5	A286 BM
A286/713C	107,300	93,700	9.5	18.2	A286 BM
A286/713C	106,000	92,000	6.5	9.4	A286 BM
A286/713C	110,000	84,300	8.0	11.6	713C BM
A286/713C	106,200	87,200	10.5	24.2	A286 BM

*Specimen extracted from weld overlap

**BM - Base Metal
 HAZ - Heat-Affected Zone

Test Results

Some mechanical property test results are given in the previous section; therefore only several metals are listed below and consideration is given to varied mechanical properties.

STEEL AND SUPERALLOYS

A comparison of hardnesses in the base metal, the weld joint, and heat-affected zone on three alloy steels, AMS 6265, 6470, and 6415, reveals little variation across the weld joints and in some instances the hardness was higher than the base metal.

The potential advantages of electron-beam welding have been evaluated for the fabrication of bimetal wheels. A286 is generally used as a disc alloy, 713C is used as a blade alloy, and, depending on the application, U-500 may be used for either the disc or the blades. Spin testing of electron-beam welded wheels produced failures at approximately three times service requirements.

Mechanical property tests indicated that electron-beam welds of A286 to 713C and of U-500 to 713C (all in cast form) retain excellent tensile and stress-rupture strengths. All the test results exceeded current fusion welded specification minimums by significant margins. (See Table 1.12.)

REFRACTORY METALS

The second generation of electron-beam welded columbium alloys exhibit improved ductilities; Table 1.13 summarizes the elevated weld properties of several columbium alloys.

ELECTRONICS

The reliability of electron-beam welding has been demonstrated in terms of the mechanical and electrical characteristics of interconnections in the as-welded condition and also after environmental exposure. Pull strength tests have been carried out on copper ribbons welded to the metallized edges of individual wafer assemblies simulating conditions in a microassembly stack. In this instance the pull tests were carried out on weld pairs.

Table 1.13 Weld 2200°F Tensile Properties. Values Reported Are Average of Two Tests.

Material	Thickness, in.	UTS, $\times 10^{-3}$	YS, $\times 10^{-3}$	Elong, %	Weld process[a]	Location of fracture[b]
FS-85	0.060	41.2	32.1	5.0	TIG	FZ
		43.3	35.0	4.0	EB	FZ
FS-85	0.020	35.8	31.3	2.0	TIG	FZ
		42.9	33.1	2.0	EB	HAZ
B-66	0.060	44.8	32.3	32.5	TIG	BM
		45.9	32.5	33.5	EB	BM
B-66	0.020	39.8	29.4	30.5	TIG	BM
		41.3	31.2	43.5	EB	BM
D-36	0.050	13.1	13.1	56.5	TIG	BM & FZ
		14.5	14.4	11.5	EB	BM
D-36	0.020	11.7	11.4	100+	TIG	BM
		12.1	11.3	100+	EB	BM
				16.8	TIG	BM[c]
Cb-752	0.060	34.9	25.1	22.5	TIG	BM
		34.5	24.9	19.2	EB	BM
Cb-752	0.030	27.4	24.1	29.4	TIG	BM
		26.3	22.0	31.9	EB	BM

[a] TIG—gas tungesten-arc; EB—electron beam. [b] FZ—fusion zone; HAZ—heat-affected zone; BM—base metal. [c] Pulled at 10 times loading speed for comparison

Joint Design—Test Results

The salient points of electron-beam welding electronic components show the following from the standpoint of mechanical and electrical characteristics:

1. Pull strength. None of the welds failed in the fusion zone or at a load less than 200 g. Statistical analysis of a relatively large sample population demonstrated at a 90% confidence level that less than 2.6 welds out of 200,000 would fail below 200 g.
2. Weld resistance. It was demonstrated at a 90% confidence level that less than three out of 1000 weld pairs would exhibit a resistance of 0.010 ohm or greater.
3. Thermal shock. No degradation of pull strength was observed. From the statistical evaluation of the resistance measurements before and after exposure, it was demonstrated that no more than 4.5 out of 1,000,000 would exhibit an increase of 0.001 ohm or greater.
4. Extreme ambient temperature tests. All weld pair resistance measurements from -55 to $200°C$ were below 0.010 ohm.
5. Vibration test. Microassembly stacks vibrated according to Method 204B of MIL Std 202A did not exhibit any visible degradation.

APPLICATIONS AND FUTURE POTENTIAL

Characteristics of electron-beam welds as shown in numerous previously cited figures suggest even more exciting applications for the process. The uses, needs, and applications range from nuclear, aircraft, missiles, space, electronics, underseas to commercial cars, saw blades, etc. Electron-beam welding has also proven to be an excellent salvage tool for expensive machine components and production items. With all these applications and they are continually increasing, electron-beam welding is not the complete panacea for all the welding industry. Just as there are applications for high and low voltage machines, so the same is true for electron beam and all the other welding processes.

Nuclear

Electron-beam welding has been applied to the fabrication of fuel elements for pressurized water reactors. The distinct advantages of the electron beam weld are the following:

1. Vacuum environment prevents contamination of the weld joint.
2. Vacuum environment insures that the weld metal possesses good corrosion resistance and mechanical properties.
3. High concentration of weld heat. The extreme concentration of ther-

mal energy is an important advantage of electron-beam welding of plate-type fuel elements. It results in a deep, narrow weld compared with the wider fusion zone that characterizes the tungsten-arc process. Complete weld penetration is required, but excess penetration is undesirable, especially if it extends over the heat transfer area of the fuel. With the high depth-to-width ratios obtained during electron-beam welding obliteration of adjacent seams is avoided. Welds made by the tungsten-arc process may overlap adjacent seams; accurate tracking then becomes difficult.

4. High concentration of heat reduces shrinkage and thermal distortion. For example, in fuel element welding, total transverse shrinkage per weld seam is only 0.001 to 0.0015 in. for electron-beam welds when compared with 0.010 to 0.012 in. for tungsten-arc welds.

The potential advantages of electron-beam welding for fabricating PWR-type fuel elements have been extrapolated into cost estimates. By taking advantage of the various benefits of the process, including reduced weld shrinkage and distortion, it appears possible to reduce direct labor charges by 35 to 40% and overall costs of a typical fuel assembly by 8%.

Ion propulsion systems with their need for refractory metals has made use of the electron-beam process. Contact ionizers with a molybdenum jacket have been welded to a heavy-walled tungsten cylinder.

Electron-beam welding is ideally suited for fabrication of ion propulsion systems that includes joining of not only similar refractory materials but also dissimilar refractory materials. Complex, high-restraint configurations have been successfully welded without cracks or leaks; distortion was held to 0.002 in., although some cases required as much as 100 in. of welding for fabrication. Engine life tests of ion propulsion systems have demonstrated that the weld and porous tungsten in the weld vicinity are capable of withstanding numerous thermal cycles and long usage without any evidence of failure or degradation.

Underseas

Experimental shells with electron-beam welded inner ribs of HY-80 steel (Figure 1.35) are being actively considered and pursued in tests as a submarine hull structure for underwater uses.

Motors–Wheels–Gears–Engines

Electron-beam welding of critical aircraft engine parts such as highly loaded power train and accessory gears have provided the engineer with a

Figure 1.35 Full penetration through electron–beam welds in HY-80 steel cylinder.

new design concept to reduce cost, decrease weight, increase reliability, and solve complicated design problems.

Current design stress levels for double helical aircraft gears require that the gear teeth be hardened after cutting. The desirability of maintaining the smallest possible gap between helices for minimum size and weight precludes finish grinding of the gear teeth after hardening because of the large gap required for grinding wheel clearance. The designer is limited in his choice of the gear material to a nitriding steel to control hardening distortion. When the gear must have integral bearing raceways, experience with nitrided steel as a bearing raceway material is limited as compared with the more conventional through-hardened or carburized bearing steels.

By using the electron-beam welding process, the design of a double helical gear may be improved in one of two ways:

1. The double helical gear teeth can be manufactured in two pieces from carburizing steel, cut, case-carburized, hardened, and ground separately, and electron-beam welded at the center.
2. The gear teeth can be made from nitriding steel, cut, shaved, nitrided, and the bearing raceways made for carburizing steel and electron-beam welded to the gear tooth section.

In the first choice the necessity of a large gap for grinding wheel clearance is eliminated and in the second choice the desirable properties of two different steels are utilized on an integrated piece.

On large double helical gears (20 to 22-in. diameter) feasibility electron-beam welding tests have indicated the potential for joining the two finished ring gear details in one operation and subsequently welding this subassembly to the web detail.

Frequently, accessory cluster gears requiring high levels of accuracy are joined together with splines, locknuts, or bolt flanges in order to avoid the necessity of providing space for grinding wheel clearance. Electron-beam weldment design has eliminated the splines and bolts in engine accessory gears by finishing the gears separately and then electron-beam welding the finished gears to their shafts. The cluster gear for a modern turbine engine accessory gearbox is shown in Figure 1.36.

Electron-beam welding is ideally suited to manufacturing simple shaft-gears. Electron-beam welding the gear blank to the shaft results in savings of material. A costly one-piece die forging has been replaced by a two-piece

Figure 1.36 Accessory cluster gear. Electron beam welded at webs. (Arrows point to electron–beam welds) [44].

Applications and Future Potential

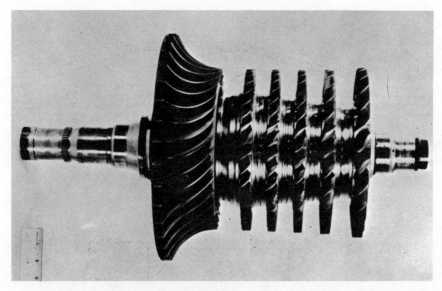

Figure 1.37 Bladed electron–beam welded titanium roter [45].

construction consisting of a simple pancake forging and a piece of bar stock. Further cost reduction can be attained by making the shaft from a less expensive material than the more costly gear material.

The properties of bimetal turbine wheels were shown previously in Table 1.12. In addition, however, the principal advantage of any bimetal design is the capability to use materials of superior elevated-temperature strength for the blade section of the rotor while manufacturing the disc from materials of good ductility and strength at the lower operating temperature of the disc. Thus a bimetal rotor has optimum material properties throughout, and cost is reduced by usage of the expensive, high temperature alloys only in the blading.

Jet engine components, rotor/shaft assemblies, engine nozzles, and labyrinth seal assemblies are currently being electron-beam welded. Electron-beam welding has provided a better end product made in less time than with previous joining methods.

Two new applications for electron-beam welding are the compressor rotor and power shaft which are components of a rotating gas turbine engine. The bladed rotor, 6AL-4V titanium, is shown in Figure 1.37.

Silicon-iron laminates are stacked together to make up the stator core assembly for electrical actuator motors. Conventional welding techniques for holding the laminates together at a high rate per minute were unsuccessful. The uneven and inconsistent weld penetration and width resulted in nonuniform magnetic paths. The electron-beam process allowed the penetration to be controlled to within 0.001 to 0.002 in. and resulted in small uniform welds that did not adversely affect the magnetic properties of the core.

Aircraft–Missiles–Re-Entry Vehicles–Boosters

Development programs for electron-beam welding encompassing a variety of diverse shapes, including tees, crosses, channels, angles, and combinations thereof for the SST (Supersonic Transport Airplane) are currently under-

Gage pressure: 2830 psi
Hoop stress $(Pr)/t = 240,000$ psi
Tensile Test Results:

Tensile strength, psi	Yield strength, psi	Elongation, % in 2 in.
206,000	164,500	7
214,500	173,400	7

Avg. UTS (210,000) × 115% = 240,000 psi

Figure 1.38 Electron–beam welded H-11 steel 9.5-in. diameter pressure vessel after burst [23].

Applications and Future Potential 89

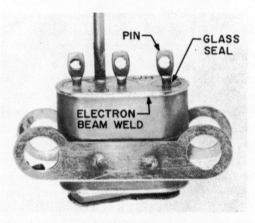

Figure 1.39 Electron–beam welded high temperature switch [25].

way with the results to be utilized in designs and manufacturing within the next two years.

Titanium forgings have been joined by electron-beam welding to form the rotor hub 10 ft in diameter for the new armed helicopter-airplane.

Pressure vessels are a constant companion to any space rocket, booster, re-entry vehicle, or orbiting laboratory. Electron-beam welding has entered this field and done exceptionally well. Figure 1.38 shows the results of burst tests on an H-11 pressure vessel. Other metals such as 17-7PH steel, 6061 aluminum, and 6AL-4V titanium have been electron-beam welded as pressure vessels.

A 5083 aluminum rocket engine injector was recently converted from GTA welding to electron beam. The 23-in. diameter part was designed so that the twenty-one fuel and oxidizer channels did not have a common joint. Any interchannel leakage could be disastrously explosive if the propellants mixed internally. Electron-beam welds hermetically closed the oxidizer channels of the inner face.

Electronics

Wave guide antennas for the X-20 airplane and the Scimitar antenna for the Apollo command module have required the use of electron beams.

Figure 1.6 illustrates the tooling used in the application of electron-beam welding for electron tubes and Reference 21 is an excellent source for any further information on this subject.

Electron beams are useful in hermetically sealing transistors, welding transistor junctions, and replacing soldered connections. In this application electron-beam welding provides a mechanically and electrically sound joint containing no residual corrosive flux and does not affect the spring characteristics.

Increasing reliability requirements demand that electronic circuits and other electrical devices, such as microswitches, be hermetically sealed without the danger of corrosive impurities remaining within the capsule. Electron-beam welding accomplished in a vacuum provides an ideal process for such encapsulation. A high temperature switch (Figure 1.39), used in a high-performance aircraft, requires an all-welded assembly because flux residues associated with soldering causes deterioration of internal components and subsequent changes to the dielectric properties of the switch itself. In this instance the 17-7 PH cover assembly was electron-beam welded to a cold-rolled steel base in a vacuum and the unit subsequently backfilled with an inert gas. The welds were 0.012 in. wide and were located within 0.030 in. of glass hermetic seals that were not damaged during welding. Hundreds of these switches have been successfully processed.

Repair Welding

Electron-beam welding is accumulating an impressive history of cost savings through repair of castings, forgings, and fabrications. For example, a rocket engine manufacturer reported a savings of $100,000 by the repair of an assembly that had failed because of a high operating temperature. A brazed Hastelloy "C" section was successfully electron-beam repair welded without removal of the brazing alloy.

Some of the repairs requiring salvaging have included the following:

1. Oversize holes that can be saved by welding in a plug and remachining. Three similar aircraft housings (6 AL-4V titanium), Figure 1.40, which contained a good deal of costly and complex machining were involved. Unfortunately, one of the bored holes, approximately 0.200 in. in diameter, was oversize. A close-fitting titanium plug of the same alloy was inserted in the hole, electron-beam welded, with the results as seen on the right side of Figure 1.40.
2. Some cracks can be repaired by scanning over the joint with a small piece of filler wire. An example of this was a magnesium turbine nose section that experienced fatigue cracking during operation of the engine. Sound repair of these welds was accomplished by electron-beam welding. Here the capability of electron beam to make a very narrow weld 12 in. below the surface of the part was a necessity. Furthermore, it was vitally necessary to locate precisely the position of the weld before turning on the beam. Otherwise, extensive damage of the part would have resulted.

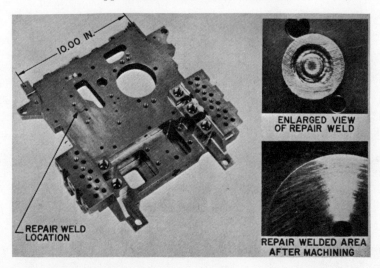

Figure 1.40 Repair of C120AV titanium housing by electron–beam welding [23].

Simpler, less costly items, such as tooling and fixturing, also can be repaired using electron-beam welding. An example is a fabricated aluminum drilling fixture which is quickly and easily repair welded without distortion. Here the fixture was fabricated by normal arc-welding techniques. However, on completion it was found that one of the welds had cracked. At that time the locating holes had been positioned to close tolerances, and repair by normal welding would have resulted in extensive distortion. Electron-beam welding solved the problem.

3. Lugs which have been forgotten from a drawing during the design stages of a program can be welded onto finished parts without danger of distortion. Precisely machined AMS 5616 control-spider housings had been manufactured on a limited production basis before complete development testing that revealed the need for improved structural rigidity. Unfortunately, the area requiring stiffening was immediately adjacent to a hole finish-machined to very close tolerances. However, with electron-beam welding it was possible to add a stiffening member without affecting significantly the dimensions of the hole. During development testing it was found necessary to add two bosses to a large complex 431 stainless steel casting. Because of casting problems, schedule considerations, and costs involved, it was extremely important that a means be found to add these bosses to existing semifinished and finish-machined parts. Efforts to add the bosses by arc welding resulted

in cracking and excessive distortion. Fortunately, electron-beam welding solved the problem.
4. Wear surfaces that can be built up by melting shim stock onto machined areas. Some electron beam welding installations claim they save $30,000 a month by salvaging items that would ordinarily have to be scrapped. These claims are very valid as evidenced by Table 1.14, which indicates the savings realized from June through Decem-

Table 1.14 Savings Through Repair Electron–Beam Welding for a 6-Month Period [22].

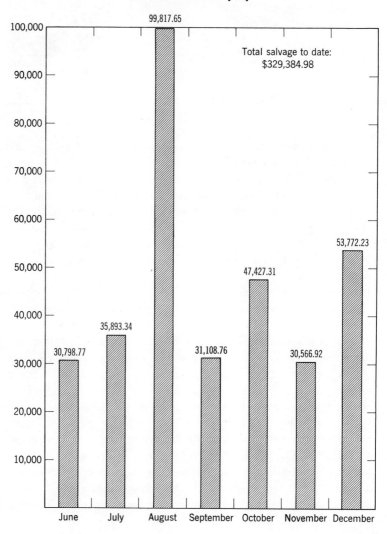

Applications and Future Potential

ber, 1962, of an aerospace manufacturer. These numbers are based on the value of the part at the time of rework minus the cost of the rework.

Since hundreds of case histories on repair and salvage welding could be listed here, the best source if further information is needed are the electron-beam equipment manufacturers.

Miscellaneous

1. Electrical connections for an entire computer memory array have been joined by an electron beam in a single pass. In a continuous pass precise welds connect terminals of ferrite core planes with those immediately above and beneath in a memory array. Conventionally, each terminal is welded separately.

    A ferrite core plane contains thousands of tiny magnetic doughnuts wired together. A number of these planes is interconnected in stacks to make a computer memory array.

    In welding the tips of a column of electrical terminals in the stack pass through the beam of electrons. The beam strikes each terminal and all four sides of the memory array are handled the same way.
2. A hammer unit that has two leaf springs is being joined to a hammer socket and hammer assembly today for the computing industry. Heat buildup is at a minimum and efficiency of the beam is well over 70%, and with the beam focused to a spot only 0.001 in. in diameter, the hammer and its socket are not distorted and the temper for the spring material is not affected by electron-beam welding.
3. Electron-beam welding has been used successfully in welding a thin foil seal over an explosive charge in the top cap of a liquid ammonia battery. Hermetically sealing the heat sensitive battery compartment by electron-beam welding provides a smaller battery than would be possible if sealing were done by other, more conventional welding processes.
4. Bimetal band saw blades fabricated by electron-beam welding cut faster and last longer than their more costly single-metal counterparts.

    Traditionally, heavy-duty band saw blades are machined from strips of hardened high-speed tool steel. Because of the rigidity of the tool steel, bending stresses in the blade are concentrated at the root of the saw teeth as the band wraps around drive wheels of the saw machine. It is in this area that blade failures usually originate. In addition, the tool steel bands cannot be drawn as taut as would be desired for high cutting speeds because of the material's rigidity.

These problems have been eased by using composite blades consisting of a narrow strip of a high-speed tool steel, M_2, which is electron-beam welded to a wide strip of low-alloy carbon steel, AISI 6150. The strip of tool steel is just wide enough so that the roots of the saw teeth are machined into the low-alloy carbon steel. Thus stresses caused by bending of the blade and impact loading of the teeth are applied to the more flexible metal and less expensive 6150 that handles flexing and tension and distributes the stresses more evenly. Again, high energy concentrating in a narrow weld zone leaves most of the steel unaffected metallurgically.

5. Shown in Figure 1.41 is an electron-beam welded diaphragm section made of 0.001 in. thick titanium, which is part of a hydraulic fluid flow control system in a missile; a similar type is used in the Apollo spacecraft. On command either pressure or an explosive charge ruptures the diaphragm to permit fluid flow in the line.
6. Electron beams have been used to weld components of infrared detector dewars, tantalum stings, and wind tunnel models for mach 15 and 20 simulations, thermionic heating devices (Figure 1.42), and the thirteen-layered columbium bellows in Figure 1.43.
7. Electron beams have recently been used as a heating and vaporizing source to improve the wettability of beryllium by braze alloys. Titanium was chosen as a surface activator to promote wetting. Vacuum

Figure 1.41 Electron–beam welded diaphragms.

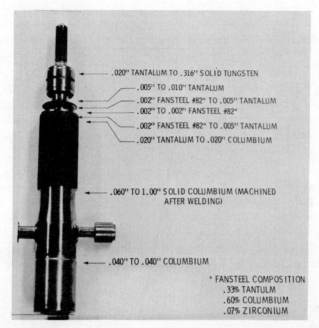

Figure 1.42 Electron–beam welded thermionic heating device.

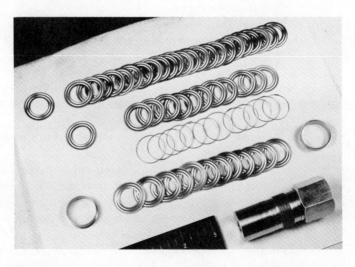

Figure 1.43 Details, welded columbium bellows, and holding fixture for thermionic generator.

evaporation of titanium onto the surfaces to be brazed was chosen as the most desirable method for applying the activator. Silver alloys produced relatively good joints with this method because the titanium layer and the short heating cycle prevented extensive reaction between the braze and the base metal. However, exceedingly good joints were also achieved using 1100 aluminum as the braze alloy. In addition, titanium has been successfully used to wet graphite and tungsten carbide. Wetting and flowing of the brazing alloy on titanium-coated base material occur by liquid metal penetrating, displacing, and tunneling under the vapor-deposited film.

FUTURE

What about the future of electron-beam welding? Future electron-beam welds will be x-rayed without removing the work from the vacuum chamber. A simple fixture positions the weld specimen and film in relation to a supplementary tungsten target which produces x-rays when hit by an electron beam. The method provides operators a fast check while parts are being welded—10 minutes versus several hours normally. Its principal advantage, however, is that it will control quality. The system will operate in the following manner. A specimen is placed in a fixture on one side of the vacuum chamber. Film packets will be placed immediately behind the specimen. When the beam is turned on the tungsten target, x-rays pass through the weld and expose the film. The film can be developed in 10 sec after removal from the chamber.

Once weld settings and position have been determined, the rest of the run can be finished with a high degree of assurance that quality will match that of the sample. To date the film does not provide ideal sensitivity. In time a 2% sensitivity will be reached by improved film.

The electron-beam space gun (Figure 1.9), however, now with its own power-pack mounted on the back of an astronaut will be in use as a space tool.

The other future trends on partial and nonvacuum welds are discussed in the special section which follows.

In conclusion, with high-vacuum systems large enough to accept automobiles, portable clamp-on systems for still larger work, and electron-beam welders capable of penetrating over 5-in. geometric limitations are rapidly disappearing for the designer considering electron-beam welding. He can now design weldments to a precision once achieved only by machining. Here is a tool that puts new perspective into his designs and broadens the horizons of weld applications. But with this increased sophistication comes the need for intelligent appraisal. Electron-beam welding is not for the unskilled; it is a precision tool. The weld design moves out of the hands of the artisan and into the hands of the technician. Joint detail becomes a problem of the designer supported with a development program. Continued success in production depends on rigid control. Despite this seemingly complex approach to weld problems, the low distortion of electron-beam welds can lead to great rewards through lower cost, faster production, or new products not possible by other joining techniques.

REFERENCES

[1] K. H., Steigerwald, "Electron Beam Milling," Third Electron Beam Symposium, 269–290 (March 1961).
[2] J. A., Stohr, "Electron Beam Welding in France," Third Electron Beam Symposium, 102–115 (March 1961).
[3] G. Burton, Jr., and .. L. Matchett, "Electrons Shot from Guns Make High Purity Welds," *Am. Mach.,* **103**(4), 95–98 (1959).
[4] G. Burton, Jr., and W. L. Frankhouser, "Electron Beam Welding," *Welding J.,* Research Supplement, 401s–409s (1959).
[5] J. R. Pierce, *Theory and Design of Electron Beams.* Van Nostrand, Princeton N. J., 1954.
[6] H. H. Schwartz, "Electron Beam Processes at Different Voltages," Second International Vacuum Congress, Washington, D. C., 1961.
[7] M. J. Druyvesteyn, and F. M. Penning, *Rev. Mod. Phys.* **12,** 87 (1940).
[8] O. C. Wells, Third Electron Beam Symposium, 291–321, (March 1961).
[9] H. H. Hoffman, "Electron Beam Welding at High Voltages," Third Electron Beam Symposium, 116–144 (March 1961).
[10] K. J. Miller, and T. Takenaka, "Electron Beam Welding," Welding Research Council Interpretive Report No. 100.
[11] *Welding Handbook,* Section III and Section IV, Fifth Edition, pp. 54.21, 54.6, 69.59 (1963).
[12] W. J. Farrell, "Electron Beam and Laser Techniques for Joining," ASM Golden Gate Metals Conference, 451–510 (January 1964).
[13] W. J. Farrell, "The Use of Electron Beam to Fabricate Structural Members," Creative Mfg. Seminars, ASTME Paper SP63–208 (1962–1963).
[14] United Aircraft Quarterly Beehive, Fall 1961.
[15] BTI "Production Designed" Electron Beam Welders, Bulletin 11-1 (September 1966).
[16] J. S. Stohr, *Fuel Elements Conference,* Paris, pp. 18–23, November 1957; and Le Vide **75,** 163 (1958).
[17] J. W. Meier, "High Power Density Electron Beam Welding of Several Materials," Second International Vacuum Congress, TP61-02, 1–25 (October 1961).
[18] J. W. Meier, "New Developments in Electron Beam Technology," *Welding J.,* 925–931 (November 1964).
[19] J. C. Sterling, Jr., "Electron Beams," Steel, S-1–S-5 (July 1966).
[20] G. V. Anderson, "Portable Electron Beam Vacuum Chambers Weld Large Structures," SAE-ASNE National Aero-Nautical Mtg., 676A, 1–6 (April 1963).

[21] M. L. Kohn, F. R. Schollhamer, and J. W. Meier, "Electron Beam Techniques for Fabrication and Assembly of Parts for Electron Tubes," Final Report 21067-pp 63-81-B, Contract No. DA-36-039-AMC-03625 (E), July 1, 1963 to March 31, 1966.
[22] M. M. Schwartz, "Electron Beam Welding of Refractory Metals," SAE 1962 Aero-Space Mfg. Forum, Los Angeles, Calif., October 1962.
[23] H. A. Hokanson, and J. W. Meier, "Electron Beam Welding of Aircraft Materials and Components," *Welding J.*, 999–1008 (November 1962).
[24] F. R. Schollhamer, "Microelectronic Joining Techniques by Electron Beam Processes," TP61-06, Fifth Symposium on Welded Electronic Packaging, Sunnyvale, Calif., August 1961.
[25] J. W. Meier, and F. R. Schollhamer, "Application of Electron Beam Techniques for Electronics," National Electronics Conference, TP62-15, 1–10 (October 1962).
[26] M. T. Groves, and J. M. Gerken, "Evaluation of Electron Beam Welds in Thick Materials," Technical Report AFML-TR-66-22, 1–324 (February 1966).
[27] J. F. Hinrichs, and P. W. Ramsey, "Electron Beam Butt Welding of 5254 Aluminum Alloy Pressure Vessel," *Welding J.*, 39–46 (January 1967).
[28] M. W. Brennecke, "Electron Beam Welded Heavy Gage Aluminum Alloy 2219," *Welding J.*, 27s–39s (January 1965).
[29] J. Van Audenhove, M. Meulemans, and D. Tytgat, "Electron Beam Welding of Sintered Aluminum," Fourth Electron Beam Symposium, 431–446, (March 1962).
[30] B. M. MacPherson, and W. W. Beaver, "Fusion Welding of Beryllium," WADD Technical Report 60-917, pp. 12, 24–29, 38, 54 (1960).
[31] W. T. Hess, H. J. Lander, and S. S. White, "Electron Beam Welding of Beryllium," Third Electron Beam Symposium, 185–187 (March 1961).
[32] R. E. Roth, and N. F. Bratkovich, "Characteristics and Strength Data of Electron Beam Welds in Four Representative Materials," *Welding J.*, 229s–240s, (May 1962).
[33] W. Schwenk, and A. F. Trabold, "Weldability of René 41," *Welding J.*, 460s–465-s (October 1963).
[34] R. E. Travis, V. P. Ardito, and C. M. Adams Jr., "Comparison of Processes for Welding Ultra-High Strength Sheet Steels," *Welding J.*, 9s–17s (January 1963).
[35] H. I. McHenry, J. C. Collins, and R. E. Key, "Electron Beam Welding of D6AC Steel," *Welding J.*, 419s–425s (September 1966).
[36] G. G. Lessman, and D. R. Stoner, "Welding Refractory Metal Alloys for Space Power System Applications," Ninth SAMPE Symposium, Paper #1-5, WANL SP-009 (November 1965).
[37] A. Rice, "Metallurgy and Properties of Thoria–Strengthened Nickel," DMIC Memorandum 210, October 1, 1965 (AD474854).
[38] R. E. Monroe, D. C. Martin, R. V. Wolfe, and N. Nagler, "Recent Developments in Welding Thick Titanium Plate," DMIC Memorandum 211, November 24, 1965 (AD477403).
[39] W. Schwenk, W. A. Kaehler, Jr., and J. R. Kennedy, "Weldability of Titanium Alloy sheets 6AL-6V-6Sn and 8AL-1Mo-1V," *Welding J.*, 64s–73s (February 1967).
[40] R. E. Monroe, and R. M. Evans, "Electron Beam Welding of Tungsten," DMIC Memorandum 152, 3–15 (May 21, 1962).

References

[41] S. Robelotto, "Electron Beam Welding of Refractory Materials for Ion Propulsion Systems," Fifth Electron Beam Symposium, 218–229 (March 1963).

[42] H. A. Hokanson, S. L. Rogers, and W. I. Kern, "Electron Beam Welding of Alumina, Aluminum, and Beryllium," Fourth Electron Beam Symposium, TP62-06, 464–495, (March 1962).

[43] *Electron Beam Welding Data Manual*, Hamilton Standard, Division of United Aircraft Corporation, April 1964.

[44] N. F. Bratkovich, R. E. Roth, and R. E. Purdy, "Electron Beam Welding-Applications and Design Considerations for Aircraft Turbine Engine Gears," *Welding J.*, 631–640 (August 1965).

[45] S. M. Silverstein, V. Strautman, and W. R. Freeman, "Application of Electron Beam Welding to Rotating Gas Turbine Components," Ninth SAMPE Symposium, Paper 1–3 (November 1965).

[46] E. F., Westland, H. G. Lohmann, "Vapor Plating Aluminum on Beryllium for Welding," *Welding J.*, 207–210 (March 1967).

Additional Section to Electron-Beam Welding

COLD CATHODE OR PLASMA ELECTRON BEAMS (EBMD OR PEB)

A series of tests was undertaken to determine the effects of hollow cathode discharges on the cleaning of glass substrates used for vacuum-deposited, thin film circuitry. A shortage of materials necessitated the substitution of copper screening for molybdenum sheet in the construction of hollow cathodes used. A new phenomenon was noted during a test using a screen-walled hollow cathode. It was observed that this cathode produced a well-collimated electron beam, quite unlike the normal hollow cathode discharge. A number of measurements of the volt-ampere characteristics of this discharge mode, along with a literature study, demonstrated that this effect was novel.

Some radio frequency measurements were made of the electron-beam mode and some interesting kmc oscillations were noted. A high power neon sign transformer was used with which it was possible to produce electron beams of sufficient power to evaporate holes through ceramic crucibles.

Thus a program was initiated to study and develop this novel electron-beam mode gas discharge. The perforated-wall hollow cathodes have been

Cold Cathode or Plasma Electron Beams

developed to the extent that electron beams of several kilowatts of power are available.

MECHANISMS OF PROCESS AND EARLY DEVELOPMENTS

The electron-beam mode discharge (EBMD) is a high impedance cold cathode discharge which is characterized by the production of one or more well-defined electron beams. A perforated-wall hollow cathode structure formed from fine mesh wire screening is used to produce the electron-beam mode discharge (Figure A1). This cathode, which is provided with an aperture at least three times the diameter of the screen perforations, is installed in a vacuum chamber whose pressure can be maintained at a fixed value between 1 and 1000 μ Hg. An anode is also placed in this vacuum at some distance from the cathode. The anode must lie outside the region normally occupied by the cathode dark space. Operating voltage for the EBMD has been reported from 0.5 to 150 kV with beam current in excess of 1 A.

Figure A1 Cold cathode electron gun welder.

Beam Formation

Formation of the plasma electron beam is associated with the interaction of ionized regions inside and outside the cathode with electric fields between its perforated surface and the plasma boundaries. Positive ions formed in the beam path by electron collisions with gas atoms play an important role in beam collimation by neutralizing the negative space charge which would otherwise cause beam spreading.

When the gas pressure and cathode voltage are adjusted properly, the beam mode of operation sets in and the electron beam path is marked by the luminousity of excited atoms in its path. Ordinarily, the gas density is too low to cause appreciable scattering of the beam. Beams up to 30 in. in length have been produced by this method with little spreading [1]. The focusing effect of positive ions in the beam path is an important factor in maintaining collimation [2]. Typically, the beam cross section is about the size of the exit aperture, although it can be made smaller by adjustment of pressure and cathode voltage.

At pressures and voltages outside the range of the beam mode a more or less diffuse glow discharge is obtained. In argon the beam mode is most easily supported at pressures from 3 to 10 μ with applied voltages from 5 to 10 kV depending on gas pressure and cathode characteristics. In lighter gases such as hydrogen and helium the beam mode is obtained at higher pressures (10 to 100 μ)

Measurements of beam power delivered to a shielded cup have shown that from 75 to 95% of the input power resides in the electron beam [1]. Presumably, the power not accounted for in the beam is consumed in maintaining the positive glow region surrounding the cathode and the plasma inside it. Some power is, of course, dissipated as heat in the cathode, but thermionic emission does not normally contribute electrons to the beam. More recent work has suggested a method for suppressing radial electron emission from a hollow cathode. A closely fitting concentric shield provides the solution. When such a shield is applied, no ions can reach the side walls. Any discharge that occurs between the grounded shield and the cathode is forced to take the long path between the outer surface of the shield and the inner surface of the cathode or points near the beam exit aperture. This situation confines the positive ion bombardment to the inner cathode surface and to the external surface of the face plate containing the aperture. With proper spacing of the end of the shield with respect to the plane of the face plate most of the ions can be focused on the aperture and thus can be made effective in generating secondary electrons inside the cathode.

As a net result of these phenomena a body of plasma fills the cathode, and beam formation takes place. By surrounding the cathode with the anode cylinder, or shield, beam efficiencies are increased by a factor of two or more because external electron losses can now occur only at the end of the cathode. These electrons are useful in ionizing the gas surrounding the cathode, thus providing additional positive ions.

In the shielded cathode a region of high voltage gradient surrounds the end of the cathode. This region is identified by reduced light emission and a roughly spherical boundary. The highly conducting gas outside the boundary of this so-called cathode dark space acts as a virtual anode and the electrons gain all but a small fraction of their energy as they are accelerated in the space between cathode and dark space boundary. As the gas pressure is increased, the dark space shrinks and the beam current increases, and the influx of positive ions is correspondingly increased.

Cathode Construction

Several different cathode designs have been used successfully. Some of them were made of wire mesh and thin perforated sheet metal [1]. For power levels of 2 kW or less a cylindrical cathode made from perforated stainless steel sheet has been found convenient. Construction is very simple, the parts being cut to shape and assembled by spot welding. Some cathodes are supported by a coaxially shielded lead in order to prevent a discharge from the supporting high voltage rod. High power cathodes are usually made from molybdenum wire mesh using solid sheet molybdenum end plates. Assembly is by means of spot welding, using thin tantalum foil between the molybdenum parts to facilitate the welding.

The first demonstration of a spherical cathode produced an electron beam with only 77 W. This cathode was developed in four stages to produce a high power electron gun capable of delivering 4.6 kW for welding. (See Figure A1.) The important design problem which the cathode met was that it was operational at the high power levels and was capable of dispersing the positive ions created by the electron beam. These positive ions will normally travel along the path of the electron beam and initiate the hollow cathode mode. The dispersal of the positive ions can be accomplished in two ways. A negatively charged ring surrounding the electron beam and a concave cathode surface surrounding the cathode aperture both have the effect of dispersing the positive ion beam. The final cathode design uses a combination of both these ideas. The cathode shown in Figure A1 has operated in a gas atmosphere of 3 to 4 μ Hg pressure.

Grid Control

It has been found that the floating grid always assumes a positive potential (several hundred volts) with respect to the cathode. When a variable dc bias has been applied from an isolated potentiometer, the beam current could be varied over a wide range without affecting its focus.

Even though substantial grid current is drawn, the grid power is only a few tenths of a watt because of the low grid potentials employed for beam control. It should also be pointed out that the direction of current flow, in the conventional sense, is always observed to be from grid to cathode in the external circuit, which indicates a net loss of electrons from the grid to the plasma.

Beam Focusing

With no auxilliary focusing, the spot size of the electron beam has been varied from less than $\frac{1}{32}$ in. to as large as desired by adjusting the cathode potential over a range of a few percent at any given pressure. The self-focusing property of the plasma electron beam is due to the combined action of positive ion space charge near the aperture, in the beam path, and the electrostatic lensing property of an aperture separating regions of different potential gradients. A region of high potential gradient surrounding the cathode accelerates electrons from the aperture and collects the relatively slow moving positive ions which enroute form a positive space charge concentration augmenting the extraction field at the exit aperture. The focal length f of such an aperture lens is given approximately by

$$f = \frac{4V}{E_2 - E_1}, \qquad (1)$$

where E_2 and E_1 are the external and internal field strengths, respectively, and V is the cathode potential. This accounts for the voltage dependence of the beam focus. Since the thickness of the sheaths separating the internal and external plasma boundaries from the cathode walls diminishes with increasing gas pressure, E_2 and E_1 also vary with gas pressure and thereby influence beam focusing.

To achieve adjustable beam focusing without the necessity for changing the cathode voltage, and hence the electron velocity, a simple magnetic lens has been used. Because the beam is already collimated, or slightly divergent, the lens need not be a strong one. Only a few hundred ampere turns are needed to focus the beam. Spot sizes approximately 10 mils in diameter have been achieved with a short solenoid supported a few inches below the cathode.

In addition, there are several other means of controlling the focal distance of the electron beam: by varying the focusing ring voltage, by varying the diameter of the focusing ring, and by varying the distance of the focusing ring from the cathode. The electron gun used in this electron-beam welder has the focusing ring attached directly to the cathode, thereby eliminating the need for a separate focusing power supply. A set of easily replaceable focusing rings of various diameters are used with this cathode. Since the cathode is quite small, it can easily be moved in the welding chamber as shown in Figure A1.

Electron Origin

Because the cathode operates at temperatures too low for appreciable thermionic emission, other mechanisms must be examined to explain the origin of the beam. Although the plasma body inside the cathode may supply a small fraction of the electrons, most of them must come from the inner cathode surface. This is borne out by the observed suppression of the beam current on application of a negative grid potential. Most of the electrons from the inside of the cathode are suppressed when a negative bias of 8 V is applied. This suggests that low-energy secondary electrons produced by high-energy ion bombardment or photoelectrons (or both) make up the bulk of the beam current. Since the grid is open ended, it cannot exercise complete cutoff of the beam [1]. However, more complete cutoff could be obtained if the end were closed with a sheet of mesh having an exit aperture for the beam.

Cathodes with oxide films or other deposits become clean after a few minutes of operation. This indicates that positive ions drawn out of the surrounding sheath are accelerated to high velocities and, by entering the perforations and the aperture, bombard both the inside and outside surfaces of the cathode. Foreign surface layers are thus quickly removed from the cathode surfaces by the impact of ions from the discharge.

Even though the combined areas of all the perforations or mesh openings total many times the area of the beam aperture only a small fraction of the cathode current is accounted for by electrons escaping from these small openings. These electrons are accelerated outward from the cathode and by collisions ionize the surrounding gas. A continuous supply of positive ions is thus maintained. A large portion of the electrons is not stopped by gas collisions and these strike the walls. This is evidenced by fluorescence and slight heating of the glass walls (see Figure A2).

EQUIPMENT–ADVANTAGES–ECONOMICS

Commercial equipment is now on the market. It normally operates at voltages under 15 kV; these lower voltages eliminate the need for x-ray shielding. The workpiece and cathode are enclosed in a glass bell jar providing full visibility of the work area. The reduced pressure and controlled environment in the bell jar work chamber minimize workpiece contamination. A blue transparent safety cover over the bell jar provides protection for the operator's eyes and permits observation of the operation in process.

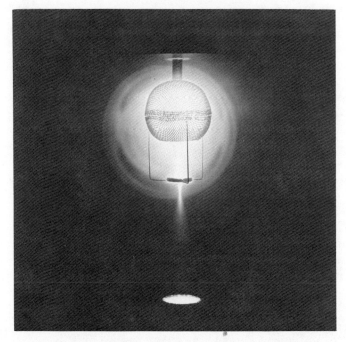

Figure A2 Welding cathode operating at low pressure.

A workpiece is mounted on a holding attachment consisting of three slides to provide motion in the x, y and z direction. Each slide is driven by a single phase, constant speed, ac motor. The machine operates from 200 V, single-phase 60 cps, 15 A service. The cathode is powered from a single-phase power supply provided with the unit. The power supply is rated at 1.75 kW.

The system can utilize any gas. The pressure is adjusted by a control knob on a panel and can be maintained at any desired level between 25 and 1000 μ. Generally, two basically different types of cathodes are available. One type, the hollow cathode, produces an unfocused beam whose thickness is related to the geometry of the emitting surface of the cathode. The second type, the contoured cathode, focuses the beam at a focal point or line. The focal distance from the cathode emitting surface is a direct function of the shape of the emitting surface.

Another machine currently available makes use of two focus coils rather than a single solenoid to produce a magnetic field. The total power available is limited only by the number of cathodes one selects.

Some advantages possessed by this unique type of electron gun are the following:

Equipment–Advantages–Economics

1. Insensitivity to contamination.
2. Long operating life.
3. Ability to operate in a partial vacuum.
4. Grid control of beam intensity with high speed of response.
5. Self-collimating beam properties.
6. Excellent depth of focus.
7. Ruggedness and simplicity.

The welder is not only simple in construction but also easy to operate. The cathode is self-cleaning and operates well even in a "dirty atmosphere." Operation of this device is not interrupted by severe outgassing, material evaporation from welds, and other normally undesirable products of high-temperature welding. This device is not limited to welding and can be used equally well for cutting, hardening, brazing, stress relieving, etc. Because a cold cathode gun requires a much less complicated vacuum system, electron gun systems have been developed for less than the full vacuum welding systems.

MATERIALS–TEST RESULTS–FUTURE

Materials that have been welded include tungsten, tantalum, molybdenum, titanium, and stainless steel. In addition, fusion of dissimilar materials such as porcelain and tantalum has also been accomplished. Figure A3

Figure A3 Edge weld made in argon at 20 torr [4].

Table A.1 Welding Parameters for Butt Welds in 2219 Aluminum Alloy[4].

Thickness Inches	Beam KW	Beam Kilovolts	Tensile PSI	Traverse Speed In./Min
.478	6.3	30	40,500	48
.478	7.0	40	42,800	46
.352	4.5	40	43,200	46
.250	3.4	40	40,700	46
.089	1.2	30	37,000	46

is an example of a columbium cup about $\frac{1}{4}$ in. in diameter and 0.006 in. thick which was edge-welded into a columbium ferrule in an argon atmosphere at 20 torr. A 30 kV, 30 μA beam was used. Average values of welding parameters and tensile strengths for butt welds of 2219 aluminum

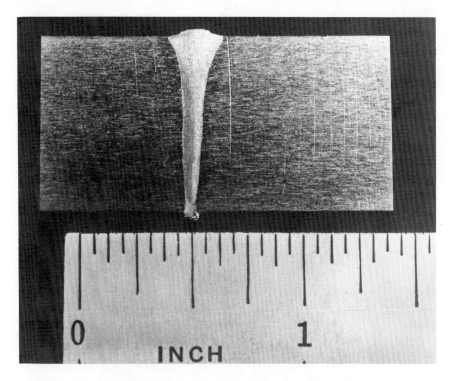

Figure A4 Plasma electron–beam bead-on-plate weld in 6061 aluminum.

alloy are given in Table A1. Depth-to-width ratios of the melt zones in these welds are at present less than those for welds made by conventional electron-beam equipment, and tensile strengths are about 75% of those of the parent material. Even though this alloy has a relatively high copper content and is considered difficult to weld, few cracks or voids have been observed in the plasma-made weldments.

A $\frac{1}{4}$ in. bead-on-plate weld section in 6061 aluminum difficult to weld by conventional electron beam due to cracking is shown in Figure A4. It was made at a traverse speed of 44 in./min. Here the larger depth-to-width ratio and narrower heat-affected zone reflect the difference in thermal properties as affected by composition.

Butt welds of good quality have been made in low carbon steel, chrome alloy steels, and stainless steels in thicknesses up to $1\frac{3}{8}$ in. Additionally, plasma welds in René 41 have been made experimentally.

In addition to the applications mentioned, cold cathode and plasma electron beams are the following:

1. Fusion welding pipes having diameters ranging from $\frac{1}{4}$ to 15 in. and wall thicknesses up to $\frac{1}{4}$ in. These have been butt welded in several seconds by focusing an annular electron beam around the pipe. Axial and planar beams have also been used to make butt and lap welds.
2. Brazing metal to metal and ceramic to metal joints. The beam heats ceramics as easily as it does metals. The beam is applied either to the metal or to the ceramic or to both, and the heat is conducted to the brazing alloy. Brazing operations have been conducted in reducing or oxidizing atmosphere as needed.
3. Diffusion bonding joints between similar or dissimilar metals or between ceramics and metal. When the beam is applied, heat is conducted to the adjoining material and under precisely controlled temperature and pressure, diffusion between the workpieces forms a high quality diffusion bond.
4. Hardening by encircling the workpiece or any local area. Sections of ferrous alloys, for example, have been austenitized. Quenching has been accomplished by direct air impingement during venting to provide metallurgical transformation.
5. Crystal growing. Single crystals have been grown by the "floating zone" method by placing the starting material in the center line of an annular beam. Floated upward slowly, at several centimeters an hour, the molten zone solidifies continuously as a single crystal.
6. Joining nonconductors. Nonconductors as well as conductors have been joined easily. The workpiece was immersed in a discharge; therefore electrical current was carried through the discharge and was not im-

peded by the presence of a nonconducting workpiece. These nonconductors included fused quartz, alumina, zirconia, and related refractory ceramics.

An untapped source of welding potential exists and future development programs producing larger machines and establishing welding capabilities will resolve the limits of cold cathode and plasma electron-beam welding.

PARTIAL (PVW) BEAM WELDING

Because electrons are of very small mass, they are deflected on impact with bodies of greater mass, for example, gas molecules. As a result an electron beam passing through a gas at atmospheric pressure very quickly becomes diffuse and its effectiveness as an intense heat source is lost. For this reason and since high voltages are involved, electron beams are usually generated, formed, and controlled in regions of very low pressure, usually in a vacuum environment of 10^{-4} torr or lower.

Until recently almost all electron-beam welding was accomplished with the workpiece also in a high-vacuum environment. However, during the past few years technological breakthroughs have resulted in a capability for electron-beam welding with the workpiece in a protective gas at atmospheric pressure. In addition, experimental work has shown that very satisfactory welding can often be accomplished at intermediate vacuum levels of about 100 to 400 μ. Thus there appear to be three welding modes: high vacuum, intermediate vacuum, and atmospheric pressure. Each has its advantages and disadvantages.

As stated most electron-beam welding to date has been accomplished with the workpiece in a vacuum environment of 10^{-4} to 10^{-5} torr or 0.1 to 0.01 μ. In this environment it is possible to obtain the most precise, highest quality welds. Electron scattering is not encountered and thus an extremely fine, high-power density beam can be produced and maintained over relatively long distances.

The major disadvantage associated with high-vacuum electron beam welding lies in the requirement for a work chamber large enough to enclose the workpiece, including any workpiece motion, and incorporating structures and seals which have extremely low leak rates. Furthermore, surfaces, materials, and designs must minimize outgassing. In order to provide reasonable chamber pumpdown times, it is necessary to incorporate relatively large mechanical and diffusion pumps, which add to the cost and the complexity of the system. In spite of these disadvantages, high-vacuum electron beam welding has proved to be a practical and economical production process providing construction and design possibilities not available when other joining techniques are used. It can be considered the most flexible and versatile electron-beam welding mode.

When the workpiece is enclosed in an environment of intermediate vacuum level, that is, 100 to 400 μ, some beam spread occurs because of electron-molecule collisions, particularly when working distances are increased. This spreading of the beam, however, is relatively small. A beam passing through 16 in. of air at 300-μ pressure shows some fringe effect. But even at this pressure and this distance beam power densities still remain high enough to produce metal fusion and thus welding.

MECHANISMS OF PROCESS AND EARLY DEVELOPMENT

The characteristics of electron beams of the welding variety have been investigated to a limited extent at intermediate vacuum levels. Basically, it consists of the modification of a standard high-vacuum electron beam welder in which a diffusion pumping port with an orifice just below it has been inserted in the electron optical column. In the upper portion of the electron column the pressure is maintained at normal high-vacuum levels, and in the lower portion, which contains the optical viewing system, magnetic lens, and deflection coil, higher pressures exist.

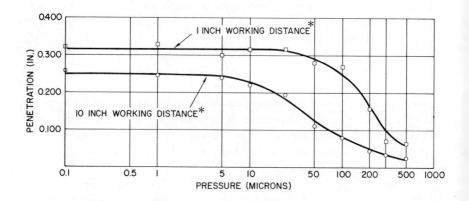

Figure A5 Effect of pressure on weld penetration [5].

Figure A5 is a plot of weld penetration versus operating pressure at two different working distances. It must be noted that the working distances indicated do not include about 15 in. of higher pressure beam path in the lower portion of the electron optical column. As might be expected, penetration falls at higher pressure levels and the point at which this occurs is somewhat lower for a longer working distance than for a shorter one. It is of interest, however, that for the shorter working distance full penetration could be maintained up to a pressure of over 20 μ. Until pressures exceed about 20 μ very little effect on penetration is encountered even at long working distances. For shorter working distances pressures in the 100 to 300 μ range still produce significant depth-of-weld penetration.

The effect of pressure on the shape of the fusion zone shows that extremely deep, narrow fusion zones, or the so-called deep weld effect, can be obtained up to pressures greater than 200 μ at the operating conditions involved. In fact, very little effect of pressure can be seen up to 50 μ. However, above this pressure the width of the top bead increases due to beam spread, and depth of penetration is reduced. At 300 μ pressure most of the fusion penetration appears to be the result of thermal conductivity from the initial beam impact surface.

EQUIPMENT–TOOLING–ADVANTAGES–ECONOMICS

Two major manufacturers of hard vacuum equipment have also equipment currently available in the partial vacuum generation of machines.

One type of machine sketched in Figure A6 consists of a basic unit to which special tooling can be attached. Here the tooling includes the small work chamber sized to fit the specific part involved. Three typical tooling configurations are shown, two for making relatively large circular welds using workpiece rotation and the third for making small circular welds using beam deflection. This tooling can be quickly and easily changed. With this machine the operating parameters are set and locked at a control station. The operator merely loads the part into the machine and pushes a button, closes a door, or perhaps moves the fixture into position. The rest of the cycle is fully automatic. Power supplies up to 25 kW at a 150,000 V rating are available.

Another machine is a high-production electron-beam welding facility that essentially involves a two-chamber machine:

(1) high vacuum gun chamber always maintained at 0.1 μ or less.
(2) medium vacuum chamber having a minimum of volume into which one part at a time is automatically loaded and welded.

The stroke of the lower tooling automatically actuates a vacuum valve mounted on the bottom of the electron-beam gun housing that alternately

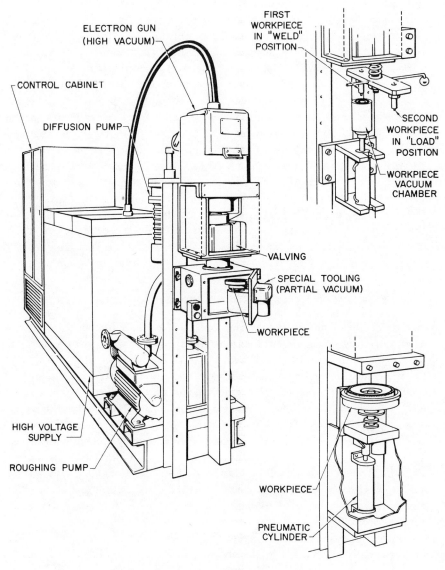

Figure A6 Production partial vacuum electron–beam welder [5].

seals or vents the work chamber. A second beam column valve opens to permit the electron beam to impinge on the piece part when the local chamber is evacuated. A small orifice, just large enough to permit passage of the electron beam, is located within the gun housing at the first crossover of the beam. This orifice enables the gun housing to be maintained at high vacuum (1×10^{-4} torr or less) even though the local chamber is only pumped to a roughing pressure of about 100 μ.

The system works because, at these low pressures, opening the valve between the gun chamber and the weld chamber does not cause the pressure in the two chambers to be equalized. At a pressure of 100 μ, the air molecules have a long mean free path, so that only a few of them find their way randomly into the gun chamber and are disposed of by the diffusion pump.

After welding is completed, the valve between the two chambers is closed before the welding chamber is filled with air and opened. This maintains the gun chamber vacuum at 0.1 μ. Power supplies up to 30 kW at a 60,000-V rating are available.

Both types of hard vacuum welders are capable of being retrofitted to produce partial or soft electron beam welds. It is important to keep weld chamber size small. Even with the fast-acting mechanical pump used to obtain the soft vacuum, pumpdown time is proportional to the volume of the welding chamber. It is therefore necessary to keep this volume as small as possible. One solution is to keep the welding chamber small. Fortunately, many mass-produced components that can be electron-beam welded are relatively small, so a small chamber will accommodate them. When the equipment is to be used on both small and large assemblies, the welding chamber must, of course, be designed to accept the largest work.

Electron-beam welding equipment can be made versatile by providing it with replaceable welding chambers. With this type of design the welding chamber is lowered for loading and unloading, and raised for evacuation and welding. O-rings between the welding chamber and the gun housing maintain the desired vacuum.

When a separate welding chamber is built for each type of product welded, the chamber becomes part of the tooling. Not only is the chamber designed to accept the part and locate it under the electron-beam gun but also tooling for moving the part so that welding occurs along a desired line can be made integral with the chamber itself.

To what extent this concept can be put in practice depends almost entirely on the ingenuity of the tool designer. Applications range from a double-drum housing for welding the tool steel cutting strip to the softer metal backing in bandsaws to a double-gun system for producing two welds 24 in. long and 12 ft apart.

Table A2. Comparison of Electron Beam Welders Capabilities [22]

	Full Vacuum	Partial Vacuum	Nonvacuum
General Operating Level	30,000 to 150,000 volts	60,000 to 150,000 volts	150,000 to 175,000 volts
Power Range	Up to 30 kw	Up to 25 kw	12 to 25 kw
Maximum Beam "Working Distance"	Over 36 in. from point of beam entry into chamber	27 in. from beam entry into workpiece enclosure	About $\frac{1}{2}$ in. from beam exit
Workpiece Pressure Level	10^{-4} mm Hg	10 to 300 microns (10^{-2} mm Hg to 3×10^{-1} mm Hg)	Atmosphere
Maximum Penetration (stainless steel at power level)	2 in. at 6 kw 5 in. at 25 kw	2 in. at 7.5 kw 4 in. at 25 kw	$\frac{1}{2}$ in. at 12 kw $\frac{7}{8}$ in. at 25 kw
Pumpdown Time	3 to 30 minutes, depending upon chamber size	Fraction to 10 seconds, depending upon enclosure size	None
Type of Work (suitability)	Batch loading, medium volume	High production rate, high volume	Continuous operation—simple configurations
Versatility (variety of application for one machine)	Excellent	Limited	Poor

Although the ultimate capabilities of electron-beam welding cannot be obtained at these higher pressures, weld characteristics remain perfectly acceptable for a wide variety of applications. The advantages of using the intermediate vacuum level are significant. They include much faster pumpdown, simpler and less costly vacuum equipment and vacuum chambers and work-handling mechanisms, and easier automation with vacuum feedthroughs. In particular, this mode of operation lends itself ideally to high-production repetitive operations in which the chamber is designed to fit closely around a specific part. In conventional EB welding, vacuum pumpdown can take 30 min or more. With the semivacuum approach, the time is in the 20-sec range. A comparison of the three modes of electron-beam welding is shown in Table A2.

MATERIALS–TEST RESULTS–FUTURE

It has been clearly demonstrated that high-vacuum electron-beam welding produces welds with superior mechanical properties. In many materials these properties approach or equal those of the base material, for example, this is true of heat-treated high-strength alloy steels.

As might be expected, electron-beam welds made at intermediate vacuum levels also exhibit excellent mechanical properties. Table A3 lists the tensile properties of butt welds made in annealed AISI 304 stainless steel, rimmed AISI 1010 low-carbon steel, 6061-T4 aluminum alloy, and 6 AL 4V titanium alloy. It can be seen that in three of the four materials failure occurred in the base metal, whereas in the fourth, the aluminum alloy, failure occurred at the edge of the weld, but at strengths exceeding the established minimum for this alloy in the solution-treated condition. Furthermore, porosity in gas-producing materials such as rimmed steels appears to be minimized. With regard to chemical contamination, it can quite readily be shown that the concentration of foreign atoms at a vacuum level of

Table A.3 Mechanical Properties of Partial Vacuum Electron–Beam Welds[5].

	UTS (PSI)	YS (PSI)	Elongation (% in 2 inches)	Fracture Location
Stainless Steel AISI 304 - Annealed 1/4 inch - Butt Weld				
Weld 1	89,000	31,800	74	base metal
Weld 2	86,100	33,700	66	base metal
Base Metal - Nominal	85,000	30,000	62	-
Rimmed Steel - AISI 1010 3/16 inch - Butt Weld				
Weld 1	49,000	38,700	24	base metal
Weld 2	48,400	37,700	25	base metal
Aluminum - 6061 (weld in T4 - no post weld treatment) 1/4 inch - Butt Weld				
Weld 1	34,400	31,000	5	edge of weld
Weld 2	35,500	28,000	5	edge of weld
Base Metal - Minimum (T4)	30,000	16,000	18	-
Titanium 6A1 - 4V - Annealed 1/4 inch - Butt Weld				
Weld 1	137,500	-	9	base metal
Weld 2	131,250	129,000	10	base metal
Base Metal	135,000	130,000	15	-

Note: All welds made at pressure of 200 microns.

100 to 200 μ is about the same as that found in inert shielding gases. Thus with most materials chemical contamination will not be a problem at these vacuum levels. Steel, aluminum, and titanium up to 1 in. thick has now been welded.

Current and future applications include gears, saw blades, flywheels, slip yokes, and pipe fittings. The 14-in. outside diameter ring gears and counterweights as well as flywheel stampings were electron beam-welded in 1968 automobiles.

Double helical aircraft engine gears assembled from two pieces are a natural application for intermediate vacuum electron-beam welding. It also is quite practical to weld single gears onto relatively long shafts. Band saw blade manufacturing has been revolutionized with the advent of electron-beam welding.

These applications have been accomplished in a high-vacuum environment, but have been now satisfactorily welded under intermediate or partial vacuum levels.

Figure A7 illustrates the partial vacuum-welded automotive flywheel. Here an SAE 1050 carbon steel gear was welded to a rimmed SAE 1010 steel sheet-metal disc or hub. Excellent welds were obtained at a welding speed of 300 ipm. Using a specially designed but very simple single spindle machine, these parts can be welded, one at a time, at a rate of three per minute. In other words, loading, pumpdown, welding, and unloading all can be accomplished in about 15 sec. Thus practical high-production rates can be realized.

Another intermediate vacuum application consists of another automotive part, the universal slip yoke. Significant cost reduction can be realized by using internally splined tubing welded to a simple forged yoke. Again,

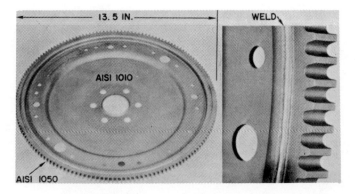

Figure A7 Automotive flywheel [5].

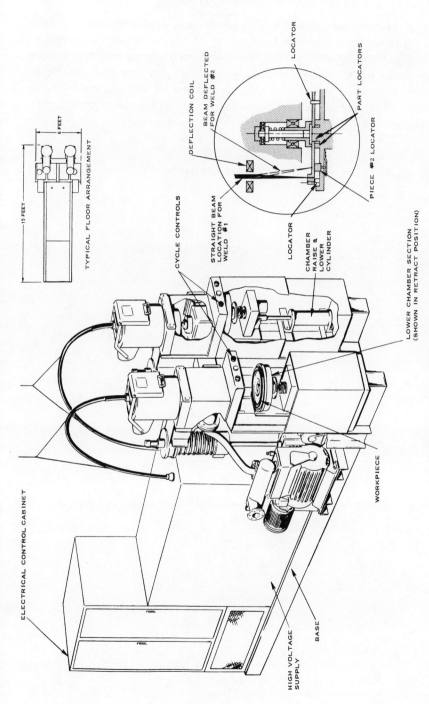

Figure A8 Typical dual station PVW machine.

using a relatively inexpensive single-purpose unit, production rates of about four per minute can be realized.

Finally, the next step will be the power train components, particularly transmission parts, for they seem to be most complex and costly. In Figure A8 is a conceptual sketch of a dual station partial vacuum machine that could weld the power train components mentioned previously or some new application.

NONVACUUM (NVW) ELECTRON BEAM WELDING

Today, most electron-beam welding machines contain the workpiece in the vacuum environment. At least two inherent advantages are to be gained from this configuration. First, a vacuum environment is chemically very pure, making possible the welding of highly reactive materials without fear of chemical contamination. The second advantage is the capability of providing an electron-beam system having a long working distance. Thus it is possible to focus the electron beam at any point from 0.50 to over 30 in. from the electron optical structure, permitting welding in very inaccessible areas and, in general, adding great flexibility to the welding system. For these reasons plus technical problems associated with maintaining the workpiece at atmosphere pressure, practically all electron-beam welding to date has been conducted with the workpiece in a vacuum chamber.

However, enclosing the workpiece in a vacuum chamber imposes limitations on the size of the workpiece and on high-speed, high-production automation of the process unless large or complex chambers and vacuum systems are provided. The use of portable vacuum chambers also tends to be complex and cumbersome. Thus the development of a practical system for electron-beam welding with the workpiece at atmospheric pressure would significantly increase the scope of application of this welding technique.

Two basic technical problems must be solved in order to produce a practical nonvacuum electron-beam welding system. First, a technique must be developed that permits the electron beam to escape from the vacuum environment in which it is formed and controlled into a region of gas at atmospheric pressure without significant loss of power. Second, the electron beam must be one that provides a reasonable working distance external of the gun structure before scattering reduces its power density to levels too low for metalworking.

The first problem is the more difficult to solve. Because of the high-power density necessary for electron-beam welding, thin low-density "windows" that would transmit electron beams of low-power density are not practical. A more practical system is a technique for staging the vacuum over a very short length using an orifice system through which the electron beam can pass. To be practical this orifice system must be large enough to transmit the electron beam without significant loss of power, must be designed to minimize the pumping requirements for reasonable pressure drop across each orifice, and must contain provisions to prevent clogging during extended welding operations.

The second problem can be solved quite readily by using a beam of relatively high-speed electrons, that is, a beam generated at relatively high voltage. It is well known that the lower the accelerating voltage the more rapidly an electron beam becomes diffused when passing through a gas at atmospheric pressure. Thus, as the voltage is increased, it is possible to retain a significant power density at greater distances. At least 120,000 V are required for a practical working distance; a reasonable and practical range is 120,000 to 200,000 V.

MECHANISMS AND PRINCIPLES OF PROCESS AND EARLY DEVELOPMENT

The problem of developing an electron-beam welding system for operation in a nonvacuum environment involves a series of theoretical and practical problems, which may be broadly divided into three groupings.

1. Electron scatter. The electron beam will be degraded by scatter as it transverses the regions at near-atmospheric pressures.
2. Beam transfer unit. The electron beam must be generated under high vacuum and made to impinge on a workpiece at atmospheric pressure. Some mechanism is required to allow the transfer of the beam through that pressure difference.
3. Electron beam generation. An electron beam must be generated which can transverse the beam transfer unit, undergo degradation by scattering, and yet arrive at the workpiece with adequate residual power density to accomplish welding. Once the requirements have been established, the design of the gun is relatively straightforward.

These problems are, of course, all interrelated and in order to solve any one completely they must all be solved. The sequence in which the groupings are presented appears to be the logical one in which to discuss the problems involved.

Electron Scatter

Calculations have been performed to estimate the sum of the unscattered fraction of the incident beam and the useful portion of the scattered fraction that reaches the weld surface [7]. Relative intensity curves have been studied for several gases at different values of gas density and two levels of accelerating voltage, 100 and 150 kV. Only the 100 kV electron beam in helium calculations is discussed because it is the most significant.

Interactions Between Swift Electrons and Atoms

Classical scatter theory predicts an alteration of the beam in traversing gaseous matter. The degree of alteration is dependent on the gas atomic number, the gas density, and the distance traversed in the gas.

For an electron beam in the low relativistic range, two types of interactions with the gas predominate:

1. Inelastic scattering electrons are scattered from the incident beam with loss of energy.
2. Elastic scattering incident electrons are scattered without loss of energy.

Inelastic scattering results in a transfer of energy to the atom either by excitation or ionization. The angle of scatter is generally small, since the interactions are predominantly those of an incident electron with an atomic electron rather than with the relatively massive nucleus. On the other hand, when the scattering is elastic, the interaction is usually one with the coulomb field of the nucleus and generally results in a larger scatter angle.

When an electron passes through the gas it may experience several such interactions in succession. The average distance traversed by an incident electron before an interaction occurs is defined as the mean free path. This distance is a function of the parameters gas atomic number and gas density as well as the momentum of the incident electrons. It is a statistical average for many electrons. The free path length of individual electrons varies in a random way, but in general it is close to the mean [9].

Mean free paths have been calculated after Goudsmit and Saunderson [8] using the Born approximation to represent the atomic field. The gas is assumed to be at standard pressure and room temperature (760 torr; 30°C).

Scatter Distribution

In addition to the classification by types of interaction, scatter theory is traditionally categorized by the number of interactions:

(1) single scattering—one mean free path or less,
(2) plural scattering—several mean free paths,
(3) multiple scattering—a large number of mean free paths.

The gas thicknesses and densities for electron-beam welding cover the range from single to multiple scattering, but plural scattering applies especially to this particular problem. Paradoxically, plural scattering theory is the most complex and has been the least studied of the three types.

Scatter distribution curves have been calculated using the multiple scattering theory of Moliere [10] as modified by Bethe [11]. Bethe's modification included a term that accounts for inelastic scattering by atomic electrons. Moliere's original theory accounted primarily for elastic scattering by the coulomb field of nuclei.

Approximations

During this study an attempt was made throughout to make conservative approximating assumptions [9]:

1. A ring source equal in diameter to the beam.
2. The expression for power density at the weld surface was maximized to define the welding area.
3. Single scattering from the ring, employing the appropriate multiple scattering distribution curve, was used to account for plural scattering.
4. A path length equal to the gas thickness was assumed.
5. Accounting for energy losses was incomplete.

The first approximation has the effect of distributing the peak of the beam over a larger area at the weld surface. The error involved in the maximizing approximation can cause only a conservative estimate. The gross effect of the third assumption is to exclude a somewhat larger portion of the scattered intensity than would actually be useful. The fourth listed approximation is not conservative, but the error is least in the low range of gas densities and thicknesses where the power density remains sufficiently high for welding.

Energy losses are partly accounted for since Bethe [11] includes an approximation for inelastic scattering by the atomic electrons. Evans [12] states that, although the hard collisions between incident and atomic electrons are infrequent, they account for about one-half of the total energy

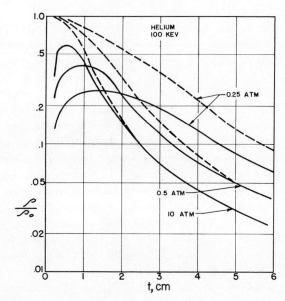

Figure A9 Ratio of scattered and total power density to the initial power density in the weld area [9].

loss because of the large energy transfer. The scatter angle for these hard collisions would generally be great enough to be excluded, by scatter theory, from the area considered in the calculations.

The curves of Figure A9 illustrate the results of the calculations. The ratio of final-to-initial power density is plotted as a function of thickness for several gas densities. The solid curves represent the contribution to the area of maximum power density of the scattered electrons. The broken lines represent the total power arriving in that area, including both the scattered and unscattered electron contributions [9].

Beam Transfer

Because the pieces to be welded required a gaseous environment at or near atmospheric pressure, some connecting link had to be found between the high vacuum of the gun chamber and the weld environment through which the beam must pass. High energy beams have normally been brought out to atmosphere through thin aluminum or beryllium foils stretched across an aperture in the chamber wall. These foils have not been satisfactory in this application because they caused dispersion of the beam and the resultant heating effect on the foil severely limited the possible power density. An alternative method for bringing a beam from the vacuum chamber into the atmosphere was to provide differentially pumped apertures.

This system limited the flow of gas into the high vacuum region by small orifices that restricted gaseous effusion and by using the secondary vacuum pumping system. The pressures to be expected were obtained with the aid of a nomograph [13]. There was a limitation to this system because the beam must traverse in the differential chamber at a distance that was large compared with the beam diameter. Since the dispersion of an electron beam was a function of gas density times the distance it traveled through a gas, excessive energy loss and scatter would occur if the traverse distance was too long. The limitation in reducing this traverse distance arose from the fact that the cross-sectional area between apertures was sufficiently large to allow a high conductance path for efficient vacuum pumping in the differential chamber. The dispersion could also be reduced by lowering the gas density along the path length. In particular, the first chamber was operated at the lowest pressure consistent with the integrated vacuum system.

The design of the nozzles and apertures included provision for cooling, for there was some dispersion of the beam in transit, with a resultant heating effect. Because the alignment of such a system was extremely accurate, heat distortion was minimized.

Controlled Atmosphere Unit

A controlled atmosphere unit was incorporated and was external to the differential pumping system proper. This was required for both the beam and the weld.

From the point of view of the beam the unit was required since the maximum attainable power density was achieved. Gas scatter was the governing factor and this was minimized since the gas through which the beam passed was exclusively helium. In order to obtain this helium environment other gases were excluded by the unit. However, certain of the materials that were welded required an inert gas seal because their properties were not to be radically altered by the process.

Since the gas content of the area in the vicinity of the weld was controlled, the chamber was integrated with the first orifice of the differential pumping chamber, one wall of which was formed or completed by the workpieces. Since gas scatter was minimized, the length along the beam axis was kept short.

In the system shown in Figure A10 the controlled atmosphere chamber was maintained at a slight positive pressure with respect to the surrounding atmosphere. The chamber was made shallow to reduce the probability of

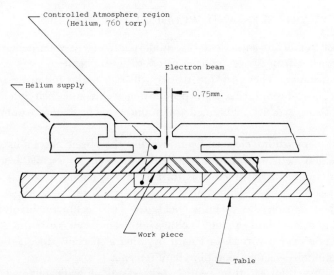

Figure A10 Detail of the controlled atmosphere chamber, showing also the workpiece, table, and backup channel [9].

electron scattering and the diameter was reduced to lower the total gas consumption. Although a smaller diameter would have decreased gas consumption even further, care was taken that the solidus material in the weld was not exposed to the possibly contaminated atmosphere outside the area covered by the chamber. Furthermore, the materials comprising the unit were not directly exposed to the molten work material in order to avoid thermal distortions.

Manufacturers continued to research the field of NVW and three years ago laboratory prototypes were in existence and intensive evaluations in progress. These laboratory units operated over a broad range of accelerating voltages and beam currents (from 2 kW at 100 kV to 12 kW at 150 kV). Initial test results from this equipment on 2219 aluminum alloy indicate tensile strengths of 41,900 psi and yield strengths of 29,000 psi. The elongation across the welds was 10% and transverse weld shrinkage was less than 0.005 in. Figure A14 shows a nonvacuum electron-beam welding machine. The manufacturer claims transmission efficiencies in excess of 90% have been obtained, pumping requirements have been significantly minimized, and orifice clogging has been avoided in regard to the major problem, that of providing an escape passage for the electrons. The unit operates very successfully at voltages up to 175,000 V and at powers up to 7 kW.

VARIABLES AND WELDING PARAMETERS

The most significant variable and welding parameter is weld penetration, which in turn affects speed, voltage, amperage, atmosphere (gas or air), and work distance. Therefore the reader will see in the following section the significance of producing fully penetrating welds and their interaction with the other variables mentioned to produce practical nonvacuum electron-beam welds.

Practical nonvacuum electron-beam welding has been accomplished by designing equipment that permits an electron beam to escape from the vacuum environment in which it is formed into a region of gas at atmospheric pressure without significant loss of power. Thus the workpiece can be positioned outside the vacuum enclosure. This technique involves the combination of a relatively high voltage electron beam with the staging of the vacuum over a very short length by using a specially designed orifice system to minimize vacuum pumping requirements.

Initial nonvacuum electron beam work was done with relatively low-powered machines limiting penetration capability to about $\frac{1}{8}$ in. Today, however, with improved electron optics and orifice systems and with output powers of about 12 kW it is possible to produce deep welds in many materials to depths exceeding $\frac{3}{4}$ in. and at relatively high speeds. Figure A11 shows this weld in AISI 4340 steel. It can be seen that, although this weld is wider than that encountered with vacuum and intermediate vacuum welding, it is still relatively deep and narrow when compared with other

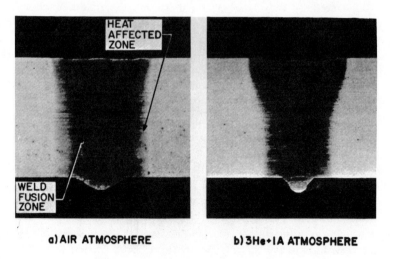

a) AIR ATMOSPHERE   b) 3He+1A ATMOSPHERE

Figure A11   Nonvacuum electron–beam butt welds—AISI 4340 [14].

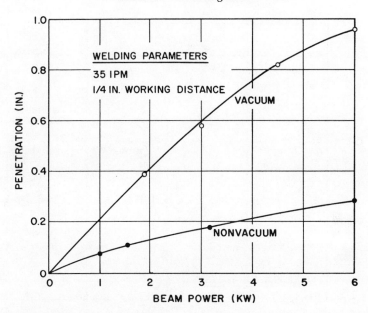

Figure A12 Vacuum and nonvacuum weld penetration versus power—AISI 304 stainless steel [14].

fusion welding techniques. Obviously, electron scatter is much greater at atmospheric pressure than at lower pressures. Tests have shown that penetration falls off very rapidly with working distance and thus, in order to produce welds with significant depth, it is necessary that the workpiece be in close proximity to the final pressure staging orifice. However, working distances up to at least ½ in. yield practical welding results. It is also found that depth of penetration is strongly dependent on accelerating voltage.

The type of gaseous atmosphere through which the electrons must pass also has a significant effect on scattering and thus on weld penetration. Again, as expected, the lower the density of the protective gas used, the greater is the depth of penetration. Thus helium gives significantly greater penetration than does argon, thus verifying the earlier work of Leonard [9].

Welding speed also has an effect on depth of penetration. However, as with vacuum electron-beam welding, the effect of welding speed is less critical as the speed increases. In other words, at very low welding speeds small changes in speed produce relatively large changes in depth of penetration. At higher welding speeds significant speed variations can occur without significantly affecting the depth of the weld.

Figure A12 compares the data on weld depth versus power in 18-8 aus-

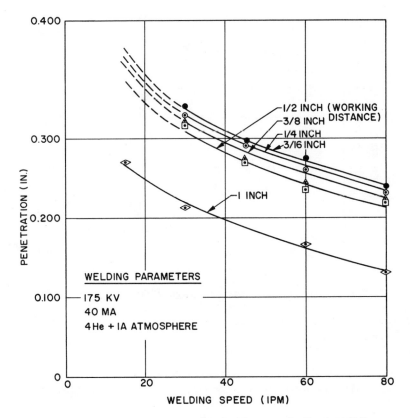

Figure A13 Weld penetration versus welding speed—René 41 [14].

tentic stainless steel with similar data obtained when welding in a vacuum under the same conditions of power and welding speed. It can be seen that for a given power the depth of penetration for nonvacuum welding is less than that obtained when welding in a vacuum. This probably represents a decrease in beam power density because of the larger beam cross-sectional area experienced with the nonvacuum technique. Figure A11 shows photographs of nonvacuum welds in AISI 4340 steel in a thickness of 0.250 in. One weld was made in air at 35 in./min and the other in a mixture of three parts of helium and one part of argon at 60 in./min. Both welds are sound, but it can be seen that the weld made in the helium-argon mixture is somewhat narrower.

Figure A13 is a plot of penetration versus welding speed for René 41 using a gas mixture of four parts of helium and one part of argon, and working distances from $\frac{3}{16}$ to 1 in. are shown.

Variables and Welding Parameters

In welding TZM molybdenum at working distances from $\frac{3}{16}$ to 1 in. in the same gas mixture as above a marked decrease in penetration (approximately 100%) of the refractory alloys versus superalloys (Figure A13) occurs.

Table A4 is a chart listing typical nonvacuum welding parameters for several different materials in different thicknesses. This shows the wide range of capabilities this process has reached at the present time.

EQUIPMENT–ADVANTAGES–DISADVANTAGES

As seen in the previous section, laboratory development equipment has been built and exhaustive evaluations conducted on the functioning of the equipment, its power supplies, vacuum systems, etc. In addition, work has been done to establish the welding parameters of various families of metals and the capability of the machine to function when welding.

Now, commercially available equipment is being marketed (See Figure A14.) This, of course, does not include any work chamber and it is apparent that a relatively clear space beneath the machine structure provides ample room for workpiece handling mechanisms with the convenience of modular construction. This modular construction permits combining the component

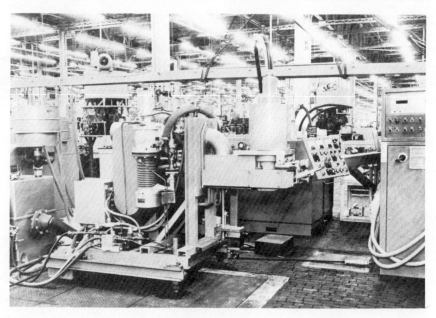

Figure A14 Nonvacuum electron–beam welder [5].

structures in a system that is most economical for a given application. Depending on the application, concepts have been evolved whereby the work can move beneath a fixed-gun column while in another concept a mechanism to traverse the gun column over a fixed workpiece is utilized.

Industry has been slow in accepting the nonvacuum unit primarily be-

Table A.4 Typical Nonvacuum Electron–Beam Welding Parameters [5,14].

Base Metal	Thickness In.	Voltage KV	Current ma	Weld Speed ipm	Work Distance In.	Atmosphere
Killed Steel	0.125	175	30	85	1/4	Air
SAE 1010 Steel	0.010	150	2.5	145	1/4	Air
AISI 4130	0.025	175	4	80	1/4	4He-1A
AISI 4340	0.250	175	40	60	1/2	3He-1A
AISI 4340 Steel	0.430	175	40	15	3/16	4He-1A
AISI 4620 Steel	0.400	175	40	26	3/16	Air
AISI 304 Stainless Steel	0.120	175	60	480	3/8	He
AISI 304 Stainless Steel	0.400	170	35	10	1/4	Air
AISI 304 Stainless Steel	0.750	175	70	10	3/8	Air
AISI 321 Stainless Steel	0.060	175	20.5	130	1/4	Air
17-7 PH Stainless Steel	0.100	175	25	108	1/4	Air
2219 Aluminum	0.500	175	70	125	1/8	He
2219 Aluminum	0.750	175	70	27	1/8	He
3003 Aluminum	0.010	175	15	3500	1/4	He
5052 Aluminum	0.0093	150	3	145	1/4	Air
Rene 41 Superalloy	0.275	175	40	15	1	4He-1A
Udimet 500 to A-286	0.200	175	40	50	1/2	Air
Inconel X	0.125	150	25	18	3/16	Air
TZM Molybdenum	0.187	175	40	15	1/2	He
Copper	0.008	150	8	145	1/4	Air

cause it has not found adequate means of handling the high volume of output. One company is using a unit to make tubewelds at high speeds. Another firm is currently incorporating a nonvacuum unit with a high-speed tube mill as a production item. Other makers of tubular products and auto accessories are closely investigating the nonvacuum units.

Of course, since extremely high voltages are being used, the problem of radiation comes up. How is the operator protected against x-ray? Two practical solutions have been proposed and are utilized. First, the equipment can be operated from a remote location. Second, if the operator is needed on location, leaded shielding has been designed to protect him against radiation. Protecting an operator against electron-beam welding radiation is no different from protecting a technician who frequently x-rays fusion welds from a remote position. There appears to be no difference if x-rays are emitted during the actual welding operation as long as proper shielding is provided.

The major disadvantage of the nonvacuum process is that it imposes a limitation on the shape of the workpiece. The contour of the workpiece must be reasonably smooth so that the beam impingement can be within 1 in. of the machine structure and at a relatively constant distance. Furthermore, in general, wider fusion zones are produced by this technique with lower depth-to-width ratios. Chemical purity is controlled by the extent of inert gas shielding in the vicinity of the weld. Obviously, nonvacuum electron-beam welding has the very significant advantage of not placing a limitation on the size of the workpiece being welded. Moreover, it simplifies high-speed work handling and high production application for parts of certain geometry.

MATERIALS–TEST RESULTS–APPLICATIONS AND FUTURE

Table A5 listed numerous materials which have been nonvacuum electron-beam welded. Welding of TZM molybdenum results in large recrystallized grains in the weld zone that diminish in size in the heat-affected zone. These recrystallized grains essentially have lost the strain-hardened structure through which wrought base metal achieves its strength. Consequently, welds made in TZM molybdenum are usually weaker than the base metal.

The representative tensile data of Table A5 for nonvacuum electron-beams welds are in agreement with the above and indicate that the wrought base metal exhibits greater tensile strengths than the weldments when tested between a temperature range from room temperature to 2200°F. The results also indicate that the wrought base metal loses its strength more rapidly than the weldments due to the partial loss in its strain-hardened structure.

Table A5 Tensile Data.

Base metal	Thickness, in.	Heat treat condition[a]	Test temperature, °F	Ultimate tensile strength, psi	Yield strength (0.2% offset), psi	Elong. in 1/2 in., %	Elong. in 1 in., %	Elong. in 2 in., %	Location of failure[a]
AISI 4340	0.400	T-BM	72	222,000	214,000	...	14	9	BM
AISI 4340	0.400	TW	72	208,000	202,000	...	4	2	HAZ
AISI 4340	0.400	TWT	72	191,000	185,000	...	9	5	HAZ
AISI 4340	0.400	TWNAT	72	219,000	208,000	...	...	11	BM
René 41	0.400	SG-BM	72	190,000	139,000	...	28	24	BM
René 41	0.400	GW	72	153,000	112,000	...	30	9	FZ
René 41	0.400	GWG	72	183,000	142,000	...	12	12	FZ
René 41	0.400	SWG	72	181,000	140,000	...	9	9	FZ
René 41	0.400	SWSG	72	181,000	142,000	...	20	11	FZ
TZM molybdenum	0.100	BM	72	138,000	123,000	10	...	...	BM
TZM molybdenum	0.100	W	72	74,000	58,000	4	...	...	FZ
TZM molybdenum	0.100	BM	1400	99,000	80,000	16	...	...	BM
TZM molybdenum	0.100	W	1400	49,000	41,000	22	...	...	FZ
TZM molybdenum	0.100	BM	1600	96,000	81,000	16	...	...	BM
TZM molybdenum	0.100	W	1600	48,000	42,000	22	...	...	FZ
TZM molybdenum	0.100	BM	1900	85,000	75,000	16	...	...	BM
TZM molybdenum	0.100	W	1900	52,000	42,000	22	...	...	FZ
TZM molybdenum	0.100	BM	2200	69,000	56,000	26	...	...	BM
TZM molybdenum	0.100	W	2200	50,000	38,000	22	...	...	FZ

[a] BM—base metal; W—weld; T—temper at 775° F for 2 hr; N—normalize at 1650° F for 1 hr; A—austerritize at 1525° F for 1 hr, oil quench; S—solution anneal at 1950° F for 4 hr; G— age at 1400° F for 16 hr; FZ—fusion zone; HAZ—heat-affected zone.

Reported data [15] indicate that the tensile strength of weldments can be even greater than that of the base metal when tested above the recrystallization temperature due to the inherent instability of the wrought structure at the higher temperatures combined with exaggerated grain growth under the influence of the applied stress [16].

The elongation values obtained for the weldments tested at room temperature are considerably lower than those of the base metal. When the test temperature is increased to 1600°F, however, the ductility of the weldments approaches that of the base metal (See Table A5.)

Figure A15 presents tensile data on nonvacuum electron-beam welded 0.250-in. thick annealed austenitic stainless steel with the weld surface machined slightly undersize to produce fracture in that region. It can be seen that these data compare very favorably with handbook nominal and guaranteed tensile strength values. Test results have been plotted statistically for 0.062-in. annealed austenic stainless steel and data on vacuum welds have been combined with data on nonvacuum welds. All data points fell on the same straight line, indicating that statistically the nonvacuum welds have strengths equal to those of vacuum welds and that both have strengths equal to the base metal.

When 0.400-in. thick welded AISI 4340 steel was tested in the tempered

Specimen	Ultimate Tensile Strength (PSI)	Yield Strength 0.2% Offset (PSI)	% Elongation in 2 in.	Loading Speed (IPM)
Welded Specimen				
1	82,600	35,300	57	0.05
2	81,900	36,200	53	0.05
3	84,000	36,400	63	0.05
4	84,100	37,100	49	0.05
Base Metal				
Nominal[17]	85,000	35,000	55	--
Guaranteed[18] Minimum	80,000	30,000	50	--

Weld reinforcement on the top and root was removed prior to testing.

Figure A15 Tensile test data—nonvacuum electron-beam welded AISI 304 stainless steel (0.250-in. thick annealed sheet) [14].

after-welding condition, the weldments retained approximately 90% of the average tensile strength of heat-treated base metal. Welded 0.400-in. AISI 4330 specimens that were tested in the as-welded condition retained 95% of the average tensile strength of the heat-treated base metal. This slight decrease in tensile strength was the consequence of overtempering in the heat-affected zone where the specimens failed.

The average tensile properties of representative AISI 4340 tensile specimens are presented in Table A5.

Optimum tensile strength and maximum ductility of René 41 weldments were obtained when the material was welded in the solution annealed condition and subsequently re-solution annealed and aged after welding. Average tensile data for representative René 41 weldments are included in Table A5.

A comparison between weldments produced by nonvacuum electron-beam welding and those produced by the gas-tungsten-arc process indicates that the tensile properties of the nonvacuum electron beam weldments are equivalent and in some cases superior to those produced by the gas-tungsten-arc process. Beemer and Mattek [19] report tensile property data for welds made in 0.040-in. thick René 41 sheet which was aged after welding. The tensile properties reported by these investigators were equivalent to René 41 nonvacuum electron-beam weldments that were similarly aged after welding. However, tensile strength data reported by Schwenk and Trabold [20] on weldments made by the gas-tungsten-arc process in $\frac{1}{16}$-in. thick René 41 and aged after welding were approximately 12% lower than the tensile strength of nonvacuum electron-beam welds that were similarly aged after welding.

A series of fatigue tests on René 41 specimens were designed to provide a maximum constant stress over a given tapered section. The weld was included in this tapered section so that the weld and base metal could be tested in fatigue at a common stress level. The specimens, which were tested at 90 cycles/sec, employed a technique of incremental stress increase that involved stressing the specimens at a level lower than their predicted endurance limits and then increasing the stress level at successive 10,000 psi increments after every 10^7 cycles until failure.

The low boundary fatigue limit of both base metal and nonvacuum electron-beam welds at 10^7 cycles was 35,000 psi. All of the specimens including the base metal failed between 40,000 and 60,000 psi. The location of fracture for all the specimens was well defined and inspection indicated that no voids or inclusions were present on the fracture surface.

It is therefore quite obvious from Tables A4 and A5 that nonvacuum electron beams can weld metals in all the major categories, ferrous as well as nonferrous.

Materials–Test Results–Applications and Future

Nonvacuum electron-beam welding is a much newer technique and therefore fewer specific applications have been developed up to now. However, interest in this technique is high and its production potential is promising. One outstanding production application for nonvacuum electron-beam welding is in the manufacture of thin-walled tubing and steel pipe where control of weld bead geometry is important. Welds on tubing have been produced at speeds up to 10 ft/min and $\frac{1}{8}$-in. wall stainless steel pipe has also been welded at a speed of about 30 ft/min. Again, in this instance the weld bead has been excellent, and the welding speed is remarkable when compared with other production techniques for this type of welding. Nonvacuum techniques make an electron beam wide enough to overcome fitup and drip-through problems associated with resistance methods, which can be eliminated through careful regulation and sensing of heat input. Undoubtedly, in the relatively near future much high quality tubing and pipe will be welded using nonvacuum electron-beam technology.

Another production application, shown in Figure A16, is a spherical pressure bottle 6 in. in diameter with a wall thickness of approximately 0.1 in. made from 17-7 PH stainless steel. As indicated, the equitorial weld was made by nonvacuum electron-beam welding. After welding, the structure was subjected to a TH 1050 heat treatment that produces an ultimate tensile strength of about 180,000 psi. Mechanical properties of the welded bottle were evaluated by a simple low cycle fatigue test that included pressurizing the bottle several times at increasingly higher pres-

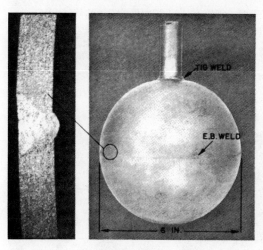

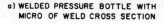

a) WELDED PRESSURE BOTTLE WITH MICRO OF WELD CROSS SECTION   b) BURST PRESSURE BOTTLE

Figure A16 Nonvacuum electron-beam welded 17-7 PH stainless steel pressure bottle.

sures. Burst finally occurred at 9500 psi. This pressure corresponds to about 93% of the base metal strength. However, as can be seen, the fracture originated at the point where the inlet tube was welded into the bottle and propagated perpendicularly through the nonvacuum electron-beam welds, which suggests the absence of notch sensitivity in these welds. Thus it can be concluded that the nonvacuum electron-beam weld exhibited a strength comparable with that of the base metal.

The turbine wheel that was previously electron-beam welded in vacuum was tack-welded in four quadrants by the vacuum electron-beam welding machine so that no special weld fixture would be necessary for nonvacuum electron-beam welding. The turbine wheel was fixtured in a standard rotary fixture as shown in Figure A17 with a copper backing ring positioned approximately 0.010 in. under the weld joint to improve weld quality. The workpiece-to-orifice distance was $\frac{1}{2}$ in. The wheel was set in motion at a predetermined weld speed and the electron beam was turned on at full power (175 kV, 40 mA) for one full revolution. An air atmosphere was used. Several welded turbine wheels successfully passed visual dye penetrant and radiographic inspection.

One nonvacuum electron-beam welded turbine wheel was balanced and spin tested to destruction. Burst occurred at 35,900 rpm, which corresponds to a tip velocity of approximately 1500 fps. Photographic techniques were employed to record the instant of burst as shown in Figure A17. From this photograph the fracture mode can be seen as a tearing away of the rim at the weld where the cross section dimension was a minimum.

The burst strength corresponding to an angular velocity of 35,900 rpm was 104,000 psi at the weld joint. The weaker of the two materials, A-286, exhibited a hardness of Rc 23, corresponding to a yield strength of 85,000 psi and ultimate tensile strength of 130,000 psi. The Udimet 500 had a yield strength of 121,000 psi and an ultimate tensile strength of 142,000

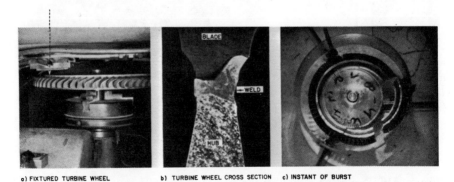

a) FIXTURED TURBINE WHEEL b) TURBINE WHEEL CROSS SECTION c) INSTANT OF BURST

Figure A17 Nonvacuum electron–beam welded bimetallic turbine wheel.

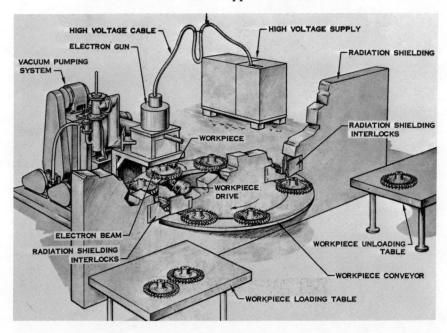

Figure A18 Production system for welding turbine wheels.

psi. Thus it can be seen that the turbine wheel exhibited a strength approaching that of the weaker base metal.

An artist's conception for producing high rates of specialized assemblies like the turbine wheel is shown in Figure A18. The turbine wheels, installed on individual rotary spindle assemblies, are mounted on a rotary conveyor system. The rotary conveyor system positions the composite turbine wheel under the electron beam where the wheel is rotated on the rotary spindle. Loading and unloading the rotor are accomplished from a separate control room that is radiation-isolated by appropriate wall structures and interlocks.

What about the future? It is expected that this technique will supplement vacuum electron-beam welding, but it will not replace it. Widespread industrial interest in this technique has already been aroused. Potential areas of application of the process include the fabrication of thin and heavy-walled tubing, metal containers, missile cases, ship structures, and automotive parts. In general, any component requiring high-quality welding at high speed and having a reasonably smooth contour that permits access of the beam exit orifice close to the weld joint is a worthy candidate for this process.

In Figure A19 is an artist's concept of future nonvacuum electron-beam

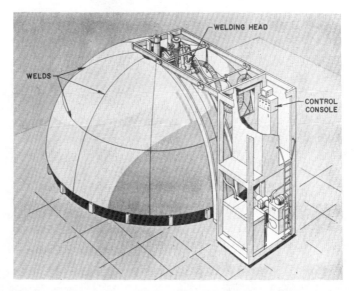

Figure A19 Concept for nonvacuum electron–beam welding of rocket structures [5].

machines, depicting how electron-beam welding can be applied to the fabrication of large structures. To date no full-scale cases have been welded using this technique, but preliminary test work has shown conclusively that high quality welds in typical rocket case materials can be produced at high speeds.

Versatility of welding large sections in both the "X" and "Y" axes is another future development. The electron-beam gum column and its supporting vacuum equipment are mounted on a common platform. By the use of a slide-rail assembly for "Y" motion and a rack and pinion drive for "X" motion, two-dimensional welding can be performed. Optimum location of the workpiece beneath the gun can be obtained by appropriate fixturing or by lowering or raising the platform base supports. This development will concurrently conceive a unique optical viewing system that permits telescopic viewing of the workpiece at any location in the applicable X-Y plane. Operation of the machine and welding process will be achieved from a radiation-isolated control booth.

In addition to equipment improvements and innovations, research and development will continue. They will be directed toward obtaining even deeper welds and greater depth of penetration with a given power rating. Work will also be concentrated in an effort to increase the present working distance to 1 in.

Much of the success of electron-beam welding in air will depend on the ability to move the process over the workpiece. Development of several guidance controls currently on the market that could be used to convey the equipment along the weld seam will require investigation and evaluation. Increased speed will be another item for development. It is anticipated that the eventual construction of a 25 kW-unit capable of welding ¾-in. wall steel pipe in a single pass at the rate of 100 ipm will also evolve.

The future for nonvacuum electron beams as well as all other types of electron beams is especially promising. It is now up to the user to select carefully the equipment and tool the component in order to realize the full potential of electron-beam welding.

REFERENCES

[1] L. H. Stauffer, and M. A. Cocca, "Grid Controlled Plasma Electron Beam," Fifth Electron Beam Symposium, 345–367 (March 1963).
[2] E. G. Linder, and K. G. Hernquist, "Electron Space Charge Neutralization by Positive Ions," *J. App. Phys.* **21,** 1088 (November 1958).
[3] H. L. L. Van Paassen, E. C. Muly, and R. V. Allen, "Cold Hollow Cathode Discharge Welding," National Electronics Conference (October 1962).
[4] L. H. Stauffer, "Plasma Electron Beam Welding," Ninth SAMPE Symposium, Paper No. IV-1, Dayton, Ohio (November 15, 1965).
[5] J. W. Meier, "Electron Beam Welding at Various Pressures," Second International Conference Electron and Ion Beam Science and Technology, New York City (April 17, 1966).
[6] J. R. King, "High Production with Electron-Beam Welding," *Tool and Mfg. Eng.,* **57,** 4 (October 1966).
[7] L. H. Leonard, et al., Development of NonVacuum Electron Beam Welding, final report on contract AF 33(657)-7237 (1963).
[8] S. Goudsmit and J. L. Saunderson, *Phys. Rev.* **57,** 24 (1940).
[9] L. H. Leonard, "Electron Beam Welding at Atmospheric Pressure," Fifth Electron Beam Symposium, 378–394 (March 1963).
[10] G. Moliere, *Naturforsch, Z.,* **32,** 78 (1948).
[11] H. A. Bethe, *Phys. Rev.,* **89,** 1256 (1953).
[12] R. D. Evans, *The Atomic Nucleus,* McGraw-Hill, New York, 1955.
[13] B. W. Schumaker, "The Designing of Dynamic Pressure Stages for High Pressure/High Vacuum Systems," USTIA Report No. 78, University of Toronto, 1961.
[14] J. W. Meier, "Recent Developments in NonVacuum Electron Beam Welding," First International Conference Electron and Ion Beam Science and Technology TP64-04, Toronto, Canada (May 6, 1964).
[15] J. A. Houck, "Physical and Mechanical Properties of Commercial Molybdenum Alloys," DMIC Report No. 140, Battelle Memorial Institute (November 1960).
[16] A. G. Ingram, and Ogden, "The Effect of Fabrication History and Microstructure on the Mechanical Properties of Refractory Metals and Alloys," DMIC Report No. 186, Battelle Memorial Institute (July 1963).

[17] *Metals Handbook,* Vol. 1, "Properties and Selection of Metals," Eighth Edition, 414, (1961).
[18] *Stainless Steel Handbook,* Allegheny Ludlum Steel Corp., 2 (1956).
[19] R. D. Beemer, and L. J. Mattek, "Welding Nickel Base Alloys," *Welding J.,* Research Suppl., 267-s to 273-s (1962).
[20] W. Schwenk, and A. F. Trabold, "Weldability of René 41," *Welding J.,* Research Suppl., 460-s to 473-s (1963).
[21] F. Schollhamer, "NonVacuum Electron Beam Welding," MMP Project No. 7-926a I-II-III-IV, Contract AF 33(615)-1052 (February 1963 to June 1965).
[22] J. C. Sterling, Jr., "Electron Beams," Steel, S-1–S-5 (July 1966).

Chapter 2

PLASMA-ARC WELDING

Since 1955 plasma-arc technology has made rapid advances in utilizing the advantages of high heat energy of an ionized gas stream. Testing of re-entry vehicles, cutting of metals, and spraying of powder compositions have been developed and applied. Chemical flames cannot approach the temperature possible with constricted arc plasma; only electron-beam energy concentration is higher. Plasma is being recognized as a source of heat for fusion of metals in a welding process and the fundamental principles of application are being developed. Basic torch designs follow those used in plasma-arc cutting, which include generation of the plasma stream and torch cooling. Gas velocities are lowered for a given orifice size to prevent expelling the melted metal from the joint to be formed.

As a heat source tool for welding plasma arc falls between the unconstricted electric arc used in the gas-tungsten-arc welding process and the high energy concentration of an electron stream in electron-beam welding. Thus, because collisions are greater between the electrons of the arc plasma and the molecules of the plasma gas, higher temperatures are achieved in the plasma jet than in conventional arc welding. For conventional shielded metal-arc welding, temperatures up to approximately 10,000°F (6000°K) may be obtained in the arc, for the gas-tungsten-arc, temperatures up to 35,000°F (20,000°K), and in the nontransferred arc plasma jet, temperatures up to 90,000° (50,000°K). This subject is discussed in the next section.

Many engineers know that plasma welding has been under investigation for several years. What they do not know is its capability. Is it in the same class as the electron beam? What alloys will this new process weld? Is it a production process?

Some investigators feel that this process is the most important development to evolve from the welding industry during the past ten years. Once industry has an opportunity to experiment with the process, many new and unusual applications will surely emerge.

THEORY AND PRINCIPLES OF THE PROCESS

The basic principle of this process consists of heating a plasma-forming gas such as nitrogen or hydrogen to a high enough temperature so that the gas molecules become ionized atoms. The process produces coalescence from the heat of a constricted electric arc-gas mixture called a plasma. In recent years the term "plasma" as applied to arc processes has been associated with constricted arc techniques.

Arc Modes

The columnar arc, often called the plasma arc or arc jet, exists in two common forms, the transferred and nontransferred modes. These two arc modes are used to generate a plasma for welding. With a transferred arc the arc is established between the workpiece and an electrode within the torch. With a nontransferred arc the workpiece is not in the arc circuit

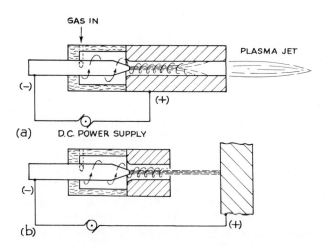

Figure 2.1 Basic modes of plasma torch operation: (a) nontransferred—nozzle the anode; (b) transferred—workpiece the anode. Note the use of vortex gas flow which gives added stability to the arc column [1].

and the arc is established between the constricting orifice and the center electrode inside the torch. In the two modes illustrated in Figure 2.1, assume the pointed electrode to be the cathode. In Figure 2.1a the nozzle piece serves as the anode and in Figure 2.1b the workpiece anode lies beyond the nozzle. The nontransferred geometry of Figure 2.1a was first introduced by Schönherr [2] in 1906 for the production of chemicals. This design was further developed in Germany during World War II for the manufacture of acetylene. Multiple 8000 kW units were installed for this purpose at Huls [3]. The Schönherr arc column is long and is stabilized in its path along the duct by a vortex flow of gas. For acetylene production the gas was natural gas with an arc voltage of about 8000 V.

In the 1930's renewed interest in the nontransferred columnar arc led to a succession of interesting torch designs and applications. Metal spraying was first described by Reinicke [4] in 1939 and further developments were made by Rava, [5] which led to our modern torch designs.

In most of these earlier references little interest was paid to the phenomenon in which the arc column extended through long ducts contained in metal. In fact, when attempts were made to describe this phenomenon, completely conflicting statements may be found. For example, Wist [6] describes a situation in which the arc column entirely fills the duct. A similar geometry [7] states that the arc column at all times is well separated from the duct walls by a cool sheath of gas.

Arc Column

Detailed studies of arc column stabilization have recently been made by several investigators. Significant advances in the theory started with the interesting work of Maecker, Peters and their associates in Germany. Maecker [8] describes the unique "wall-stabilized" arc. In the absence of a forced gas flow this arc appears to be in contact with the duct walls. However, gas aspirated into the duct is thought to provide a thin tubular sheath of cool gas which prevents the arc from actually contacting the wall.

Maecker was the first investigator to describe accurately the mechanism of magnetic pumping. The diverging electric flow lines along the arc in the vicinity of the cathode produce a force capable of accelerating the plasma to high velocity. A similar situation exists when the arc contracts to pass through the narrow passage. The magnetic field in the vicinity of the hole is distorted sufficiently to cause a reverse flow of plasma toward the cathode. These flows meet to form a plasma disk. Extending from the duct toward the anode is a stream of plasma and gas possessing high momentum.

The phenomenon of "double arcing," as first described by Maecker [8], places an upper limit on the length of the arc duct. For the wall-stabilized situation, "double arcing" occurs when the voltage drop along that portion of the arc column contained within the duct becomes greater than the combined anode and cathode voltage drops. The current flow switches from the duct to an alternate path through the conducting metal nozzle, thus establishing two arcs—one between the cathode and "A," the second from "A" to the anode. For forced gas flow considerably longer ducts are necessary to produce double arcing. In the absence of such adverse effects as momentary gas stoppage, double arcing can occur only when the arc column is fully developed, that is, in the wall-stabilized condition when essentially all heat transfer from the arc flows radially into the duct walls. Until this situation is established, the arc column is "gas stabilized," the arc column being positioned well away from the duct walls by a relatively cool sheath of gas.

The fully developed arc column has received a considerable amount of attention recently and will continue to be investigated until all known variables are verified. The numerous investigations are primarily because of the simplification of the arc model where axial properties remain unchanged. The arc in the gas-stabilized or undeveloped region of the duct continues to expand, thus presenting a much more complex situation.

Stine [9] and Weber [10] recently described the nontransferred arc in detail. The models which they proposed show that the arc column grows larger with distance along the duct, becoming fully developed or wall stabilized only after traveling a substantial distance through the duct. For example, Weber states that the arc column in helium remains undeveloped in $\frac{3}{8}$-in. diameter ducts ranging from 2 to 6 in. in length.

Determining the location along the duct where the flow becomes fully developed is obviously important. Any practical arc device for heating a gas should have a duct shorter than that required to develop the wall-stabilized condition, for beyond this critical duct length the efficiency of electric-to-gas power energy conversion drops rapidly.

All these recent investigations have resulted in a better understanding of the mechanisms of arc stabilization for the nontransferred arc. Unfortunately, little attention has been directed toward verifying whether the same situation exists for the transferred-arc, particularly under laminar gas flow conditions when used in plasma welding. Rava [5] discusses a transferred-arc welding device that is gas stabilized by a vortex gas flow. On the other hand, Gage [11], using a similar device with nonwhirling flow, reports that the arc column is wall-stabilized. In a later publication [12] the same author describes a gas-stabilized transferred-arc.

In an effort to determine the actual situation, Cooper, Palermo, and

Browning [1] conducted a series of tests on transferred-arc stabilization. The undeveloped transferred arc is quite complex. The inner arc column is small in relation to the duct opening and the arc plasma within it has fantastic temperature. The arc column is immediately surrounded by an inner sheath of hot gas. (This flow geometry was first reported by John [13] for the nontransferred mode.) The inner sheath gases are not sufficiently ionized to sustain an electric current flow. They are brought to high temperature by heat conduction from the central arc core. The major flow of gas, however, comprises the outer sheath which is not subject to significant heating and thus remains invisible. It is this outer flow that positions, or stabilizes, the arc column well away from the duct walls.

KEY VARIABLES AND MECHANISMS

The detailed discussion of the previous section was necessary as development of plasma welding has evolved. In welding, when using the plasma process, the arc gases actually cut or "keyhole" entirely through the piece, thereby producing a small hole which is carried along the seam.

Keyhole Action

The term "keyhole" has been applied to a hole that is produced at the leading edge of the weld puddle where the plasma jet displaces the molten metal, allowing the arc to pass completely through the workpiece. Presence of the keyhole, which can be observed during welding, gives a positive indication of complete penetration. Figure 2.2 shows a top view of a "keyhole" with weld bead in $\frac{1}{2}$ in. titanium. The metal that has been melted during this cutting action is drawn together immediately behind this hole by surface tension forces to produce the weld bead. It is, of course, necessary that the arc gases possess sufficient momentum to create the "keyhole." Undue jet diameter and momentum, however, lead to too large a "keyhole" and the molten metal may collapse from the workpiece.

The keyholing capability of plasma arc is related directly to the gas flow rates used. The rates commonly quoted in plasma-welding schedules are not substantially different from some gas-tungsten-arc welding. This is misleading because such rates are on a cfh volume usage basis. The difference between gas tungsten-arc and plasma-arc welding is only obvious if the gas rates are computed relative to the cross-sectional area of their respective torches. This produces gas flow rates in terms of linear feet per hour, that is, linear speed. This computation shows a very great momentum factor in favor of plasma as well as a great difference in heat transfer by gas particles.

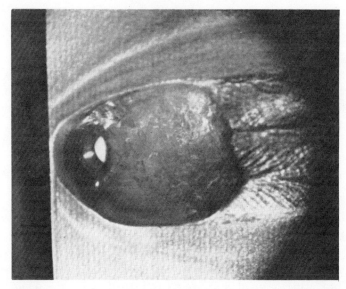

Figure 2.2 Top-view of "keyhole" with bead in ½-in. thick plate of titanium [1].

It does not appear to be practical to keyhole thin gages, that is, material less than $\frac{1}{8}$ in. thick. For material thicknesses less than this the parameters of the weld schedule are too critical for keyholing; the difference between a weld and a long, wide hole is a very sensitive relation. This limiting factor on material thickness may be only temporary since further development of the process, particularly in the area of focusing the arc energy, may extend both the upper and lower limits for keyholing.

Another mode besides keyhole utilized in welding, although infrequently, is the normal mode. This mode generally utilizes reduced orifice gas pressures so that the plasma stream does not go completely through the material; it is used most often in multiple pass welds.

Focusing

The outer sheath of cool gas has some adverse effects on the plasma welds. Primarily this major mass flow of gas is contained within this outer sheath and these gases are too cool to contribute significantly to the welding process. Their presence serves to spread the arc diameter and to influence adversely the proper "keyholing" action. One method for reducing this adverse influence is sketched in Figure 2.3. The torch geometry used a $\frac{3}{32}$-in. duct diameter. (Ducts as small as $\frac{1}{32}$ in. are useful in thinner materials.) Three separate gas flows are used. An inner, relatively low flow of gas passes in conventional manner through the arc duct. This gas is given a rotary motion to maximize "gas-stabilizing" effects. A second flow of gas, the "focusing" gas, is introduced into the plasma-arc stream through several holes steeply inclined to the axis of the arc column. It is necessary that the "focusing" gas intersect the plasma-arc stream before the latter strikes the workpiece. Given sufficient momentum, the cool impinging gases "focus" and concentrate the hot gases and arc column into a much narrower stream. The "keyhole" can be carried to greater depth and the overall stream momentum is reduced.

The "focusing" action of the impinging secondary flow is not well understood. It may partly result from some of the outer sheath gases actually being separated, or blown, from the plasma-arc stream, thus exposing the arc column to increased cooling action. When the plasma and "focusing" gas streams are of the same type gas (for example, argon), it is probably this cooling action which concentrates the arc. When the gases are different (argon plasma with helium the "focusing" flow), an increased contraction of the arc is evident. On the other hand, when helium is the plasma with argon the "focusing" flow, an increase in the arc diameter may result. These effects are probably a result of the different ionization potentials of these gases.

The third flow of gas is a conventional shield gas to protect further the weld area from atmospheric effects. The three flows may be different types of gases, each adjusted to give optimum results, or, to simplify the

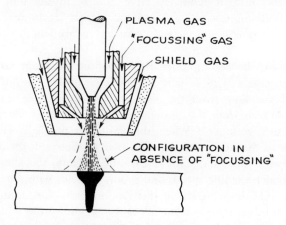

Figure 2.3 Sketch of plasma-welding torch providing "focusing" action of arc column [1].

torch structure, the "focusing" and shield gases are generally fed from a common manifold.

Plasma Gases

The plasma-forming gas used has a marked effect on the depth of penetration at a given amperage. Pure argon can be used; however, the addition of small amounts of hydrogen to the argon increases the penetration at a given amperage level. Plots of amperage for keyholing versus gas compositions for nickel alloy 200 and nickel-chromium alloy 600 have shown that an addition from approximately 5 to 7% hydrogen in argon is optimum.

Helium can also be used, but the arc-starting characteristics do not appear to be as good as the use of either argon or argon-hydrogen mixtures. Also, the penetration pattern at a given amperage is comparatively shallow with helium. This shallow penetration pattern could be an advantage for some applications, for example, when filler metal is being used.

A gas mixture of helium-argon has been recently tried but found to be not as effective as straight argon. One last factor yet to be discussed is the plasma flow rate and its effect on joint penetration.

Arc Shaping

The columnar nature of the constricted arc makes the plasma-arc process less sensitive than the gas tungsten-arc process to variations in arc length. Since the unconstricted gas tungsten-arc has a conical shape, the area of heat input varies as the square of the arc length. A small change in arc length therefore causes a relatively large change in the unit area heat transfer rate. With the essentially cylindrical plasma jet, however, the area of heat input and intensity is virtually constant as arc length varies within normal limits.

Arc shaping can be accomplished by bracketing the main orifice of the arc-constricting nozzle with a pair of auxiliary ports. The effect of the relatively cold gas flow from the side ports is to squeeze the normally circular heat pattern from the plasma jet into an oval or elongated configuration on the workpiece. When the multiport nozzle is aligned with the center line of the side ports perpendicular to the weld, the arc is elongated to line with the joint. This is desirable because the elongated heat input yields greater welding speeds and produces welds with narrower heat-affected zones. The improved heat distribution from the multiport nozzle has increased welding speeds from 50 to 100% over those obtained with single port nozzles as well as reduced the tendency for weld undercut to occur.

EQUIPMENT AND TOOLING

The welding equipment required for plasma-arc welding consists of a torch, a welding control unit, a torch mounting assembly, a remote control box, cables, and hoses. Additional requirements are a power supply, a high frequency generator, a torch travel device, and a cold wire feed unit.

Torch

Plasma-arc welding torches are available commercially for operation on either dc, straight polarity (dcsp), or dc, reverse polarity (dcrp), at arc currents to 450 A. A torch is shown in operation in Figure 2.4. Water-cooled power cables are connected at the top of the torch to supply power and cooling water to the electrode. Fittings are provided on the lower torch body for the plasma gas hose, the shielding gas hose, and cooling water for the nozzle.

Figure 2.4 Welding fixture.

Two types of electrodes are used in the plasma-arc torch. The tungsten electrode is used for straight polarity operation and is available in several diameters, depending on the current to be used. A water-cooled copper electrode is used for reverse polarity operation. The tungsten electrode is usually used for stainless steels and most other metals. The copper electrode is used for aluminum, zirconium, and other reactive metals.

One advantage of the copper-electrode is that it permits the process to operate on reverse polarity. The cleaning action of reverse polarity scours the weld zone free of harmful oxides. A second advantage involves the welding of metals like titanium and zirconium. Certain codes prohibit the use of tungsten-arc welding on these metals because of the risk of tungsten inclusions in the weld. Obviously, a copper electrode eliminates this problem.

A variety of multiport nozzle designs are available for different welding applications. The diameter of the nozzle's central port depends on the welding current and gas flow rate. The spacing between the side gas ports is influenced by the thickness of the workpiece. In general, relatively low gas flow rates (1 to 20 cfh) through the arc-constricting nozzle are preferred. This avoids excessive arc force, which produces a turbulent weld puddle and promotes weld undercut. Since the gas flow through the torch nozzle is usually too low to provide adequate shielding of the weld, auxiliary gas shielding must be supplied through a separate outer gas cup assembly on the torch. (See Figure 2.4.)

Controls

The electrical circuit for plasma-arc welding consists of a power supply, a high frequency generator, and a starting circuit. The high frequency generator is used to ionize the gas between the torch electrode and nozzle, thus permitting a high, transient, dc pilot arc current to pass through the condenser. This current provides sufficient gas ionization to establish the main arc to the workpiece.

There are two types of controls for plasma-arc welding. Both contain relays and solenoid valves to control the flow of plasma and shielding gases, to turn on and off the high frequency power for arc starting, to open and close the welding power contactor, and to control the flow of cooling water. Welding sequence is started and stopped with push buttons on the remote control box. Plasma and shielding gas flow rates are metered through the flowmeters on the side of the control box.

In addition to the features of the small control larger control units contain electronic governors for wire feed and travel carriage motors. The large control also contains gas preflow and postflow timers and a third flowmeter for the gas that protects the underside of the weld. If circumferential keyholed welds are to be made, the welding control is modified to provide also gas and current slope control.

Power Supply

A conventional dc power supply with a drooping volt-ampere output characteristic and 70 open-circuit volts is suitable for most applications, especially aluminum welding, in which argon or a mixture of argon and up to 5% hydrogen is used. Although both rectifiers and motor generators are satisfactory for the process, rectifiers are preferred because of better current stability as the power supply warms up to operating temperature. If a mixture of argon and more than 5% hydrogen is to be used, two power supplies must be connected in series to obtain the open-circuit voltage necessary for arc ignition.

Arc starting is accomplished with a high frequency spark in a pilot arc circuit consisting of a resistor-condenser combination connecting the torch nozzle to ground. A high frequency voltage applied between the electrode and the torch nozzle ionizes a path through the plasma gas inside the torch. Gas ions blown outside the torch nozzle establish an arc from the tip of the electrode through the torch nozzle to the workpiece. Reliable arc ignition can be obtained by striking the arc in pure argon and then switching over to the desired argon-hydrogen gas mixture for the welding operations.

Because of the relative insensitivity of the plasma arc process to arc length variations, arc voltage control equipment has not been used on any industrial applications. Arc voltage control, however, can be used with this process to follow contoured welded joints as long as precautions are taken to lock out the height control when current or gas sloping is employed. A plasma-arc welding machine complete with controls, power supply, etc., is shown in operation in Figure 2.5.

Tooling

Fixturing is similar to that used for gas tungsten-arc welding as shown in Figure 2.4. The fixture is made from thick common steel. The upper plates are adjustable for varying gap width beneath the joint and must be thick enough to produce adequate underbead coverage to accommodate the keyholing plasma. The lower sections of the fixture can be made movable to permit flat and "T" type welds such as burn-down and burn-through. For keyholing conditions the piercing arc is a factor in tooling design for plasma-arc welding. The underbead gas chamber must be ap-

Figure 2.5 Plasma-arc welding in the flat position.

proximately $\frac{3}{4}$ in. deep to prevent weld contamination and excessive heating of the tooling. It also makes it unnecessary therefore to employ expensive close-fitting backing bars with their attendant problems of varying chill and mechanical tolerances.

A backing bar with a simple rectangular groove is all that is required for plasma-arc welding. This type of backing bar supports the weldment, contains underbead shielding gas, and provides a vent space for the plasma jet.

Specialized Equipment

The recent development of the plasma needle arc process for the welding of foil thicknesses up to $\frac{1}{32}$-in. thick will now offer the advantages of plasma welding to foil thicknesses.

The needle arc jet emerges from a tiny orifice in collimated form, that is, as a cylindrical, nondiverging stream extending more than $\frac{1}{2}$ in. beyond the nozzle. This is substantially longer than is possible with gas-tungsten-arc welding. Plasma needle arc employs the fundamental principle of plasma-arc processes. With plasma needle arc useful arc lengths can be as great as $\frac{1}{4}$ in. thereby providing great latitude to the welder and broadening the scope of utility for welding foil and thin gage metal assemblies.

The maximum current for this needle arc system is 10 A, dcsp. Arc stability can be maintained down to extremely low currents, that is, to less than $\frac{1}{10}$ A. Although the operating currents are small with needle arc, water cooling of the torch is employed to dissipate the heat generated by the pilot arc. Two modes of needle arc are usable. The transferred arc mode can be used when the workpiece can be connected into the arc circuit. The nontransferred mode, using the pilot arc only and without electrical connection to the workpiece, can deliver a hot argon gas jet suitable for heating metallic or nonconducting materials.

One unusual faculty of the new process is the fact that the welder can adjust the shielding gas mixture to attain optimum heat and puddle behavior in the weld. The plasma-forming gas from the orifice is argon. The shielding gas is a mixture, typically, of 95% argon with 5% hydrogen, but can be varied within limits. (In fact, some metals require gases other than hydrogen; titanium for example, requires helium.) Using two flowmeters built into the control, the operator can, in effect, "fine-tune" the mixture to optimum weld conditions rather than change electrical variables.

Although plasma needle arc is a capability process rather than a speed process as such, there are advance indications that ipm (inches per minute) figures for some comparable jobs will substantially exceed those of GTA welding. In one early test, a needle arc edge weld proceeded at 100 ipm; the same job with GTA went at only 30 ipm. Arc current requirements too are somewhat lower.

ADVANTAGES–DISADVANTAGES–ECONOMICS

The state-of-the-art is such that the full potential advantages and economics of plasma arc are not known today. Since applications are continually being discovered and utilized, the following encompasses only those advantages over other welding processes proven to date or proven experimentally.

The process is fast, up to three times as fast as tungsten-arc welding. The plasma arc is also much less sensitive to welding variables than are most other methods. Included here are torch-to-work distance, electrode wear, and the height at which filler wire is added to the weld puddle. Additionally, the extent of the heat-affected zones has been reduced to the point that generous depth-to-width ratios can be produced. These ratios, in some cases, reach as high as three to one. Successful welds have been made on square butt joints in $\frac{1}{4}$-in. thick titanium plate with a bead width of $\frac{9}{32}$ in. without benefit of edge preparation by plasma welding. As mentioned previously, the bead width would exceed three times the depth in a similar weld by either gas-tungsten-arc or gas-metal-arc and would require edge preparation. It is both interesting and significant to note that best plasma welding results are obtained with *tight butt joints* rather than a beveled edge. For many welding applications, significant savings can be made by eliminating costly edge preparations. Furthermore, the nature of the "keyholing" effect in puddle formation is such that the expense of encumbersome chills can often be completely eliminated.

The aforementioned titanium weld was made at 12 ipm with a relatively low current value of 160 A, indicating a high arc efficiency. Thus it can be said at this time that the distinct advantage in using plasma-arc welding for automatic welding lies in its ability to produce extremely narrow (depth-to-width ratio) weld beads and correspondingly higher weld speed. These two factors result in a much lower degree of distortion.

Another innate advantage of the plasma welding process concerns its insensitivity to arc length distance. Whereas in an open arc the voltage will vary with the slightest change in arc length, the plasma torch will indicate only a slight change with a $\frac{1}{8}$-in. increase in arc stand-off distance. This factor can be extremely important when welding very thin materials where imperfection in mating will cause the gas-tungsten-arc to wander to the high spots and cause an imperfection in the weld.

In general, any metal capable of being welded with the gas-tungsten-arc process can also be welded with the plasma-arc process. Advantages in either weld economics or reliability are obtained on metals and thicknesses that can be keyhole welded, with the exception of aluminum and magnesium. Development is continuing on these two metals. To enumerate, the process offers the following advantages over conventional gas-tungsten-arc welding.

1. Reduced overall welding time. Typical welding conditions for gas-tungsten-arc welding of a 0.235-in. thick nickel-iron-chromium alloy would be 8 ipm and 300 to 350 A. Typical plasma-arc welding conditions for the same alloy and gage would be 16 ipm and 180 to 200 A. This is beside any weld quality considerations that certainly favor heavily the plasma-arc process in the heavy gages. In general, the welding conditions used for a particular alloy are adaptable for other alloys in the same family. For example, the same welding conditions, with possibly minor adjustments, could be used for other nickel-chromium alloys.
2. Reduced filler metal consumption and joint preparation.
3. Uniform penetration.
4. More tolerance for joint mismatch and arc length variations.
5. Elimination of the danger of tungsten inclusions.
6. Simplified fixturing.

Some of these advantages apply when welding continuously formed stainless steel tubing and circumferential pipe joints.

For some large structures involving thick joints plasma-arc welding may be an optimum choice between conventional gas shielded-arc and electron-beam methods. In these instances the plasma method can result in less distortion, lower welding and related costs, and higher joint strength than gas shielded-arc welding. Moreover, plasma can avoid the chamber restrictions of the electron-beam process with some sacrifice in travel speed, depth-to-width ratio, and consequent distortion.

The recent substitution of plasma arc for gas-tungsten-arc welding in the fabrication of a 120-in. diameter missile case showed the following savings for ⅜-in. thick D6AC (a high strength alloy steel) material. The principal advantage was a reduction in welding hours. This is illustrated by comparing the typical gas-tungsten-arc and plasma-arc conditions used to weld the rocket case girth seam shown in Table 2.1. The total weld cycle times for the two processes are compared in Table 2.2 and total weld cycle time was reduced from over 17 to 9 hr.

In a production situation the actual hours required to complete a weld were not as important as the number of work shifts involved. With the gas-tungsten-arc process, two work shifts were required to complete a weld, excluding preheat and postheat treatments. With the plasma-arc process, the same portion of the total welding cycle was easily completed in a single work shift. Maintenance costs of equipment were proportionate to the operating hours. Therefore a reduction in welding time from 7 to 1.5 hr. resulted in proportionately reduced maintenance costs on welding equipment and controls. The reduction in total weld cycle time from 17.5 to 9 hr. resulted in proportionately reduced maintenance of fixtures, heating equipment, and associated controls.

Table 2.1 Comparison of Gas–Tungsten–Arc and Plasma–Arc Welding Conditions for a 120–In. Diameter Girth Weld in ⅜ In. D6AC Steel Rocket Motor Case.

	Gas tungsten-arc		Plasma arc	
	Root pass	Fill pass	Root pass	Cover pass
Current, amp	260	275	280	200
Arc voltage, v	12	12	40	35
Travel speed, ipm	3.5	5.0	10	7
Plasma gas, cfh	...	...	18 Ar	30 He
Shielding gas, cfh	60 He-75[a]	60 He-75	60 Ar	150 He
Filler metal feed rate,[b] ipm	...	18	...	36
Number of passes	1	3 or 4	1	1
Total arc time, hr	7		1.5	

[a] He-75: 75% helium—25% argon.
[b] ³/₃₂ in. diam.
[c] 600–650° F preheat used with both processes.

Table 2.2 Comparison of Total Weld Cycle Times for the Plasma–Arc and Gas–Tungsten–Arc Processes.

Operation	Hours required	
	Plasma arc	Gas tungsten-arc
Preheating time	2	2
Mismatch measurement	1	1
Weld procedure check	1	1
Welding time	1.5	7.0
Weld cleaning	1	4
Postheating time	2.5	2.5
Total weld cycle time	9	17.5

A comparison of total production man hours and total quantities of filler metal and inert gas required to make a 120-in. diameter girth weld with the two processes showed that the plasma-arc weld required about one-third as many man hours, less than one-half the filler metal, and approximately three-fifths as much inert gas as a gas-tungsten-arc weld.

Certain disadvantages and limitations currently exist. The plasma-arc process is more complex than the gas-tungsten-arc process because two gas flows must be controlled as compared with the one shielding gas flow of the gas-tungsten-arc welding process. Thus, in order to suitably start girth welds and also eliminate the keyhole in the area of the weld overlap, slope control of both nozzle gas flow and current are required. With the gas-tungsten-arc welding process only current slope control is required for starting and overlapping girth welds.

Automatic arc length control equipment has not been applied generally to the plasma-arc process because of the relative insensitivity of the process to variations in arc length. However, if the use of such equipment were deemed necessary to compensate for excessive arc length variations, precautions would have to be taken to lock out the arc length control equipment during any period of change in nozzle gas flow or current. With the plasma-arc process the arc voltage is a function of arc length, nozzle gas flow, nozzle size, and welding current; therefore for the same arc length, voltage variations occur with variations in nozzle size, gas flow, or welding current.

Visibility at the weld is another important limitation. The torch, as presently designed, is rather bulky in the nozzle area and does not have a projecting tungsten electrode to line up with the seam in the workpiece. Alignment therefore becomes more difficult and at the same time more important because of narrower weld beads and higher speeds. Newer torch designs will probably take this factor into consideration.

Complete inert gas shielding of the weld is also difficult to maintain. Low gas flow at the nozzle and relatively high travel speeds are characteristics of the process. Therefore, auxiliary equipment such as gas trailers must sometimes be used to provide sufficient protection. This additional equipment complicates further the problem of visibility.

The addition of filler wire to the weld is accomplished by conventional means. However, besides further reducing visibility in the weld area it also introduces an important side effect. The plasma-arc process is subject to secondary arcing, which is an extraneous arc struck between the torch nozzle and the workpiece. It is caused by a voltage on the torch electrode that is high enough to strike an arc across the gas stream to the orifice insert and then to the work. The insertion of filler wire or a guide tip in the vicinity of the nozzle shortens effectively the electrical path and increases the sensitivity to secondary arcing. The remedy is found in increasing the torch gas volume, reducing the current value, increasing orifice diameter, or reducing tungsten setback. Any one or a combination of these adjustments usually eliminates the secondary arcing.

MATERIALS AND THEIR PROPERTIES

Aluminum and Copper

The plasma-arc welding process has been applied to numerous materials. Reverse polarity dc has been used in plasma-arc welding of aluminum. Only a limited amount of work has been done to evaluate the plasma-arc welding of copper. Best results thus far have been obtained on deoxidized copper up to $\frac{3}{32}$ in. thick. On this thickness the "keyhole effect" is obtained in the same manner as on stainless steel. Plasma-arc welds have also been made on electrolytic copper; however, as with the gas-tungsten-arc process, weld porosity is a problem. Sufficient work has not been done to extend the process on material thicknesses greater than $\frac{3}{32}$ in.

Nickel and Nickel Alloys

Plasma-arc welding of nickel and high nickel alloys has produced consistently high quality weld joints in thicknesses up to approximately 0.300 in. The alloys include 70-30 copper-nickel, 70-30 nickel-copper, nickel-iron-chromium and nickel-chromium. The first two alloys have been used in the sea water systems, the high pressure air and oil systems, and in the high pressure steam system of nuclear powered submarines. The weld properties of the latter alloys indicate that nearly 100% joint efficiency can be expected when welding annealed base metal.

When welding cold-worked base metal, the weld and heat-affected zone are in the annealed condition, and joint efficiencies of less than 100% are to be expected.

Steel (Hot Work Die and Stainless)

D6AC steel has been very successfully plasma-arc welded. Typical results obtained from flat tensile specimens show 209 ksi tensile and 194 ksi yield strengths which, compared with welding D6AC by gas-tungsten-arc, show 201 ksi tensile and 191 ksi yield.

Successful butt welds of type-304 stainless steel with sheared edges have been made. The sheared edges usually form a mismatch; this, however, apparently causes little difficulty when plasma welding. A butt joint would have been difficult to produce using conventional gas-tungsten-arc techniques, for the edge nearer the electrode holder would have received most of the arc heat.

Titanium and its Alloys

Titanium and its alloys are particularly suited to plasma welding. The "keyhole" is readily formed. Figure 2.6 illustrates the results of titanium tube butt welds.

Miscellaneous

Numerous metals and alloys have been experimentally welded by the plasma-arc process. Included among these are galvanized steel, René 41, and Alloy 625.

JOINT DESIGN

In general, plasma-arc welds can be used with many of the common joint types, including "T," plug, slot, fillet, and longitudinal or circumferential butt welds. The plasma-arc process can weld square butt joints of

Figure 2.6 A tube butt weld in 0.015-in. titanium.

Joint Design

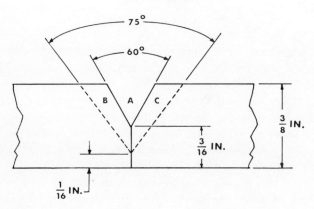

Figure 2.7 Typical joint preparation for welding of $\frac{3}{8}$-in. thick high-strength steel. Weld metal area of gas-tungsten-arc welding is $A + B + C$; weld metal area for plasma-arc welding is A.

most materials up to $\frac{1}{4}$ in. thickness in a single pass with or without filler wire addition. The use of filler wire is dependent on the applicable weld specifications.

Butt Welds

Plasma-arc butt welds are often made in one pass without the addition of filler metal. For such a weld a tight butt joint having square edges is recommended. When a multiple pass weld is needed, filler wire must be added. This technique has been used successfully on most materials ranging up to 0.250 in.

Material with thicknesses greater than 0.250 in. can be plasma-arc welded using prepared joint designs and multipass welding techniques. For materials in this thickness range the plasma-arc process requires fewer passes and offers greater weld reliability than the gas-tungsten-arc welding process. Prepared edges, U-grooves, V-grooves, etc., are normally used. Figure 2.7 shows a typical joint preparation used to weld $\frac{3}{8}$-in. thick high-strength steel. Superimposed is the outline of a conventional gas-tungsten-arc joint preparation. It is evident that the plasma-arc weld requires about one-third as much filler metal as does a gas-tungsten-arc weld. The first pass in material greater than 0.250 in. thickness may be welded with or without a filler metal, depending on the application. Subsequent filler passes must be made with reduced nozzle gas flow. The normal gas flow rates cannot be used because, in the absence of a keyhole, agitation of the molten metal becomes excessive.

"T," Slot, or Plug Welds

Plasma welding of "T" joints has been accomplished by penetrating through the top plate into the upright member to provide equally distributed fillets on each side of the upright member. A uniform bead is generated on the welded side, either flush or slightly convex, depending on the amount of filler wire added to the molten puddle. Plug or slot welding requires tooling and fixturing to provide adequate interface contact at the weld area and positive hold-down clamping adjacent to the weld. The upper sheet is then maintained in position where it cannot be lifted by thermal expansion. Spot timers and current decay control can be used to eliminate crater-cracking of plug-type welds. Wire feed timers used in conjunction with current decay rate and postgas flow controls can also be used to provide still greater assurance of consistent weld quality [15].

Fillet Welds

Nearly all plasma welding has until recently been limited to flat work because of the limited space usually available to the plasma torch. Now joints as thick as $\frac{5}{16}$ in. can be satisfactorily welded by the plasma-arc method in flat, vertical, and horizontal positions. Three services must be fed into the torch: plasma gas, shielding gas, and orifice water cooling. This requirement has resulted in somewhat larger torch diameters than are used for gas-metal-arc or gas-tungsten-arc welding. Recent refinements in the design of the plasma torch front end configuration have broadened its scope to include fillet weld capability. A cone-shaped nozzle with gas shield not only permits welding of fillet joints but also makes the introduction of filler wire easier and improves visibility. Shielding of the weld zone and arc stability are also improved by introducing the shielding gas tightly around the plasma orifice, thus reducing the aspiration effect of the plasma gas.

Since the amount of fillet welding that has been performed is limited, the conclusions thus far are only tentative. It is apparent that common fillet welding practices apply. The plasma torch should be positioned at a 45 degree angle or normal to the included angle of the weld joint to provide equal distribution of the weld reinforcement on both members and to prevent edge undercut. Fillet-type plasma welds are narrower and have deeper root penetration than those obtained with the gas-tungsten-arc welding process. Higher travel speeds are required to control excessive root penetration. The deeper penetrating arc results in fillet configurations more commonly concave than convex [15].

Plasma Needle Arc Weld Joints

Metals in the mil thickness range and in thicknesses up to $\frac{1}{32}$ in. have been welded with butt, edge, and lap-type joints.

Mismatch

Thickness mismatch normally will result in undercutting of the high side of the joint. Along with this observation it is interesting to note that widely different cross sections, for example, $\frac{1}{8}$- and $\frac{1}{4}$-in. thick material, can be joined easily, but the $\frac{1}{4}$-in. side of the joint will be undercut. However, because of the relative insensitivity of the plasma-arc process to variations in joint fit-up and mismatch, the tensile strength of the joint is comparable to the base metal strength.

APPLICATIONS AND FUTURE POTENTIAL

The plasma-arc process is suited especially to the continuous type of welding operation on uniformly thick material. As a result some of the first applications of the process have been on tube mills.

Significant increases in welding speed and weld uniformity, when compared to gas-tungsten-arc welding, have resulted from using the plasma-arc welding apparatus on a production basis in tube mills. In one tube mill, 0.225 in. wall by $3\frac{1}{2}$ in. OD Type-304 stainless steel tubing was welded with gas-tungsten-arc equipment and helium at a speed of 6 ipm. The welding speed was increased to 14 ipm by changing to the plasma arc process using a gas mixture containing 7.5% hydrogen in argon. In addition to achieving more than a 100% increase in welding speed, the plasma-arc process produced more consistent welds. A section of the 0.225-in. wall tubing is shown in Figure 2.8. The almost machinelike uniformity of the underbead typical of keyholing can be noted.

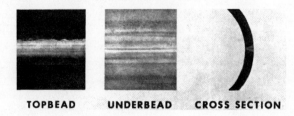

TOPBEAD   **UNDERBEAD**   **CROSS SECTION**

Figure 2.8 Plasma-arc weld in 0.225-in. wall × $3\frac{1}{2}$-in. OD Type 304 stainless steel tubing. Welding speed 14 ipm using 7.5% hydrogen in argon.

As mentioned previously in the previous section, the plasma-arc process is used for making circumferential joints in 70-30 copper-nickel, Monel, and stainless steel pipe for applications in which the pipe can be rolled. The advantages of plasma-arc welding over gas-tungsten-arc welding for this application are simplified joint preparation, increased welding speed, and elimination of backing rings. Procedures have been developed for the copper-nickel and nickel-copper piping ranging from 2- to 14-in. diameter and from 0.100- to 0.500-in. wall thickness. At present the equipment is set up to weld production joints in copper-nickel submarine piping.

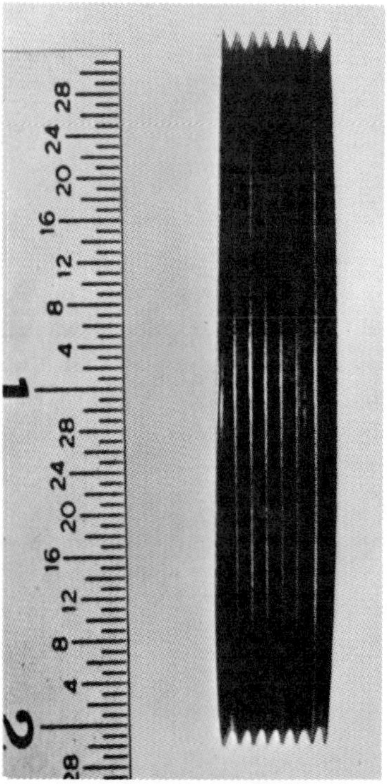

Constricted arc welding is used to join sponge compacts of reactive metals to form furnace electrodes for consumable electrode furnaces.

Since gas-tungsten-arc welding is not suitable for joining sponge compacts for manufacture of nuclear grade zirconium because of the possibility of tungsten inclusions in the base metal, gas-metal-arc welding and electron-beam welding have been used for this application. More recently plasma-arc welding with dcrp and a copper electrode has shown advantages over both former methods. Sizable savings have been made by eliminating the need for costly zirconium filler metal where the plasma process replaced gas-metal-arc welding. In comparison with electron-beam welding both the initial equipment cost and the operating costs favor the plasma-arc process. The welds are strong enough to support the weight of the electrode and have sufficient electrical conductivity to carry the remelting current. Titanium sponge compacts are also welded with the plasma-arc process. A stand-off distance of about 1 in. is used when welding sponge compact material to protect the torch from spatter caused by "popping" in the weld puddle.

Figure 2.9 A bellows edge weld in 0.007-in. stainless steel.

Needle arc welding has produced miniature bellows as shown in Figure 2.9, instrument parts, metal mesh, and light gage assemblies used in precision and aerospace equipment.

Developmental welding programs have been conducted to determine the utility and advantages of plasma-arc welding for missile case fabrication. One program required joining girth seams of 52-in. diameter, 0.175-in. thick Minuteman ICBM chamber cases of 6AL-4V titanium forgings. A five-pass procedure was required with the gas-tungsten-arc method. Three passes with filler metal addition were made in a single-vee joint from the outside of the cylinder and two additional weld beads were made on the inside of the cylinder to complete the weld at 6 ipm. With the plasma-arc process a keyhole-type root pass was made on a square butt joint at 13 ipm travel speed. The second weld was a fusion cover pass which would wash out any undercutting at 8 ipm. After a complete testing program, the following conclusions were determined:

1. Plasma weld mechanical properties are equal to or better than gas-tungsten-arc weld properties.
2. The arc time is reduced by a factor of 4.
3. Internal routing is not necessary.
4. Rewelding the root of the joint is eliminated.
5. Filler metal is eliminated.
6. Weld quality is excellent.

The second program involved the joining of girth seams on Titan III-C missile cases and the material was D6AC steel plate. Fabricators are presently using gas-tungsten-arc welding to meet quality requirements, but welding cycles are extremely long. For example, four passes with filler metal addition are required to complete either a longitudinal or circumferential weld. Maximum travel speed for any pass is about 5 ipm. The welding operation is complicated further by a 600°F preheat requirement, joint tolerance and fitup problems with this size weldment, deterioration of ceramic-coated backing bars, and interpass cleaning.

Plasma-arc techniques have been developed for making both the longitudinal and circumferential welds. Figure 2.10 shows a circumferential weld being made on a production fixture. Plasma-arc welds are completed in two passes. Total plasma arc time for an entire cylinder is about one-third of that required for gas-tungsten-arc welding. Filler metal requirements are reduced and interpass cleaning time is one-third that required for gas-tungsten-arc welding; penetration is consistent and backing bar problems are eliminated. Preliminary test results indicate that plasma-arc weld quality and mechanical properties are equivalent to gas-tungsten-arc welds.

In the field of cryogenic pipe fabrication the plasma-arc process has been used to weld preformed lengths of stainless steel pipe of various wall thicknesses and alloys. One specific application involved the use of the

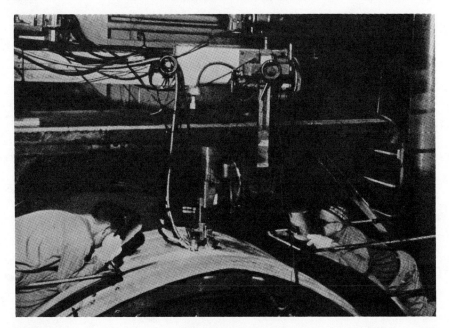

Figure 2.10 Test cylinder being plasma-arc welded on Titan III-C production fixture.

plasma-arc process in conjunction with the submerged-arc process for welding 10-ft lengths of Type-310 stainless steel pipe.

The ability of the plasma-arc process to produce x-ray quality welds consistently at high speeds led to its use in the fabrication of Inconel 600 panels. In this application the panels are used to construct large bins in the chemical processing industry.

What about the future? Plasma arc's reproducibility and ease of control should make the process most adaptable to total automation including the use of numerical control. Development work is also under way to design a manual torch. Another important aspect of the future is cost. Since many fabricators already have power supplies currently in use with gas tungsten-arc welding, the cost of a plasma-arc setup is substantially reduced. Since commercially available equipment is now coming on the market, a company will be able to purchase the entire plasma-arc system for less than $5000.

The future will bring further development and evolution of different torch designs. These designs will be particularly suitable for individual applications. For example, reversing the usual direction of the shield gas flow may well reduce bead width in the welding of thin metals. Elimination

Applications and Future Potential

of the outer sheath flow by vacuum action would produce an arc jet of high current density.

The dramatic results obtained at 300 A using single-pass welds in $\frac{1}{2}$ in. metal suggest that much thicker sections can be welded in similar fashion by simply raising the current level. Only the future can tell us what results will be obtained when the current is raised an order of magnitude. It is not improbable that thicknesses of 2 in. and more may be plasma welded from one side at a single pass with no joint preparation or filler metal required.

REFERENCES

[1] G. Cooper, J. Palermo, and J. A. Browning, "Recent Developments in Plasma Welding," *Welding J.*, 268–276 (1965).
[2] O. Schönherr, *Elektrotechnische Zeitschrift*, Berlin, p. 365, April 22, 1909.
[3] H. Gladisch, "How Fuels Make Acetylene by DC Arc," *Hydrocarbon Proc. Petrol. Refiner*, **159** (June 1962).
[4] R. Reinecke, "Apparatus for the Production of Metallic Coatings," U. S. Patent 2,157,498, May 1939.
[5] A. Rava, "Electric Arc Torch Apparatus," U. S. Patent 2,768,279, October 1956.
[6] E. Wist, "Means for Converting the Electric Arc into an Elongated Flame," U. S. Patent 1,892, 325, December 1932.
[7] German Patent 685,455.
[8] H. Maecker, "Plasma Streams in Electric Arcs as a Result of Self-Magnetic Compression," *Z. Physik*, **141**, 198 (1955).
[9] H. A. Stine, V. R. Watson, and C. E. Shepard, "Effect of Axial Flow on the Behavior of the Wall-Constricted Arc," AGARDograph 84, p. 451 (September 1964).
[10] H. E. Weber, "Growth of an Arc Column in Flow and Pressure Fields," AGARDograph 84, p. 845 (September 1964).
[11] R. M. Gage, "Arc Torch and Process," U. S. Patent 2,806,124.
[12] R. M. Gage, "The Principles of the Modern Arc Torch," *Welding J.*, 959–962 (1959).
[13] R. R. John, S. Bennett, L. A. Cass, M. M. Chen, and J. F. Conners, "Energy Addition and Loss Mechanisms in the Thermal Arc Jet Engine," Paper 63–022, AIAA Electric Propulsion Conference, (March 11–13, 1963).
[14] S. Filipski, "Plasma Arc Welding," *Welding J.*, 937 (1964).
[15] *Welding Handbook*, Section Three, Fifth Edition, P. 53.8, 1964.

Chapter 3

ELECTRIC BLANKET BRAZING

Large furnaces have made necessary prior art brazing processes in the event that the components to be brazed are sizable, which has resulted in brazed honeycomb panels being exceedingly expensive. Conventionally, an upper and lower form block accurately contoured to match the respective surfaces of the part to be brazed have been employed to hold the contour of the assembly within desired limits and prevent distortion of the assembly during the brazing operation in a furnace.

Furnace brazing is subject to several undesirable limitations. Furnaces are necessarily large in order to be able to accommodate the workpieces. They are slow to heat and cool because heating of the assembly to be brazed depends largely on air convection and radiation which is less efficient than conduction heating. Moreover, contoured form blocks are expensive and sometimes cause rippling and buckling in the surface of the assembly being brazed as a result of uneven distribution of the bonding pressure over the panel surface. Engineers can no longer wait for designs to materialize before establishing possible fabrication techniques. The design-to-hardware pace is too swift. To meet the demand "production" has been forced to anticipate "design," which has resulted in the growth and importance of production research. Consequently, the invention of the electric blanket came about.

This technique was to provide an electrothermal method of brazing metal skins to metal honeycomb core which obviates to a large extent the above-mentioned limitations of furnace brazing. Heating of the assembly to be joined was accomplished primarily by conduction, with the assembly in compression relative to the thermal source so that brazing was effected rapidly and efficiently. Means were also provided for distributing bonding pressure uniformly over the surface of the assembly; at the same time the number of contoured forms required was reduced from two to one.

THEORY AND PRINCIPLES OF PROCESS

The virtues of sandwich construction have been known for over a decade. As early as 1942 adhesive bonding received considerable attention from the Martin Company and U. S. Plywood. The limitations of the conventional material in construction, namely aluminum, and the process of fabrication, adhesive bonding, have become apparent more recently. The constant redefinition of operating environment has been responsible for this discovery. When environments exceed 330°F, both the material and the process are inadequate. This has resulted in a second generation of sandwich structure brazed honeycomb mainly precipitation—hardened stainless steels, superalloys and titanium alloys.

The advantages of honeycomb in the construction of supersonic aircraft, missiles, truck bodies, building structures, and space vehicles are so great that it is worth the tremendous effort necessary to determine the answers to the many processing problems that have risen and are continuing to arise. High strength, lightness, thermal insulation, compression, and torsion are some of the characteristics in which honeycomb sandwich structures excel.

Recognizing certain limitations and extravagances in the approach of fabricating sandwich-type structures through the use of large furnaces, the Martin Company devised and is generally given credit for developing early electric blanket brazing methods. The initial idea for this type of brazing, in place of the retort and furnace techniques, was evolved from studies conducted at the University of Colorado. Personnel there were doing simulated aerodynamic heating tests on a model wing with heavy pads. The pads brought a conduction heating blanket in firm contact with the specimen. Electric current passing through the heating strips caused heat to be conducted into the specimen. The strips used were Kovar (an iron-nickel alloy). It was necessary to insulate them from the test piece with separate anodized sheets.

It was learned later that if zirconium heating strips were used, an insulating oxide (ZrO_2) film could be deposited on the strips by heating them in air. The insulating oxide was effective to test limits of 700°F. This meant zirconium strips could be laid directly against the piece to be heated.

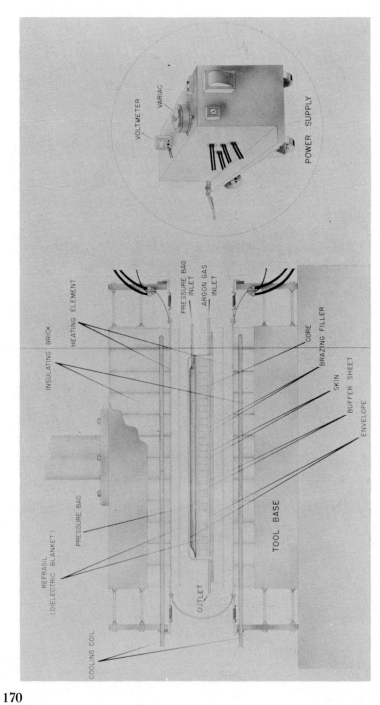

Figure 3.1 Schematic of the first electric-blanket brazing tools and portable power unit.

With this basic information the first electric blanket brazing fixture was designed (Figure 3.1), constructed, and evaluated. The first attempts were marred by lack of a good dielectric between the heating strips and the workpiece. Zirconium was no longer effective since operating temperatures were now in the 2000°F range. Stainless steel and ceramic combinations were also ineffective. A suitable dielectric was finally found in Refrasil (a leached silica cloth). Almost simultaneously nichrome V was found to be a more suitable heating element material.

Thus the early electric blanket brazing fixture evolved as a unit in which the heating elements, the cooling system, and the accessory tooling were all self-contained. It was capable not only of brazing but also subsequent heat treating.

MECHANISMS AND KEY VARIABLES

The main factors in the blanket process are uniform control of temperature and flatness of the complete assembly. In the early development the brazing heat was supplied by a portable electrical unit as shown in Figure 3.1.

The fixture consists of an upper and lower blanket (platen) of insulating firebrick in an aluminum framework. A network of steel water pipes is interplaced through the bricks so that the fixture can be cooled rapidly after an assembly is brazed. The nichrome V heating elements runs lengthwise across the face of the firebrick. The elements produce 14 W./sq. in. at 5 to 35 V. The voltage and temperature are regulated from a central control panel utilizing a 44 V, ac, three-phase power supply.

Thermocouples embedded in the surface of the fixture and workpiece indicate temperatures within 25°F for the brazing cycle and within 10°F for any subsequent heat treating. A developmental blanket and brazed panel is shown in Figure 3.2. Steel tubes inserted at each corner of the bag (Figure 3.1) are used for the removal of air and replacement with argon. Argon is circulated under pressure and provides an antioxidation environment throughout the entire brazing cycle. Pressure is provided by the dead weight of the top platen. Figure 3.3 shows an opened envelope after brazing and its contents.

The alternative method of pressure application shown in Figure 3.1 is to use a stainless steel pressure bag inside the atmospheric envelope. Argon is circulated into the pressure bag and thus provides a gaseous cushion between the top platen and the completed assembly. This provides a more uniform application of platen pressure. In both instances the outer steel envelope is separated from the heating elements by a layer of Refrasil to prevent any electrical short.

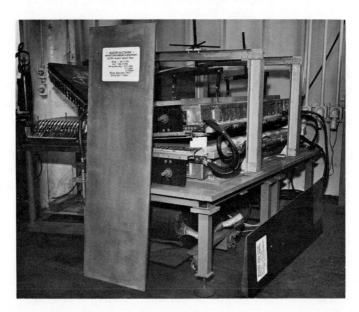

Figure 3.2 A developmental blanket fixture capable of brazing panels 6 ft by 3 ft.

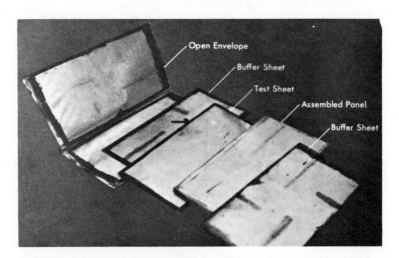

Figure 3.3 Contents of a sealed envelope. The test sheets are used to check physical properties and the buffer sheets have slits to compensate for thermal expansion within the envelope.

Mechanisms and Key Variables

When refinements were continuously made sizes increased until production-oriented-type blankets evolved, which are discussed in the next section.

TYPES OF EQUIPMENT AND TOOLING

Since the early days of experimentation and development numerous strides have been taken in the development of blanket-brazing techniques. Improved processes, mainly in the aerospace industry, have brought the all-important goal of efficient and relatively cheap sandwich fabrication much closer.

Flexible Electric Blanket

This type of blanket shown in Figure 3.4, which utilizes a high-temperature wire mesh heater in a blanket of high-temperature silica insulation, is not too different in principle from the electric blankets used in many homes. In fact, the collection of silica insulation pads used around the brazed part and the electric blanket makes the present test layout look very much like a pile of feather mattresses and pillows. The flexible electric blanket with resistance-heating elements sandwiched between the refractory

Figure 3.4 Blanket unit and controls.

cloth covers has the part enclosed in a pan or envelope with a rigid form block. The pan is then placed on an electric blanket underlaid with a refractory insulating pillow and then covered with a second electric blanket. The lay-up is completed with a second insulating pillow. This new blanket-brazing method has produced production size panels with a minimum of equipment and relatively little space.

Vacuum Blanket Brazing

A vacuum blanket without the usual steel envelope containing argon has been developed. Heating is accomplished with nichrome strips. Since there is no vacuum envelope, the components of the assembly cannot be held together in the customary way. Instead, a pressurized bladder is used between the part and the upper tool surface. Since the tooling remains relatively cool, there is little trouble with outgassing during the pump-down, and the tooling can be made from low temperature materials. For rapid cooling liquid and gaseous coolants are circulated through a cored plate sandwiched between the heating elements and the part.

Ceramic or Platen Blanket

This type of blanket brazing uses modular, cast ceramic platens with integral heating elements and cooling holes. The holes slightly larger than the heating wire are cast in the platens and resistance wiring run through. The large holes make it easy to replace wiring. The platens, finished to close tolerances, are used as contour forms in brazing. After cleaning, sandwich assemblies are enclosed in a steel envelope and placed between the platens. To assure close contact of parts to be brazed, a vacuum is maintained in the envelope and external pressure is applied to the platens.

The ceramic material is then heated electrically to the brazing temperature and held there for a short, controlled period. The platens and the honeycomb sandwich are cooled rapidly by air forced through the holes in the ceramic material. Even subzero temperatures are possible while the honeycomb assembly is still in the tool.

Single Retort Blanket

The components to be brazed are laid up in a single retort as shown in Figure 3.5, Copper is used as a temperature-gradient equalizer and air is circulated through cooling grooves in the copper for rapid, controlled cool-down. The retort and the copper are placed between the platens shown

Types of Equipment and Tooling

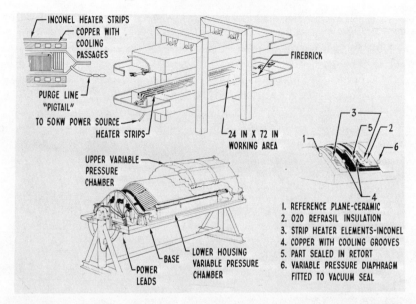

Figure 3.5 Schematic of production-type blanket fixtures—contour and flat.

and the lid is lowered into position. The platen tool shape is built up out of foamed fused-silica blocks with a fused-silica binder. The largest tool sections made by several manufacturers measure 8 by 12 ft. Up to three of these have been fitted together to maximum size of 24 by 36 ft.

Before brazing, electric heating elements are placed on the grid pattern of the bottom tool first. The retort is then welded tight and laid up in the electric blanket-brazing fixture. (See Figure 3.6) Both upper and lower dies, between which the panels are brazed, contain plenum chambers that permit rapid and uniform heating and cooling. Glasrock cement and foam bricks, mentioned previously, are used to form plenum chambers. This increased use of Glasrock material has resulted from its following properties:

(1) excellent thermal shock;
(2) practically no expansion or shrinkage;
(3) tools can be cast very close to finished dimensions (0.003 in.);
(4) no limit on the sizes or thickness of Glasrock tools.

Heat is supplied through the strip heating elements by two power supplies—one for the top heating elements and another for the bottom ones. At present these supplies can be either a manual system or automatically programmed.

Electric Blanket Brazing

All the tools and panels discussed previously were flat. In the lower sketch of Figure 3.5, an experimental dual-chamber test tool is shown that was built for a severely contoured section of a duct. It is similar to the photo on the left, which is a tool for a flat panel except that a sheet metal diaphragm is placed between the retort and the upper chamber. When this upper chamber or lid is placed in position, rubber seals create an airtight upper and lower chamber. This tool can be operated using positive pressures or modest vacuum for brazing. It can also be used at extremely low vacuums, which eliminates the need for argon, and is trunnion-mounted. The production version of the tool has produced curved panels to enclose the pods for the J-93 turbojet engines that are installed in the RS-70 airplane.

With the successful fabrication of production parts for various purposes, developments continued to determine the feasibility of electric-blanket brazing at 2300°F and use of molybdenum-coated heating elements for the higher temperature applications.

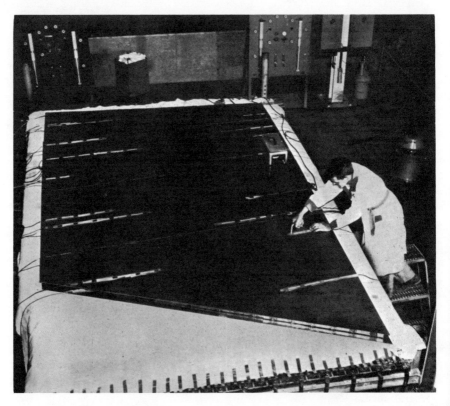

Figure 3.6 Layup of lower half of an electric blanket brazing fixture.

Types of Equipment and Tooling

Figure 3.7 Developmental blanket fixture showing upper and lower Kanthal heating elements.

Figure 3.7 shows the laboratory tool including the aforementioned heating elements that show the following.

1. Brazing temperature of 2300°F using electric blanket heating is feasible.
2. Time rate of heating is limited by the current density capabilities of the heating elements.
3. Power input cannot exceed 15 W/sq. in. of heating surface.
4. Glasrock was the most suitable refractory for electric-blanket brazing from the viewpoint of thermal conductivity, rate of expansion, thermal shock, and time rate of cooling.

In conclusion, it is obvious that savings in heat energy, time, cost of furnaces, and other factors add up to major economies in the manufacturing primarily of honeycomb panels by blanket brazing.

ADVANTAGES AND ECONOMICS

Advantages

The electric blanket was developed to eliminate the inherent disadvantages normally encountered with other specialized heating equipment such as furnace brazing. Some of the important advantages gained through the adaptation and utilization of the electric blanket method are the following:

1. The greatly increased heating efficiency results in substantial power requirement reduction—65% efficiency as compared to 10% with furnace brazing.
2. Elimination of specialized furnace equipment.
3. More efficient utilization of assembly floor area and a substantial reduction in attendant support areas.
4. Substantial reduction of occupational hazards and elimination of specialized journeyman talents.
5. Decreased brazing cycle when compared with conventional brazing techniques—an average of 6 hr as compared to 23 hr.
6. Functional design of blanket-brazing control equipment permit extreme mobility. One set of control equipment powered several blanket brazing fixtures at different intervals. This produces a reduction of power supply and temperature control units as well as a reduction in overall power requirements.
7. A substantial increase in tooling service life as compared with conventional brazing techniques.
8. Capability of rapid and economical expansion as required in event of a national emergency.
9. The amount of handling is reduced since the panel is brazed and heat-treated in the same fixture.
10. Distortion during brazing and heat treatment is controlled with less difficulty as the panel is held in the jig throughout the cycle.
11. Greater controllability is attained since heat inputs are varied on different parts of the same panel.

Economics

The price of brazed panels has decreased considerably over the past eight years with the introduction of newer methods for brazing and with improvements in general technique and control. Initial brazed panel costs were extremely high. However, the fact that these costs have been intermingled with individual research and development program costs makes them difficult to identify. Brazing costs vary considerably according to the size and configuration of the panel. Since no two panels present exactly the same type of manufacturing problem, it is difficult to compare manufacturing costs or to arrive at valid average cost figures. Without some general analysis of the cost structure it is difficult to determine the presence or absence of any price improvement trend.

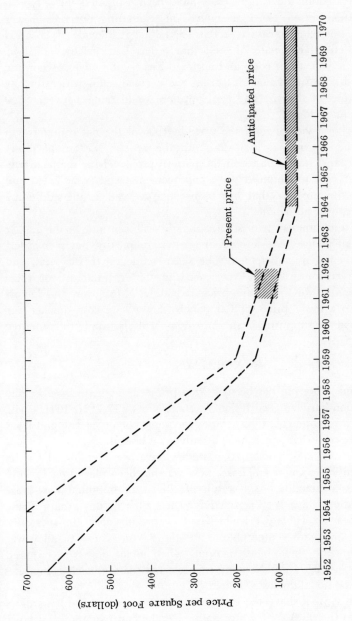

Figure 3.8 Brazed honeycomb core panel price trend [1].

Based on flat 2- x 3-ft panel with 1/4-in. cell, 0.0015-in. foil, 17-7 pH resistance-welded cres core, 0.010-in. face sheets, normal edge attachments and no inserts.

To give some indication of production performance, both at present and as anticipated in the future, costs have been estimated for a typical flat configuration-brazed panel assembly. For estimating purposes it was assumed that this panel measures 2 ft by 3 ft, that it is made of 17-7 PH resistance-welded CRES material, and that it has $\frac{1}{4}$-in. cell, a 0.0015-in. foil thickness, a 1-in. thick core, and enclosed Zee section edges. According to experienced core panel manufacturers, the average selling price for any quantity of panel of this size and configuration would range between $150 and $200 a square foot.

Figure 3.8 shows the range of the price in 1960 of the flat configuration panel described above and indicates both the general pattern of prices for this same panel in the past and the pattern projected for the next-five-year period. A very pronounced price improvement trend is indicated and clearly shows the progress that has been made as well as the additional progress that can be anticipated.

Contoured and tapered panels undoubtedly will continue to be higher priced than flat panels because of their greater complexity, but a considerable price reduction is anticipated. The sharp reduction in the scrap and rejection rate currently being experienced in the manufacture of these panels is an indication of future price reductions. Moreover, the factors contributing to a lower price for flat panels—lower core cost, quicker and cheaper methods of supplying heat, and so on—will also apply to contoured and tapered panels.

MATERIALS

Electric-blanket brazing has been utilized primarily with steels, especially the precipitation-hardening corrosion resistant alloys (17-7PH, PH15-7Mo, PH14-8Mo, AM350 and AM355). However, some cobalt, nickel, and iron-base superalloy sandwich panels have been blanket brazed.

Base metal-braze alloy combination studies have been conducted on the five alloys mentioned; most emphasis, however, has been placed on 17-7PH because of its availability. As a result of this work optimum base-braze alloy combinations have been selected for each alloy in this group. These selections appear in Table 3.1 when based on brazing characteristics only, the five base alloys are comparable. Thus the 93 Ag-7Cu + Li alloy was selected for brazing these base materials when nodal flow was required and the 62.5 Ag-32.5 Cu-5 Ni alloy was selected for sandwich composites when little or no nodal flow was desired.

AM350 and AM355 have the best oxidation resistance of the base alloys tested, whereas the 99 Ag + Li braze alloy has the best oxidation resistance of the filler alloys evaluated at temperatures up to 1000°F.

In salt spray tests a number of base-braze alloy combinations failed after

Table 3.1 Optimum Base-Braze Alloy Combinations for Brazed Sandwich Structures.

Optimum Base-Braze Alloy Combinations for Brazed Sandwich Structures

Skin and Core Material	Operational Range - °F	Braze Alloy	Recommended Braze Temperature - °F	Holding Time-Min.	Atmosphere	Remarks
A286	800-1200	Ni-Cr-Si	1915-1925	15-20	Argon or Vacuum	Dew point -100°F is required
422	RT-800	90Ag-10Cu Ni Clad	1850	15	Argon	Where no flow up nodes is desired
422	RT-800	99Ag-Li	1810	15	Argon	Where nodal flow is required
Haynes 25	1200-1800	Ni-Cr-Si	1915-1925	15-20	Argon	Dew Point -80°F is required
Haynes 25	1200-1800	Ni-Cr-Si	1915-1925	5	Vacuum	--
Haynes 25	1200-1800	Ni-Cr-Si	2150	2-5	Argon	Dew Point -80°F is required
B120VCA	RT-600	97Ag-3Li	1375-1400	10	Argon	Dew Point -100°F is required
17-7PH PH15-7Mo AM350 AM355 PH14-8Mo	RT-800	Ag-Cu 93-7+Li or Ag-Cu-Ni 62-32-5	1650-1700 1650-1700	10 5	Argon Argon	Where nodal flow is required Where no flow up nodes is desired

Figure 3.9 The inside of a honeycomb panel indicates the flow of braze alloy along and up the cell walls.

150 hr exposure. These failures were in the form of complete core-skin separation (delamination) and are attributed to a type of crevice corrosion. This form of attack and separation was originally noted with 17-7PH brazed with silver base alloy 85Mn-15Ag-Li. In general, the 93Ag-7Cu-Li type filler alloys show excellent resistance to salt spray corrosion.

Figure 3.9 is indicative of the flow and uniformity of filleting achieved with blanket brazing. The effectiveness of the 62.5 Ag-32.5 Cu and 5 Ni braze alloy is due to the infiltration of the silver-copper alloy into a matrix of a highly porous pure nickel "sponge." The sponge holds the molten alloy in place and prevents runoff. It is sufficiently porous to enable the alloy to form a solid bond between the joint surfaces. The alloy is used in all brazed honeycomb areas of the RS-70 airplane except in the wing section panels where a silver-copper-palladium-indium composition is used. This alloy was chosen because of its lower thermal conductivity. Low conductivity is important to retard transmission of heat (from skin friction in flight) to the integral fuel cells in the wings.

JOINT DESIGN AND TEST RESULTS

Honeycomb sandwich structures are composed of thin, high-strength facings integrally brazed to lightweight core material. The selection of materials

Joint Design and Test Results

to be used in this structure is of primary importance in arriving at a satisfactory design providing optimum performance. The primary parts of this composite structure are the following.

Facing Selection

Generally, the facings in a sandwich structure are selected in order to provide the highest compressive strength-to-weight ratio economically available in the size range under consideration. Figure 3.10 illustrates an 0.010-in. PH 15-7Mo brazed facing for a closure section surrounding a jet engine. In every instance the facing selection must meet the multiple criteria of economy, structural efficiency, and process compatibility.

Braze Alloy Criteria

There are now available to designers any range of brazing systems for attaching honeycomb core materials to high-strength facings. The choice of most high-strength structural brazing alloys is based on metallurgical compatibility with the base metal, which provides structural integrity throughout the environmental conditions encountered.

Figure 3.10 Compound contoured sandwich panel for jet engine closure.

Honeycomb Core

Core materials including cell size, gage, and height are selected on the basis of structural performance, environmental requirements, and economy.

With proper selection of facing, braze alloy and honeycomb core, the designer can achieve optimum structural performance with minimum weight. For example, a beam of steel plate weighing 68.6 lb can be replaced by a structural honeycomb sandwich beam of 7.8 lb with the same deflection and strength. Therefore the design opportunities for honeycomb sandwich structures are apparent.

Edge Member Details

Normally the edges of a sandwich structure are not required to be sealed off for environmental purposes since most core materials are resistant to normal atmospheric exposure. It usually is imperative, however, to close the edge off to prevent incidental damage to the core and to provide strengthening of the sandwich to carry edge attachment loads. Sketches of typical edge details are shown in Figure 3.11. A simple inspection of the various means shown should provide an idea for a method suitable for a design capable of meeting any specific edge loading.

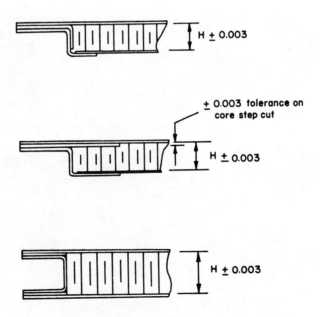

Figure 3.11 Typical edge-member configurations for sandwich panels.

Attachments and Critical Areas

In most actual honeycomb sandwich applications the success of the design is a direct measure of the ingenuity of the designer. The reason for the heavy dependence on the designer's resourcefulness is that the details of such lightweight composite materials require many unusual innovations in order to carry in-loads that are higher than can be normally sustained by the overall structure. Typical of these problems is the design of attachments, for example, two methods of attachment of frame caps. Densified core in the heavily loaded areas is attached to the light core adjacent, and welding and brazing of the frame cap to the inner skins can be performed. In addition, there are typical splice methods; when access to both sides is available, welding can be from both sides; for access from one side only, however, there is a welding approach in which the inner skin is welded first and then a cover plate is added to the outer skin.

In many designs it is desirable to have the overall lightweight of a sandwich structure, but it is still necessary to meet extremely high load requirements in some localized areas. These conditions can be met by using a high density honeycomb in the local area, by adding doublers under the skin (see Figure 3.11), or by the use of various composites of inserts.

The final determining factor of a satisfactory design is the strength of the panel when subjected to the loading conditions most likely to be encountered in service. Destructive tests offer the only known means of determining the actual strength of sandwich structures or materials.

Among the more generally significant properties of the basic materials used to form sandwich construction are the ultimate strength, the yield strength, and the modulus of elasticity in tension and compression and in shear. Standard materials-testing machines are used to make simple tension tests on samples of core foil and facing sheets.

Although bare core is tested occasionally in flatwise tension or compression or in shear, more frequently samples of composite sandwich are used to determine these properties. The most important test of bare core is the one made to determine the node-weld strength, which is done either by the core delamination test or the node-weld shear strength test. The core delamination test is performed by pulling a small section of core blanket in a direction that separates the individual strips of foil material that have been joined usually by welding at the nodes; several nodes are tested in parallel by tension-loading them. The node-weld shear strength is evaluated by sectioning a piece of core blanket so that tension on the sample applies shear loading to a minimum of nine single node-welds in series.

The brazing alloy contributes a number of important characteristics to the quality of a brazed joint. In addition to the destructive tests performed on sections of the composite structure, brazing alloys are also evaluated on the basis of trial samples made in the course of process development to determine the type of bond formed and other properties affecting bond quality. The quality of such braze alloy characteristics as filleting, wettability, warpage, flow, ductility, and undercutting usually can be determined by visual examination of the test samples. The exact nature of the bonding and the metallurgical structure are determined by metallographic tests.

In the following types of tests specimens are subjected to calibrated forces in standard-materials testing machines to evaluate various aspects of the strength of composite sandwich structure:

(1) flatwise tension;
(2) edgewise and flatwise compression;
(3) core shear strength (direct and short beam);
(4) flexure (long beam);
(5) peel;
(6) fatigue (edge compression, end moment and free resonance).

Although the fatigue of materials is of paramount concern in structural design, especially aircraft, and although sandwich structures generally offer an improvement in fatigue resistance over that obtained with conventional structures, fatigue test requirements or specifications for sandwich materials have yet to be well standardized. Three methods have been proposed as being the most promising for further use in industry. These are based on edgewise compression loading, the imposition of end-bending moments, and the development of cyclical stresses by vibration of the specimen at its resonant frequency. Considerable experience remains to be acquired before fatigue test results obtained by these or any other methods can be correlated with service conditions.

APPLICATIONS AND FUTURE POTENTIAL

Electric-blanket brazing is a joining process of considerable utility and in a honeycomb panel it is a structure with promise for the future. Several uses of the last five years include complex components that were required on the RS-70 "Valkyrie" and the B-58 "Hustler" airplanes.

A current use of blanket-brazed sandwich panels is shown in Figure 3.12 where PH14-8 Mo panels are installed into the inner crew compartment of the Apollo command module. Another current blanket-brazing application is the fabrication of various types of sandwich thermal conditioning panels and heat exchangers for aircraft and missiles.

For the future Figure 3.13 lists numerous potential assemblies and parts

Applications and Future Potential

Figure 3.12 Blanket-brazed sandwich for Apollo space vehicle.

AIRCRAFT SKINS	COMPRESSOR AND TURBINE BLADES	MISSILE SKINS
LEADING EDGES	GUIDE VANES	MISSILE FINS
EMPENNAGE	COMPRESSOR CASINGS	MISSILE NOSE SECTIONS
FLIGHT CONTROL SURFACES	TAIL CONES	ENGINE COMPARTMENTS
WIND TIPS	THRUST REVERSAL COMPONENTS	
DIVE BRAKES		
BELLY LANDING PADS		
EXHAUST BLAST PANELS		
HYDRO SKIS	FLUID HEAT EXCHANGERS	NUCLEAR REACTOR STRUCTURES
HELICOPTER BLADES	INSULATED STIFF DUCTING	VACUUM FLASKS
PRESSURIZATION DOORS	INSULATION SHROUDS	ACOUSTICAL INSULATION
BOUNDARY LAYER BLEEDS	FIRE WALLS	EXPEDITIONARY STRUCTURES
INTEGRAL FUEL TANK STRUCTURES	ANTI-ICING STRUCTURES	
SHOCK MOUNTS		
GAS PRESSURE SEALS		

Figure 3.13 Current and future blanket-brazing products.

capable of being blanket brazed. In addition to this list could be added the following:

(1) jet shrouds;
(2) jet blast deflectors;
(3) combustion chambers;
(4) turbine casings;

(5) trucktrailers;
(6) inner refrigerator cabinets (refrigerant could circulate inside the sandwich);
(7) pressure vessels;
(8) radiators;
(9) condensers and evaporators.

From the discussion of the foregoing developments it can be seen that the properties of all-metal sandwich structures have found broad application in industry.

During the development of the state-of-the-art, electric-blanket brazing has become the process best suited for production of sandwich structures. The combination of high structural efficiency (honeycomb) combined with reasonable manufacturing cost (blanket) make an attractive and potent duo for future production.

REFERENCES

[1] W. A. Tweedie, P. D. Tilton, and R. C. Rollins, *Present Status of and Future Outlook for All-Metal Sandwich Construction for Air Vehicles,* SRI Project No. 1U-2932 (August 1959).

Chapter 4

LASER WELDING

The advent of high intensity, low momentum surface heating devices is having a natural impact on the field of materials processing, particularly fusion welding. The laser and electron beams offer potential capability for precise control of energy and location that cannot be approached by older sources such as arcs and flames, also an important advantage is that under certain operating conditions they transmit little or no thrust to the material being worked. However, the extreme intensity of these sources, expressed in dimensions of power per unit area, presents problems as well as advantages when considered with the objective of local melting as in fusion welding.

With more conventional welding heat sources there was seldom, if ever, any question of too high an intensity. Indeed the problem has generally been quite the reverse in that there is a tendency for metals having relatively high thermal conductivity to conduct the heat away from the weld region almost as fast as it is supplied. In this sense the term "melting efficiency" has been used to express the fraction of the total heat delivered to the metal that is actually used for melting, which, with a typical open arc, generally falls below 50%. Melting efficiency is related directly to the intensity of the heat source, and, as the intensity increases, the melting efficiency also increases. In fact, with the intensities available from laser or electron beams they approach 100%.

The newest and most exciting of these two mentioned processes is the laser (sometimes called optical maser). It represents a virtually completely new scientific development. The possibility of a laser was first suggested in 1958 and the first experimental model was made in 1960. Since that time advances in laser technology have been rapid and numerous laboratories are performing laser research. Although much of this research is of a basic scientific nature, many of the newer developments can be applied to welding. Experimental laser machines are already in use. (See Figure 4.1.) No basic scientific discovery in history has been applied so fast to metalworking.

Many scientists consider the laser the wonder tool that emits the most intense radiant energy known to man and will revolutionize metal processing techniques within five years.

THEORY AND PRINCIPLES OF PROCESS

A laser's most important characteristic is its ability to generate coherent electromagnetic radiation at either infrared or light frequencies. A laser also can amplify light, a function from which its name is obtained: light amplification by stimulated emission of radiation; this phase accurately describes what a laser does—it amplifies light. To appreciate the potential of lasers, one must understand the nature of coherent electromagnetic energy, particularly, coherent infrared and light. The waves of coherent energy are uniform in length, extremely well collimated, and are "in phase" in the same direction over the width of a beam of energy. Uniform wavelength and complete phase alignment make a beam of coherent energy somewhat analogous to a strip of corrugated sheet metal. By contrast incoherent energy is analogous to a gushing brook with waves of varying lengths tumbling over one another in wild disorder. The sun and the stars are the most powerful sources of incoherent energy. They emit radiation of wavelengths ranging from kilometers to fractions of angstroms.

Figure 4.1 Experimental unit for laser welding. Large, shiny cylinder is the laser head. Operator uses microscope for precise positioning of workpiece for welding.

Despite the abundance of natural (incoherent) electromagnetic energy, much of recent technological progress is based on use of man-made coherent energy. To produce this coherent energy the action of a laser depends on well-known phenomenon—the ability of some materials to absorb energy in the form of electromagnetic radiation, then re-emit electromagnetic radiation. (Light is one form of electromagnetic radiation.) This phenomenon is utilized in fluorescent lamps and in television picture tubes. In the television picture tube, for example, the back face of the screen is coated with a fluorescent material. A beam of electrons sweeps over this face, creating a series of fluorescent spots that form a picture visible to the human eye.

Fluorescence is an atomic reaction. When the electronic structure of an atom of a fluorescent material is stimulated by radiation, the atom is said to be in the excited state, one analogous to a spring that has been compressed. When the stimulus is cut off, the atom returns to its normal energy state, emitting energy in the form of electromagnetic radiation, usually light. Each fluorescent material is stimulated by light of certain colors (wavelengths) and emits light of the same or other wavelengths.

The search and development of this coherent energy began essentially in 1958 when A. L. Schawlow and C. H. Townes published their theoretical paper that predicted that laser action should be possible. The material they chose for sample calculations was potassium. They showed that its spectroscopic properties made it a promising candidate and, since it was to be used in vapor form, extremely good optical quality was assured. Although considerable effort has been expended on this material, it has never lived up to expectations. Laser oscillations in potassium have not yet been reported.

Since the gain per unit length resulting from stimulated emission is proportional to the number of excess electrons in the upper level of the laser transition when compared with the number in the lower level, one would suspect that larger gain per unit length could be attained in materials with high concentrations of the active ions or molecules (such as solids or liquids) than in materials such as gases in which the concentration is lower. Although there are some exceptions, this has generally been true. It was for this reason that ruby (aluminum oxide doped with Cr^{+3}) was chosen as a likely candidate by several investigators, including T. H. Maiman [2,3]. This material has been prepared artificially in single-crystal form since the late nineteenth century. Techniques had been developed to such a point that large single-crystal rubies of good optical quality were readily available commercially. In 1960 Maiman succeeded in producing laser action in a rod of this material, and the new laser technology was born. Dr. Maiman's success was duplicated by Collins, Nelson, Bond, Garrett, Kaiser and Schawlow [4].

Theory

Atomic theory explains that atoms and ions exist only in states with certain allowed ranges of energy. These are the narrow energy levels of an atom in a gas or of an ion in a dielectric crystal. An example of an ion would be ruby: chromium Cr^{+3} ions in an aluminum oxide lattice. It is also known that the electrons in a semiconductor can exist only in certain well-defined energy bands. Although the discussion of the phenomenon of stimulated emission presented refers only to atomic energy levels, it applies as well to energy levels of an ion in a dielectric solid as to energy bands of a semiconductor. An atom with two energy levels might be compared with a switch that offers either one of two positions. If an atom shifts from a state with energy E_1 to another state with energy E_2, it can make up the difference in energy by absorbing or emitting a photon. The transition frequency of the photon may then be given by

$$v_{12} = \frac{(E_2 - E_1)}{h},$$

where h is Planck's constant.

The essence of operation lies in stimulated emission and absorption of radiation from the atoms or ions. In such processes an atom is made to jump from a state of energy E_1 to that of E_2 by the application of electromagnetic waves of frequency v_{12}. As shown in Figure 4.2, if the atom or ion is initially in the state of lower energy, a photon is absorbed from the electromagnetic wave. This accounts for the absorption of light by objects such as colored glass. Furthermore, if the atom or ion is in the excited state (i.e., with greater energy), then it is stimulated by electromagnetic fields to make the transition to the state of lower energy, emitting a photon in the process. In the laser the fields in the resonator, which

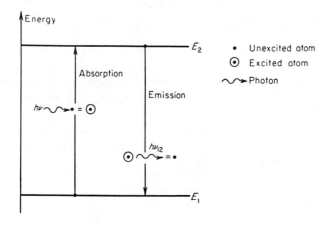

Figure 4.2 The two energy levels of an atom or ion.

is formed by two parallel mirrors, stimulate the atoms or ions to emit photons, adding $h\nu$ of energy to the resonator microwave cavity for each photon emitted.

Stimulated emission is also significant because of the photons emitted are in-phase with the wave that induces the emission [5]. In the laser the atoms or ions emit photons in-phase with the exciting signal, thus reinforcing it.

The rate of addition of photons to the resonator is proportional to the number of atoms or ions in the state of higher energy. That is, if there is a large excess population of atoms or ions in that state subject to the inducing electromagnetic wave, a large number of photons are emitted. It is possible to have a sufficient density of active atoms or ions to emit enough photons to augment considerably the wave that induces the emission. In fact, in suitable cases, energy can be supplied to the resonator rapidly enough to overcome losses, resulting in spontaneous oscillation. This mechanism is the basis for laser operation.

It may be difficult to understand how a continuous wave can be made up of discrete particles such as photons, even if they all are in-phase. The situation is complicated; nevertheless quantum field theory suggests that this is actually so. One clue is that the number of photons is sufficiently great (3.5×10^{15} photons/sec for 1 mW of red light) so that the granular structure becomes insignificant.

Briefly, then, the laser works because the atoms or ions in a laser cavity can be induced to emit photons so that they reinforce the electromagnetic fields in high-order modes of an optical cavity. When energy is fed into the cavity fast enough to overcome the losses, the laser can be made to amplify and oscillate.

Inversion

If there is a large excess population of atoms in the upper of two energy states (a situation referred to as inversion or inverted populations), then there will be a net increase in the electromagnetic fields in the laser cavity through stimulated emission of photons. The existence of an excess number of atoms in an excited state is not an equilibrium situation since spontaneous transitions from the state with higher energy to that with lower energy reduce the population of the former.

These transitions create a tendency to place the atoms in the lower energy states. Actually, if, because of random thermal motion, the atoms are in thermal equilibrium of an absolute temperature T, with no cavity radiation present, the ratio of the number of atoms N_2 in the state with higher energy E_2 to the number N_1 in the state with lower energy E_1 is given by the Boltzmann ratio

$$\frac{N_2}{N_1} = \exp\frac{E_1 - E_2}{kT},$$

where k is the Boltzmann constant.

The condition where N_2 is greater than N_1 is therefore a nonequilibirum situation and some efforts must be expended to create and maintain it. By using suitable techniques, it is indeed possible to attain inverted populations. Five basic schemes have been successfully used so far. These all put the gas atoms or ions in the dielectric crystal into a state of higher energy or put electrons in a semiconductor into a higher energy band. The process occurs faster than the atoms, ions, or electrons, respectively, can leave the excited state by spontaneous processes. These various schemes can make use of ruby, a solid material, neon and helium atoms, a gaseous mixture, or benzoylacetonate with methyl-ethyl alcohol, a liquid. These mentioned lasers are discussed in the next section on key variables.

Principles

The basic principles of the laser as mentioned previously are the same as those of the microwave maser. The first requirement is a resonant cavity. (See Figure 4.3.) In a laser this is formed by two precisely oriented mirrors, one of which is slightly transparent. Resonant modes exist between the mirrors at frequencies for which the spacing is an integral number of half wavelengths. An active medium, which may be either a gas or a crystal doped by certain atoms (chromium in the ruby), is placed between the mirrors.

The medium must possess two atomic states separated in energy by an

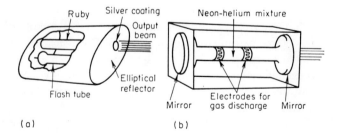

Figure 4.3 (*a*) Essential elements of solid-state laser: mirror, material with active ions (ruby), and means of exciting the material (flash tube). (*b*) Neon-helium gas laser. The helium excites the neon atoms which supply the resonator made by parallel mirrors [6].

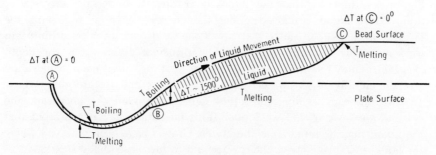

Figure 4.4 Liquid movement during welding [7].

amount corresponding to the frequency desired, and it must be possible to overpopulate the upper of these states with respect to the lower. This is done by "pumping" the atoms from a ground state to a higher energy state either electrically or optically.

From this higher energy state the atoms usually decay nonradioactively to the upper of the two energy states involved in the desired transition. From the upper state some atoms decay spontaneously to the lower state and emit light the way it occurs in a neon sign. The light caused by spontaneous emission is incoherent and is radiated in all directions at random. In the presence of the resonant cavity, however, some of this spontaneous emission excites one of the resonant modes of the cavity, and the field associated with the resonance induces emission in the medium.

This induced emission is phase coherent with the field that induces it and, if the interaction is strong enough, a coherent electromagnetic wave builds up that corresponds to one of the modes of the resonant cavity. Some of this energy leaks through the partially transparent mirror forming one end of the cavity and emerges as a sharply defined beam of coherent light. The significant thing is that this beam of light is a plane coherent electromagnetic wave, just as would be produced by a radio transmitter, but of vastly higher frequency.

The availability of coherent light greatly increases the scope of things that can be done with optical systems. Having coherent light is tantamount to having a point source. With this point source, laser now become suitable as a source of energy for welding.

In order to place laser welding in its proper frame of reference with respect to, say, arc welding and electron-beam welding, we must compare thermal events that occur in these processes. During the formation of an arc weld, liquid is moved from an area directly under the arc toward the rear of the pool as shown in Figure 4.4. This movement is rapid and

Figure 4.4 indicates that the liquid layer directly under the arc is very thin. When liquid is moved under the energy source, it transports the latent heat and superheat to the rear of the weld pool. In an electron-beam weld a similar situation develops. The principal difference is the depth of penetration. The concentrated energy spot produces deep penetration and the liquid moves up the side of the crater and toward the surface. In electron-beam welding the liquid is moved around the hot beam and close at the rear of the weld pool. This movement can be so rapid that the liquid may be lifted from the crater. Craters have been observed where the liquid metal rose above the surface and formed a solidified spout.

Laser welding is compatible with the events observed in arc welds and electron-beam welds where liquid is moved away very rapidly from the area of energy impingement. These events occur because of violent convective movement in the weld pool. The amount of metal removed by the laser beam can be rationalized by a relatively simple calculation. It has been found that only one-third of the available beam energy is required to raise all the material in a hole in a piece of steel to the melting point, provide latent heat, and then superheat it to the boiling point. An 8-J beam operating under the conditions indicated provides about 1.8 cal to the plate. The actual energy requirement for melting and superheating is only 0.6 cal. By a coincidence 1.8 cal are required to vaporize all the superheated liquid; thus sufficient energy is available to vaporize two-thirds of the metal in the hole. Vaporization would account for the violent expulsion of liquid from the crater.

One can readily see an analogy among the three welding processes. In the low energy density arc-weld process, liquid is moved toward the rear of the pool at a relatively slow rate and acts as a coolant to remove heat from beneath the arc source. In the electron beam the motion is more violent. Similar conditions exist with a laser where concentrated energy in a single overpowering pulse causes the liquid metal to move away from the point of energy impingement.

KEY VARIABLES AND MECHANISMS

Three things are essential to achieve laser action: suitable materials, appropriate energy (pump) sources, and designs that promote stimulated emission.

The prime requisite of a laser material is its ability to undergo stimulated emission. The material must also have the following properties:

1. It must luminesce when excited by optical, electrical, radio frequency, or some other type of energy.

2. The emission transition must have a long spontaneous emission decay time so that a population inversion can be achieved.
3. The absorption properties of the material must be such that a large fraction of the exciting energy is absorbed and is effective in producing a population excess in the upper level of the emission transition.
4. The spectral width of the emission transition must be relatively small so that sufficient gain per pass can be obtained to overcome the losses through the partially reflecting end surfaces.
5. The optical quality of the crystal must be sufficiently good so that scattering caused by imperfections, strains, or inhomogeneities does not result in excessive loss of the emitted radiation.

A number of solids and several gases meet these requirements. Among solids, chromium, uranium, and several rare earths have turned out to be best suited for stimulated emission. These materials are used only in the form of dopants, which are minute quantities dispersed in a variety of host materials. The active substance generally accounts for less than $\frac{1}{2}\%$ of the total doped material. The host materials, single crystals, glasses, or liquids, serve primarily to keep the active atoms apart from one another and to remove unwanted heat.

Single crystal ruby—aluminum oxide in which some aluminum atoms are replaced by chromium atoms—is the most widely used solid-state laser material. This should not imply that laser action of solids is limited to single crystals. Glass, a noncrystalline substance, also is used as a host for active rare-earth elements. Glass even has potential advantages over single crystals. For instance, glass is hardly subject to limitations in size or shape. By contrast most single crystals are limited to relatively small sizes. Argon, helium, krypton, neon, oxygen, and xenon have turned out to be active laser substances. There is hope that eventually hydrogen and nitrogen may also be made to lase. Oxygen is used in mixtures of argon-oxygen or neon-oxygen. Neon produces laser action either as a single gas or as a constituent of a helium-neon mixture. The other gases—argon, krypton, neon, and xenon—produce laser action when used in their pure form. In Tables 4.1 and 4.2 are listed some of the solid, liquid, and gaseous laser materials with some of their properties.

These requirements for laser materials are deceptively simple. The list actually must be increased greatly when a more rigorous analysis of the material requirements is made. In practice, the search for new laser materials is an extremely difficult task requiring complex material-preparation techniques and apparatus; detailed measurements of absorption and emission spectra of quantum efficiencies and of emission decay times over a wide range of temperatures; theoretical analysis of the spectroscopic data;

Table 4.1 Solid and Liquid Lasers [6].

Crystals

Material	Temperature, deg K	Output wavelength, microns	Pulsed or continuous operation
$Al_2O_3:Cr^{3+}$ (0.05 wt. %)	4.2–300	0.69	p*
$Al_2O_3:Cr^{3+}$ (0.01 wt. %)	77	0.6934	c
$Al_2O_3:Cr^{3+}$ (0.5 wt. %)	77	0.701	p
		0.704	p
$Al_2O_3:Cr^{3+}$ (0.05 wt. %)	300	0.6929	p
$CaF_2:U^{3+}$	4.2–300	2.5	p
		2.6	p, c
		2.24	p
$BaF_2:U^{3+}$	77	2.6	p
$SrF_2:U^{3+}$	4.2–90	2.4	p
$CaWO_4:Nd^{3+}$	77–300	1.06	p, c
$SrWO_4:Nd^{3+}$	77–300	1.06	p
$SrMoO_4:Nd^{3+}$	77–300	1.06	p
$CaMoO_4:Nd^{3+}$	77–300	1.06	p
$PbMoO_4:Nd^{3+}$	300	1.06	p
$CaF_2:Nd^{3+}$	77	1.05	p
$SrF_2:Nd^{3+}$	77–300	1.04	p
$BaF_2:Nd^{3+}$	77	1.06	p
$LaF_3:Nd^{3+}$	77–300	1.06	p
		1.04	p
$CaWO_4:Ho^{3+}$	77	2.05	p
		2.06	p
$CaF_2:Ho^{3+}$	77	2.09	p
$CaWO_4:Tm^{3+}$	77	1.91	p
$SrF_2:Tm^{3+}$	77	1.97	p
$CaWO_4:Er^{3+}$	77	1.61	p
$CaWO_4:Pr^{3+}$	77	1.05	p
$SrMoO_4:Pr^{3+}$	20	1.0	p
$CaF_2:Sm^{2+}$	4.2	0.7083	p
$SrF_2:Sm^{2+}$	4.2	0.6967	p
$CaF_2:Dy^{2+}$	77	2.36	p, c
$CaF_2:Tm^{2+}$	4.2	1.12	p

Glasses

Ion	Host glass	Temperature, deg K	Output wavelength, microns	Pulsed or continuous operation
Nd^{3+}	American Optical Barium Crown	300	1.06	p**, c
Nd^{3+}	Eastman Kodak "Optical Glass"	300	1.06	p
Nd^{3+}	SiNaCaAlSb	77	0.918	p
Yb^{3+}	LiMgAlSi	77	1.02	p
Gd^{3+}	LiMgAlSi	77	0.3125	p
Ho^{3+}	LiMgAlSi	77	1.95	p
Nd^{3+}-Yb^{3+}	LiMgAlSi	77	1.06 and 1.02	p

Organic Compounds

Material	Temperature, deg K	Output wavelength, microns	Pulsed or continuous operation
Benzophenone-naphthalene	77	0.47	p
Eu^{3+}-TTA Chelate	77	0.6130	p

Liquid

Material	Solvent	Temperature, deg K	Output wavelength, microns	Pulsed or continuous operation
Eu^{3+}-Benzoylacetonate chelate	Methyl-ethyl alcohol	100	0.6130	p

* Output power as high as 500×10^6 watts has been reported for ruby, operated in the giant spike mode. Energy outputs in excess of 50 joules have been reported for normal pulsed operation.
**Output energies in excess of 50 joules and efficiencies in excess of 2 per cent have been reported for pulsed operation of Nd-glass.

Table 4.2 Gaseous Lasers [6].

Gas composition	Output wavelength, microns	Output power, milliwatts	Pulsed or continuous operation
He-Ne	0.6328	4	c
	0.6401		c
	0.6293		c
	0.6118		c
	1.0798		p
	1.0845		p
	1.1143		p
	1.1177		p
	1.1390		p
	1.1409		p
	1.1523	20	c
	1.1601		
	1.1614	1	c
	1.1767		
	1.1985		
	1.2066		
	1.4276		c
	1.4304		c
	1.4321		c
	1.4330		c
	1.4346		c
	1.4368		c
	1.5231	3	c
	3.3913	10	c
He	2.0603	3	c
Ne	1.1523	1	c
	2.1019	1	c
	5.40		
	17.88		
A	1.6180		
	1.6941	0.5	c
	1.793		
	2.0616	1	c
	12.14		
Kr	1.6900		
	1.6936		
	1.7843		
	1.8185		
	1.9211		
	2.1165	1	c
	2.1902	1	c
	2.5234		
	7.06		
Xe	2.0261	5	c
	5.5738	very strong	c
He-Xe	2.0261	10	c
	2.3193	weak	c
	2.6269		c
	2.6511	strong	c
	3.1069	weak	c
	3.3667	strong	c
	3.5070	very strong	c
	3.6788	weak	c
	3.6849	strong	c
	3.8940	weak	c
	3.9955	strong	c
	7.3147		c
	9.0040		c
	9.7002		c
	12.263		c
	12.913		c
A-O$_2$	0.8446	1	
Ne-O$_2$	0.8446	1	
Cs	7.1821	0.025	c

patience and luck. The search for laser substances and suitable materials now also covers semiconductor and organic materials. Scientists have reported that gallium arsenide might lase and organic materials are now under investigation.

There are advantages and disadvantages to the various types of lasers. The gaseous type is superior in frequency stability and monochromaticity, but has low efficiency (less than 0.1%). The solid state are more efficient (up to 10% for a perfect ruby), but they will not transmit continuous wave and they give rise to temperature problems. Glass, fabricated in fiber form (more surface area per volume), should ease the heat removal problem and perhaps make continuous-wave transmission possible. Researchers have given least attention to liquids and little has been documented about them. They would probably have at least one advantage—there would be no apparent limitation in their size.

LASER MATERIAL

Of all the laser devices with sufficient energy output for welding currently, only the pink ruby-chromium system has been actively used. The ruby is a rod ranging from $\frac{1}{8}$ to $\frac{1}{2}$ in. in diameter and from 3 to 8 in. in length. It is a matrix of aluminum oxide (Al_2O_3) and 0.05% chromium (Cr^{+3}) added as the active ion. The chromium gives the ruby its characteristic red color. The greater the amount of chromium in the ruby, the darker its color. Chromium ions have the property of emitting red light when stimulated by green light. The aluminum oxide transmits green and red light and is inert insofar as fluorescent action is concerned. It merely serves as a lattice or holder for the chromium ions, which are in what solid-state physicists call the "valence^{+3}" state. (An ion is an atom that has gained or lost one or more electrons.)

In the ruby-chromium system, electrons are normally in the ground state or the E_1 level as shown in Figure 4.5. The electrons are then "excited" by absorbing green light from a "pump," usually a xenon flash lamp. When the electrons absorb light, they are excited to a higher energy state, E_3 level. This is an unstable state with an average lifetime of about 0.05

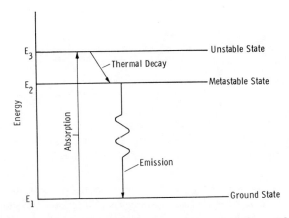

Figure 4.5 Energy level diagram of three-level laser [7].

Key Variables and Mechanisms

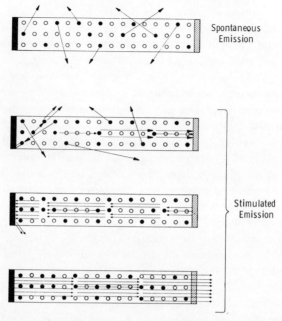

Figure 4.6 Amplification of photon wave by stimulated emission [8]. ○—excited atoms; ●—ground state atoms.

microsec. The electrons quickly decay to a metastable state in level E_2. Under normal conditions the electrons remain at this level for about 1 msec and then fall back to the ground state, spontaneously emitting photon energy as they decay. In the laser the first photons emitted by chromium ions returning to the ground state serve as triggers that stimulate other excited chromium ions to emit photons. The result is an intense flash of red light.

Laser output is greatly improved by cooling the ruby rod. The probability of chromium atoms falling directly to the lowest energy level is greatest at high temperatures. In a room-temperature laser, fewer ions will remain in the metastable state which is necessary for "lasering." For this reason ruby lasers are often cooled to liquid nitrogen temperatures ($-321°F$) during their operation. This doubles the output.

For laser action to occur the emission must be stimulated, not spontaneous, and there must be a means provided to direct and amplify the intensity of the light wave. Stimulated emission requires that there be a population inversion of electrons in the levels E_1 and E_2 of Figure 4.6. Normally, the higher the energy level, the lower the population of electrons. With a population inversion, E_2 will contain more electrons than E_1. If this condition prevails when the ruby is being pumped with a strong green light, a chromium atom may emit a photon of red light. When the photon is propagating, it may strike another chromium atom still in the excited state and cause it in turn to emit a photon in the same direction and in-phase with the colliding photon. This is the principle of stimulated emission.

Once stimulated emission is achieved it is necessary to provide for extensive amplification of the wave and an opportunity to propagate the wave from the laser in a coherent manner. This may be accomplished by polishing both ends of a circular laser rod optically flat and parallel to each other. One end of the rod is completely silvered so that it reflects virtually all of the light that strikes it from the interior of the ruby. If the rod is intended for use at room temperatures, the other end of the rod, the output end, is silvered so that it is about 90% reflective. If the rod is to be used at liquid nitrogen temperatures, the output end is not silvered. It does, however, reflect some of the light that strikes it.

When the chromium ions in the ruby are stimulated to emit light, this light can travel in any direction. Light that is traveling in a direction parallel to the axis of the rod bounces back and forth between the ends of the rod and stimulates additional chromium ions to emit photons. The ruby rod, then, acts as a resonator. A wave is built up parallel to the axis of the crystal as shown in Figure 4.6.

Some light leaves the partially silvered end of the rod on each reflection, but as long as the stored energy released to the beam on each double pass is greater than the part transmitted out the end, the light bouncing back and forth in the ruby continues to increase in intensity. Eventually (in a few thousandths of a second) most of the light is transmitted out the partially silvered (or, with low-temperature lasers,) clear end of the rod.

The output of the laser is coherent in both space and time. Spatial coherence means that all wave fronts are planes perpendicular to the direction of propagation. Time coherence means that the wave crests follow each other at regular intervals or, to put it another way, that the light has a single, fixed phase. The light is intense, for a great number of atoms have been stimulated to emit photons of light almost simultaneously, and it is highly directional because only the light traveling parallel to the axis of the ruby can be part of the laser beam.

Laser beams can be focused by simple lenses into small areas, giving controllable power densities that have not been previously available to man as monochromatic light. Any known material can be vaporized, melted, machined, or welded with a laser beam.

Although the ruby rod described in the preceding discussion is a circular rod with both ends flat, it is not the only configuration available. Laser rods have been developed with one end confocal, chiseled, or faceted. All rod designs, however, accomplish the same end result. The grinding of ruby crystals is a highly important factor in solid-state laser design. Tolerances for some of the physical dimensions are extremely stringent. For instance, the flatness of the end plates is expressed in fractions of a wavelength of sodium light and the rod ends generally are parallel to within 2 sec of arc. The crystals must provide for internal reflection of energy to bounce the wave back and forth between the end plates. Generally, reflection is achieved through the silvering of the flat end plates. Recently the crystals with spherical, faceted, or conical ends that have appeared on the market require no silver coating.

The gaseous laser shown in Figure 4.3 consists of two mirrors and a long enclosed glass tube filled with gas. The gas in the glass or quartz tube is usually a mixture of helium and neon gases. In lasers of this type, photons are furnished by excited neon atoms. Energy supplied from an r-f oscillator or a d-c discharge produces a glow discharge in the gas mixture, similar to the discharge of an ordinary neon sign. As with the solid ruby rod, the output is a plane wave passing through one of the mirrors. Although this output is relatively weak (milliwatts), it is a continuous, rather than a pulsed beam.

The two previously discussed lasers have been utilized and evaluated more than any others. As improvement of laser materials continues, however, other materials will be developed. (See Tables 4.1 and 2.)

Pump Energy

Another key factor is getting the pump energy into the laser material. Given the laser material and the energy source, this factor determines largely the strength and duration of the output beam. Shown in Figure 4.7 is a schematic of the models currently in use in which the crystal rod and

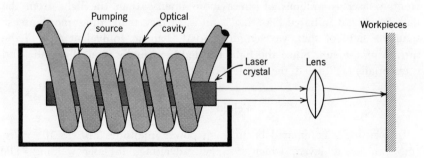

Figure 4.7 Typical laser welding system.

a straight flash tube are located at the foci of an elliptical cavity lined with a reflective material. This design has proven quite successful because the cavity wall focuses a large portion of the flash tube's output onto the crystal. The flash tube is normally xenon gas filled, using barium oxide-coated tungsten and tantalum cathodes and anodes. In place of a flash lamp with xenon scientists have substituted an exploding wire. A wire that is vaporized by an electrical discharge is one of the most brilliant sources of light known and is capable of furnishing extremely high input to a laser.

The exploding wire acts as a small-diameter line source of light. A polished elliptical cylinder of stainless steel focuses the light onto a ruby laser. The laser rod is surrounded by concentric hollow cylinders of plastic and glass to help absorb shock, to filter out extraneous radiation not useful for pumping, and to prevent shattering of the ruby.

Nichrome, aluminum, copper, and tungsten wires have been used. These wires have been exploded by electrical discharges capable of an output from 33,600 J to 20,000 V. Each in initial experiments has produced a coherent light output of 0.4 J from a ruby laser operating at room temperature.

Efficiency

Laser output is measured in joules (watt seconds). Output of the currently available lasers ranges from a fraction of a joule up to 30 to 40 J for the largest units.

Energy losses in any laser system are high. In one relatively efficient system input from the power supply is over 10,000 J; laser output is 30 J. Only a small part of the energy of the flash lamp is in the form of white light and only that portion of the spectrum in the green region initiates laser action; much of the light from the laser passes through the sides of the ruby rod and does not become part of the laser beam. Nevertheless, these energy losses are acceptable because the focused spot of light from a laser is millions of times more intense than the light from the flash lamp that initiated "lasering" and is, in fact, many times more intense than the light of that wavelength emitted from an equivalent area of the surface of the sun. Since the light emitted is virtually monochromatic and is essentially collimated, the problem of focusing is simplified.

Welding Effects

In focusing a collimated beam, high power intensities (10^5 to 10^8 W/sq cm) can be achieved, which is a key advantage to laser welding. The high intensities reduce the energy required for a typical laser spotweld on thin stock to about one-tenth that required by an electric arc. In the past the arc was used as a relatively high density heat source. More recently the electron beam has provided a more concentrated source of heat. The laser provides a similar high density heat source. For example, a conventional welding arc using 200 A and 20 V will provide a 1-cm spot with a power density of approximately 5×10^3 W/cm².

An electron beam operating at 100 kV and 8 µA on a 0.005-cm spot provides an energy density of 4×10^7 W/cm². A laser of very modest power providing 1 J and 1 msec on a 0.012-cm spot provides an energy density of 9×10^6 W/cm². Decreasing the spot size can increase this value to 10^8 W/cm². Therefore energy produced by a laser is readily comparable to the energy density produced by an electron beam.

The optimum laser output required for welding depends on the absorptivity, thermal conduction, density, heat capacity, melting point, and surface condition of the metals to be joined as well as on the duration of the laser pulse. For example, metals with high thermal conductivity, such as copper and silver, require greater energy than do less conductive metals such as iron and nickel. Greater output energy is also required for highly reflective surfaces than for dull or rough surfaces.

Thus the welding speed and plate thickness that can be welded with a laser are not limited by the available power output but by the rate of heat conduction in the metal. The maximum temperature that can be maintained at the surface of a metal is its boiling point, which limits the rate of heat conduction into a metal. Surface boiling occurs in most common metals at a power intensity of about 10^5 to 10^6 W/sq cm, which is well below the maximum capability of the laser shown earlier. If the intensity is appreciably greater, then the excess energy will be lost in vaporizing the metal. At the same time the reaction force of vaporization can expel the liquid metal, leaving a hole rather than a weld bead. Therefore care is required to avoid exceeding appreciably the intensity that produces surface boiling.

Excessive outputs can be avoided by increasing pulse durations when laser energy output is increased. The more advanced laser welders automatically select the required pulse duration for any given output.

Since the power intensity that can be utilized is limited, there is a corresponding limit to the thickness of material that can be welded for a given pulse duration. For example, theory predicts that for a 0.032-in. diameter weld on copper, the maximum allowable metal thickness is 0.014 in. for a 1 msec pulse and 0.024 in. for a 10 msec-pulse.

Selecting Weld Parameters

The important considerations in determining proper welding parameters involves basically three factors: (a) selecting the number of capacitors and corresponding voltage to obtain desired energy input level, based on $E = \frac{1}{2}CV^2$, where C = capacitance and V = voltage; (b) proper selection of optics to control size and shape of beam spot; (c) selection of the beam focal point either on or above the metal surface.

The number of capacitors that are used to obtain a given energy level is a critical consideration. Increasing the number of capacitors in the circuit achieves longer pulse cycle times. The power of the laser beam, however, is decreased due to this increase of pulse cycle time.

To obtain a sound weld with full penetration, which is free from undercutting, it is desirable that

(1) the laser beam have a power level that is sufficient to heat the surface of the metal to a temperature that is greater than the melting point of the metal but less than its vaporizing temperature and
(2) the pulse cycle time be long enough for the heat to be conducted through the thickness of the material.

TYPES OF EQUIPMENT AND TOOLING

Equipment

Laser equipment for experimental work has been available for five years. It consists basically of a power supply and a laser head assembly. An accurate positioning table supports the workpiece during operations. The power supply has an output of 800 to 8000 V, with a maximum energy storage capacity of about 25,000 J which is switched into the load with an ignitron. A pulse-rate control can be varied from 0.25 to 12 pulses/min.

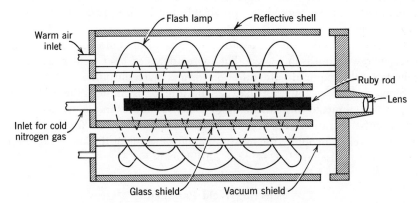

Figure 4.8 Typical laser head for welding. Cold nitrogen gas flows over ruby; warm air flows over flash lamp.

Figure 4.9 Laser beam-welding system [10].

The laser head configuration schematically illustrated in Figure 4.8 has a ruby rod at the center and is enclosed in a transparent glass tube. Cold nitrogen gas is circulated over the surface of the ruby rod and returns outside the glass tube. Between the glass tube and a helical flash lamp is a double-walled glass tube that is actually a vacuum bottle. The vacuum bottle contains liquid nitrogen that provides a supply of cold gas that is carried to the laser head by an insulated hose. The vacuum bottle prevents heat from the flash lamp from being transmitted to the ruby rod, but like the transparent glass tube it does not significantly affect the transmission of light. The entire assembly is enclosed in a reflective metal shell. A suppressor (not shown) prevents arcing between the flash lamp and the shell. Also, since the flash lamp is the most efficient when it is warm and to prevent arcing caused by humidity, hot air is circulated continuously over the flash lamp.

Tooling

As with any semiautomatic or fully automated fusion welding process, the laser welding system must have provisions for part movement, that is, indexing and/or rotation. Although the laser head assembly itself can be mounted to move and weld across a seam, the more practical approach is generally to move the work rather than the welding head. In this respect tooling to hold the object to be welded is critical in that part-rotation hesitation or out-of-roundness can seriously affect weld quality. For example, if the rotating device does not travel at uniform speeds, the weld nugget overlapping spacing can be affected; or if the object being welded is not precision tracked relative to the pinpoint light beam, the weld seam can be missed.

Production welding is being accomplished with a laser system having manual and automatic firing of the laser head coordinated with an adjustable speed positioning and indexing welding table. This arrangement makes possible continuous welding as well as automatic indexing of a series of welds or from part to part.

The weldments made with laser equipment are particularly well suited to the process in that their geometry and makeup are such that the laser system used to perform the welding is operated at 10 to 20% of its maximum energy output. The laser system used to perform the welds is capable of 10 pulses/sec at 10 J/pulse.

A government sponsored laser welding development program that in-

Table 4.3   Laser System Characteristics [10].

Item	Remarks		
Power supply	Rating: 3000 w continuous for 10,000 hr @ 30° C.		
Pulse forming network	Energy storage: 0–30,000 joules, continuously variable		
	Max voltage: 3000 v		
	Capacitors: 580 uf each, 10/bank; 2 banks total		
	Pulse duration: 0.9 to 9.0 msec		
	Pulse repetition rate; 1 firing every 6 sec (max)		
Cooling system	Laser head: recirculated distilled water		
	Heat exchanger: tap water		
Laser head	Elliptical cavity; 4 straight xenon flashlamps operated in pairs; each pair in series and charged by one capacitor bank		
	Pulse width, msec	Input, joules	Output, joules
Output energy, typical	2.7	9,600	10
	3.6	16,300	28
	4.5	21,200	42
	9.0	30,000	50
Beam spot dimensions	The beam optics control employs spherical and cylindrical lenses capable of producing either circular or elliptical spots. Circular spots (spherical): 0.016 to 0.040 in. diam; elliptical spots (spherical and cylindrical): 0.040 x 0.70 in. max.		

Types of Equipment and Tooling

cludes manufacture of a complete laser beam welding system, evaluation of the equipment, and application to numerous materials has recently been concluded [10]. An overall view of the welding system shown in Figure 4.9 has the following characteristics. (See Table 4.3.) All welding done with this system has been in the flat position, using a simple hold-down fixture with copper clamping bars and a grooved copper backing bar. The workpiece clamping device was attached to an air-driven work table, capable of providing either continuous or intermittent travel of the workpiece beneath the beam during welding. Inert gas shielding was provided on both sides of the weld surfaces by employing a gas shielding cup and a gas backing bar.

Fixture

Continuous refinements in equipment have progressed whereby a system with two laser heads is currently being evaluated. Each head delivers up to 50-J pulses that can be varied in length from 1 to 9 msec. The heads can be fired sequentially, giving an 18 msec pulse at rated output or simultaneously giving double rated output for 9 msec.

ADVANTAGES–DISADVANTAGES–ECONOMICS

Perhaps the best way to locate the niche of laser metalworking is by comparison with better known electron-beam methods. Both are capable of joining thick-to-thin metals and of welding dissimilar metals, but there are some important differences that indicate the laser will have a unique role.

1. It requires no vacuum chamber, thus removing this restriction on the size and complexity of workpieces.
2. There is no discernible temperature rise in areas adjacent to laser beam impact, so welding can be close to heat-sensitive materials or joints, including glass and ceramic seals.
3. With laser welding some preliminary test results indicate that it may not always be necessary to protect reactive metals, such as titanium or refractory alloys like tungsten and molybdenum, from atmospheric contamination.
4. The cost of acquiring, operating, and maintaining laser equipment is less than that for electron beam and most other special processes.
5. No pressures are required and surface contaminants are vaporized in welding.

Lasers have certain limitations at the present time. One important shortcoming when compared with the electron-beam process is that although electron beams can weld extremely thick sections, lasers cannot. Current capabilities in material thickness that can be welded are 0.020 to 0.030 in. maximum. Unfortunately, this material thickness capability cannot be boosted by increasing laser output because vaporization is a danger and currently pumping sources are not available for continuous operation. With continued development these limitations on thickness and equipment will be eliminated.

Current costs for welding are moderately high. The costs range from 2 to 5 cents per laser pulse and the cost of systems for laser welding is in the $2500 to $25,000 range, depending on the degree of sophistication and capacity of the power supplies. Eventually laser beam machines for production will probably cost considerably less than electron-beam machines of equivalent capacity.

MATERIALS, JOINT DESIGN, AND PROPERTIES

Laser welding has been used to produce similar and dissimilar metal joints with copper, nickel, stainless steel, aluminum alloys, iron-nickel base alloys, and the refractory metals and their alloys. Most welds made with these metals have been of the wire-to-wire type butt joint and several other types of welds.

Wire-to-Wire Type

In wire-to-wire type laser welds there are four basic configurations. Welds between metals of widely varying physical properties, such as copper and tantalum, are readily accomplished. Lap joints are commonly used in electronic micromodules; nickel is often preferred for this use because of its attractive electrical and mechanical properties as well as its good weldability.

Metals of high electrical resistance are easily welded because the laser does not depend on contact resistance to produce the required heat of fusion. An example of a high resistance material is stainless steel which has been easily welded as seen in Table 4.4.

Joint strength, joint resistance, and processing data for a variety of metals and wire joint configurations are listed in Table 4.4. The tabulated strength values are essentially pure tension for the butt joints, pure shear for the lap joints, and a combination of tension and shear for the cross and tee joints.

Generally, a laser energy output of 10 J produces a joint strength equal to that of the wire itself for 0.015-in. diameter wire in the annealed condition and about one-half as much for 0.030-in. wire. Joints made with hard

Table 4.4 Properties and Processing Data for Wire-to-Wire Laser Weld.

Material	Wire Dia,[a] in.	Joint Type	Processing Data		Joint Properties	
			Laser Output, joule	Pulse Duration, millisec	Max Load, lb	Resistance,[b] ohm
301 Stainless Steel	0.015	Butt	8	3.0	21.3	0.003
		Lap	8	3.0	22.8	0.003
		Cross	8	3.0	25.0	0.003
		Tee	8	3.0	23.2	0.003
	0.031	Butt	10	3.4	31.8	0.002
		Lap	10	3.4	34.6	0.002
		Cross	10	3.4	40.0	0.002
		Tee	11	3.6	40.1	0.002
	0.015-0.031	Butt	10	3.4	24.3	0.003
		Lap	10	3.4	24.8	0.003
		Cross	10	3.4	25.6	0.003
		Tee	11	3.6	22.6	0.003
	0.015-0.031	Tee	11	3.6	26.4	0.001
	0.031-0.016	Tee	11	3.6	19.6	0.001
Copper	0.015	Butt	10	3.4	5.1	0.001
		Lap	10	3.4	3.2	0.001
		Cross	10	3.4	4.3	0.001
		Tee	11	3.6	3.2	0.001
Nickel	0.020	Butt	10	3.4	12.2	0.001
		Lap	7	2.8	7.8	0.001
		Cross	9	3.2	6.8	0.001
		Tee	11	3.6	12.5	0.001
Tantalum	0.015	Butt	8	3.0	11.5	0.001
		Lap	8	3.0	8.8	0.001
		Cross	9	3.2	9.3	0.001
		Tee	8	3.0	10.8	0.001
	0.025	Butt	11	3.5	14.8	0.001
		Lap	11	3.5	12.9	0.001
		Tee	11	3.5	17.0	0.001
	0.015-0.025	Butt	10	3.4	11.0	0.001
		Cross	10	3.4	9.2	0.001
	0.015-0.025	Tee	11	3.6	19.3	0.001
	0.025-0.015	Tee	11	3.6	11.4	0.001
Copper-Tantalum	0.015	Butt	10	3.4	3.8	0.001
		Lap	10	3.4	5.3	0.001
		Cross	10	3.4	4.1	0.001
		Tee	10	3.4	4.0	0.001

[a] When only one size is given, both pieces are the same diameter. When two sizes are given, the first diameter is that of the top member of a cross joint or the cross bar of a tee joint. [b] Joint resistance represents the difference between the resistance of a given weldment plus wire lead, and the resistance of the wire alone.

drawn or severely cold-worked wire are never as strong as the wire itself.

Although sound joints can be produced for all four joint configurations, the lap joint is generally preferred for wire-to-wire welds. In a lap joint the laser beam is directed to the precise spot where the weld nugget is needed without stringent requirements for lead length and alignment. Only a simple fixture or clamp is needed to hold the wires together. Furthermore, multiple spot welds can be made if additional strength is required.

If the cross configuration is used, it is best to direct the laser beam at an angle to the joint so that it strikes the interface, thereby eliminating the need to melt through the top wire for welding the joint.

Butt and lap welds for joining sheet are readily produced by laser welding. Butt weld tensile strengths (0.005-in. material) have produced 123,000 psi in titanium and 46,000 psi in tantalum; in 0.010-in. material, 176,000 psi in 301-stainless steel and 88,000 psi in pure nickel.

Because welding only occurs where the light beam finds the joint area, the usual cross-wire configuration is not as amenable to laser welding as the parallel-wire joint. Instead of attempting to penetrate completely through one of the wires, the object of the parallel wire joint is to place the wires in line contact and then make the weld by fusing a portion of each wire. Surface tension exerted by the weld pool tends to draw the wires together. Several advantages of parallel-wire joints are shown in Figure 4.10.

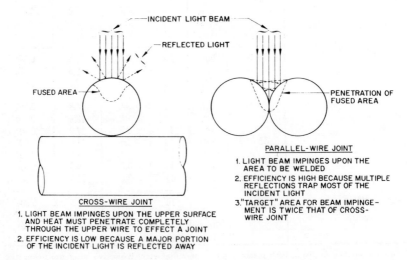

Figure 4.10 Some advantages of the parallel-wire joint over the crosswire joint when welded with a laser beam.

Wire-to-Ribbon Type

This combination has been used for electronic applications and the weld usually has penetrated through the ribbon and into the underlying wire.

When fabricating molecular semiconductor devices (molecular blocks, integrated circuits, functional electronic blocks, etc.), it is often necessary to attach very small diameter leads to small areas on the substrate of the devices. In the past this has been accomplished by several techniques, including thermocompression bonding, brazing, eutectic bonding, and soldering. Recently laser welding has attempted to attach 0.005-in. diameter gold wires to substrates of silicon and aluminum-coated silicon. In order to evaluate the strengths of these bonds and to compare them with bonds made by the conventional thermocompression bonding method, specimens were tested and the results of the tests indicate greater strength for the laser bonds, for in no instance did the lasered bond fail. These tests are preliminary and it should be noted that the laser-bonded samples had approximately half the contact areas of the thermocompression-bonded samples.

Ribbon-to-Ribbon Type

In general, the laser output required for a given sheet thickness is approximately 50% greater than that for welding wire of the same thickness. Thus, if 10 J produce a fully penetrated weld on 0.015-in. diameter wire, then 15 J will be required for 0.015-in. thick sheet. Most similar and dissimilar metal combinations that have been laser welded have been experimental. Limited strength and property data have been developed due to the newness of the process and its current limitations on material thickness. The lap weld between tungsten and aluminum alloy sheets shown in Figure 4.11 was produced by separate spot welds. The spot weld sizes are much larger on the aluminum side because of aluminum's much lower melting point. Nickel conductor ribbon of 0.002-in. thick has been welded successfully to copper-clad, glass fiber reinforced epoxy circuit boards. Sound welds can also be made between parts that vary considerably in size or mass.

Since laser welding is still in its infancy, problems exist. An experimental weld was made in 0.010-in. stainless steel. This weld was made using 0.45 J in 1 msec pulses. Spots were spaced 0.004 in. (0.01 cm) apart. A cross section of this weld is shown in Figure 4.12. It is apparent that full penetration was obtained locally. It is possible to see the very marked difference between the unfused and fused material since the heat-affected zone is almost nonexistent. Porosity is evident in one section of the weld. The heavily cold-worked area is due to an edge sheared before welding.

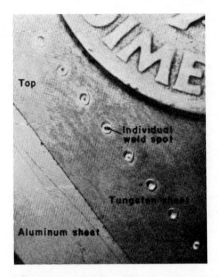 

Figure 4.11 Lap seam weld of 0.005-in. tungsten and 5052-H34 aluminum alloy sheets.

Laser welds show very narrow heat-affected zones. In addition to the porosity and incomplete penetration problems, other practical problems exist. One of these is the effect of fit between pieces before butt welding. When the welding progresses across the joint, the separation between the pieces increases. When the distance between the pieces reaches 0.001 in., welding ceases. As with other welding processes, proper fit is essential.

Figure 4.12 Cross section laser weld in 0.010-in. thick stainless steel, full penetration area, 0.45 J, 10× objective (×200).

Materials, Joint Design, and Properties

Unless the laser beam impinges on the base metal, fusion will not occur. If full advantage is to be taken of the deep penetration capabilities, it is necessary to provide excellent fit, especially for welds in thick sheets. It is difficult to envision, for example, the degree of fit required on 1 or 2 in. thick plate in structures 10 to 20 ft in diameter. It seems probable that under these conditions the laser may not be applicable. When precision fit can be readily obtained, however, the deep penetration effect could be of great value.

In determining the proper focal point and energy necessary to weld, tests are conducted and plotted. Depending on the laser unit, the energy intensity of a laser beam is mainly controlled in two ways: (a) by the power input and (b) by defocusing the position of the workpiece by moving it away from the optical focal point. The control of these two parameters will produce an optimum range for welding with the proper degree of penetration and the least amount of evaporation.

Values for the two variables, input voltage and work distance from the optical focal point, are selected in several steps with certain increments, for example, input from 3250 to 4000 V in 250 V increments, at which output energy, in terms of joules, varies from 1 to 4 J in 1-J increments. Distances differ in steps of 0.05 in. from −0.05 to +0.15 in. Figure 4.13 is an example of a plot for an 18% nickel maraging steel sheet and Figure 4.14 shows the actual photomicrograph of a butt welded specimen.

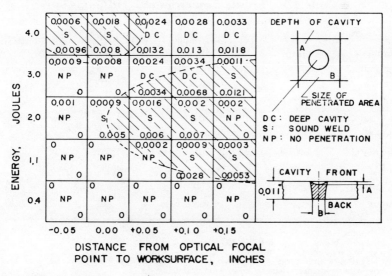

Figure 4.13 Weldability test results with 0.011-in. thick, 18% Ni maraging steel sheet; the crosshatched area indicates the conditions for sound welding.

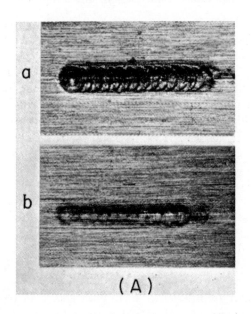

Figure 4.14 Sound laser weld obtained with optimum welding conditions: (*a*) top side of weld, (*b*) bottom side of weld.

Table 4.5 Tensile Properties of Weld Joints in 0.005 inch Cb D-36 [10].

Condition	Test temp, °F	Ultimate tensile strength, psi[a]	Yield strength, psi[a]	Elongation in 2 in., %
Base metal	RT[e]	73,400 (L)	64,600(L)	15.2
		72,000 (T)	64,600(T)	17.7
Welded	RT[e]	76,800	69,600	10.3
Base metal	1600	52,500	37,100	18.9
Welded	1600[d]	50,800	40,300	7.9
Base metal	1900[d]	37,900	34,500	35.0
Welded	1900[d]	36,000	34,400	4.7
Base metal	2200[d]	23,600	20,800	[b]
Welded	2200[d]	24,000	24,000	16.5

[a] L—longitudinal; T—Transverse.
[b] Specimens broke outside the 2 in. gage length.
[c] All values are average of three samples.
[d] At elevated temperatures, welds failed in heat-affected zone.
[e] At room temperature, welds failed in base metal.

Table 4.6 Tensile Properties of Weld Joints in 0.005 inch René 41 [10].

Conditions	Test temp, °F	Ultimate tensile strength, psi[a]	Yield strength, psi[a]	Elongations in 2 in., %[a]
Base metal	RT[e]	124,800 (L)	80,700 (L)	[b]
		110,300 (T)	71,800 (T)	13.5(T)
Welded	RT[e]	124,300	80,100	16.0
Base metal	1000[d]	103,400	72,200	14.2
Welded	1000[d]	96,900	74,700	8.5
Base metal	1400[d]	108,500	91,500	5.0
Welded	1400[d]	99,700	86,000	2.5
Base metal	1800[d]	42,600	39,000	9.8
Welded	1800[d]	42,400	37,500	3.3

[a] L—longitudinal; T—transverse.
[b] Specimens broke outside 2 in. gage length.
[c] All values are average of three samples.
[d] At elevated temperatures, welds failed in heat-affected zone.
[e] At room temperature, welds failed in base metal.

The welding potential of lasers has also been demonstrated on some of the more difficult to joint metals, namely René 41, columbium, and molybdenum. Table 4.5 reflects the tensile strength of a refractory metal alloy, D-36 columbium and Table 4.6 shows the test results of several René 41 welds. In general, Tables 4.5 and 4.6 show that the mechanical performance of welds in both alloys René 41 and D-36 compare quite favorably with their respective base metal properties. The weld joint efficiencies ranged from 90 to 100% with only small reductions in ductility. Test specimens failed usually in the base metal or heat-affected zone. The small decreases in ductility are attributed to the as-cast weld structures plus possible interstitial contamination during welding.

Continued progress in developing a continuous moving heat source capable of attaining significant increases in pulse repetition rates and pulse duration will widely broaden lasers potential usage.

APPLICATIONS AND FUTURE POTENTIAL

Where is laser welding today? Where is the goal for the future? Some people feel that future practical applications for continuous optical maser or laser welding probably will be found in the welding of minute structures such as watch parts. These same people feel that lasers will probably never lend themselves to large-scale welding operations because of their expense, not only to construct but to operate. However, fine-scale work with lasers

Figure 4.15 Hot gas bearing assembly with continuous circumferential weld in the as-welded condition. Tie-in of the weld was made in top right quarter [11].

may turn out to be competitive with other methods. The use of welding lasers in extremely fine work for exotic space age technology may be the only answer.

Other engineers seem to disagree. In Figure 4.15 is an example of a laser-welded high-temperature bearing. It is interesting to note that before laser welding of this configuration, three other joining processes had been tried without success. The first attempt involved utilizing the gas-tungsten-arc process with massive chilling techniques as heat sinks to prevent softening of the ball bearing race areas.

The second process was resistance spot welding the heat shield disc to the outer bearing ring. Resistance seam welding was also tried in an effort to obtain a seal-welded heat shield, but quality leak tight welds could not be accomplished on this configuration.

The electron-beam welding process was then attempted, first without heat sinks and later with a copper heat sink surrounding the outer bearing housing area. Although high quality welds were obtained with narrow heat-

Applications and Future Potential

affected zones, the hardness loss measured at the ball race area was considerable even though heat sinks were employed [11].

It can be seen in Figure 4.15 that the joint is a simple overlapping design. There is also an extreme difference in joint material thickness mass, with the lower heavy bearing material acting as a nearly infinite solid compared to the 0.010 in. thick heat shield material. This joint configuration caused joining problems when other fusion welding processes were used.

The laser weld consists of a series of overlapping spots around the entire periphery of the weldment. The weldment is in the as-welded condition with no postweld cleaning. The shiny surface of the weldment is normal in that the weld bursts are of such short duration that the weld or surrounding weld areas are not at temperature long enough to be oxidized by the surrounding air environment.

The small electronic sensing configuration shown in Figure 4.16 was developed into an extremely reliable laser welded configuration, again after several other joining processes (including epoxy bonds) had failed. The inner core of this assembly is wound with 0.0002-in. diameter insulated copper wire. After the proper number of windings are made on the inner

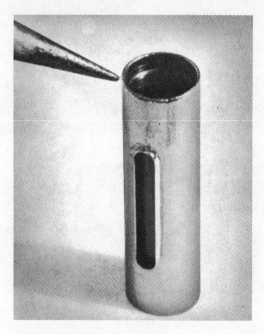

Figure 4.16 Laser weld of gold sensor package having uniform and gas-tight surface [11].

220   **Laser Welding**

core, this assembly is then slipped into an outer housing (slotted member is shown in Figure 4.16).

The configuration first laser welded consisted of 6061T6 aluminum alloy for both the inner and outer members. The material was later changed to gold for both the inner and outer members because of operating environment requirements for the sensing device. Neither material presented any problems from a metallurgical joining standpoint.

Making hermetically sealed welds in close proximity to ceramic glass seals is another fabrication application in which lasers have been utilized. Owing to package limitations, the assembly shown in Figure 4.17 developed into a serious joining problem because the joint design could not be changed to move the final closure weld away from the ceramic seal assembly.

When reviewing the design, it became obvious that a process must be

Figure 4.17 Laser weld of an electrical feedthrough in a thermistor enclosure. The ceramic is within 0.050-in. of the weld which was successfully made without heat sinks as required by gas-tungsten-arc or electron-beam welding or brazing ($\times$ 25) [11].

employed in which it is possible to avoid the expanding and contracting forces encountered with other fusion welding processes that would break and crack the ceramic glass seal. Here again the laser process was employed to do the impossible, and numerous assemblies have been weld fabricated without a single failure. Furthermore, this process is economically highly competitive with the other fusion welding processes in the time to produce the parts and no inert gas or vacuum environments are required for welding purposes.

From these examples it appears that laser welding is making progress. The practical applications developments currently underway include the following:

1. Laser rods a few years ago were not much bigger than a cigarette and they now are 3 ft in length and longer. Rods are used in experiments aimed at increasing the size, power output, and efficiency of laser systems suitable for welding. The glass from which the rod is made contains small traces of neodymium, an impurity that causes the material to amplify and orient light to form a laser beam.
2. Higher reproducibility and better distribution of energy. Now welds can be made in 1 sec which previously took 7 to 8 sec.
3. New carbon dioxide lasers with a 1000 W average output that can be operated either continuously or pulsed.
4. Improving power supplies and flash lamps through the use of coaxial or annular lamps to achieve long pulse lengths. New power supplies permit altering the duration of the laser pulse from 0 to 60 msec in 3 msec steps simply by turning a dial.

In conclusion, the laser has a definite future as a welding tool. Several areas in which the laser may assist metal workers who have conventional welding problems include

(1) control of the heat-affected area.
(2) control of weld penetration.
(3) control of metallurgical properties at the "high quality" level.

Therefore the laser has been tried and proven on special welding jobs. However, the point of standardized methods and setups has not been reached, but it is not too far distant.

REFERENCES

[1] A. L. Schawlow and C. H. Townes, "Infrared and Optical Masers," *Phys. Rev.*, **112**, 1940. (December 1960).
[2] T. H. Maiman, "Stimulated Optical Radiation in Ruby Masers," *Nature*, **187**, 493 (August 1960).

[3] A. Yariv and J. P. Gordon, "The Laser," *Proc. IEEE,* **51,** 4 (January 1963).
[4] A. L. Schawlow, "Infrared and Optical Masers," *Bell Lab. Rev.* (November 1960).
[5] G. C. Dacey, *Science,* **135,** (3498), 135 (January 1962).
[6] S. A. Collins, Jr., "Lasers," *Electro-Technology,* 64–69 (March 1963).
[7] W. N. Platte, and J. F. Smith, "Laser Techniques for Metals Joining," *Welding J., Research Suppl.,* 481-S–489-S (November 1963).
[8] A. L. Schawlow, *Sci. Amer.,* **204,** (6), 60 (June 1961).
[9] T. W. Black, "Lasers. Cast Light on Machining, Welding Problems," *The Tool and Manu. Eng.,* 85–91, (June 1962).
[10] S. Reich, Final Report on Laser (Optical Maser) Beam Fusion Welding, TRG, Inc., *Tec. Report* AFML-TR-65-33 (February 1965).
[11] K. J. Miller, and J. D. Nunnikhoven, "Production Laser Welding for Specialized Applications," *Welding J.,* 480–485 (June 1965).

Chapter 5

FUSION SPOT WELDING (GAS METAL AND TUNGSTEN-ARC)

Long before the submerged arc welding method was invented, the advantages of automatic arc welding were recognized and in many production applications unshielded base or lightly coated electrodes were used. Submerged arc welding, however, with its well-known advantages of quality and economy for wide range of applications to various materials was until recently the only automatic fusion welding process available to the industry. This situation has changed considerably during the last two decades.

The gas-tungsten-arc welding process was developed during World War II. This process has provided many opportunities and challenges for the welding industry, although the inert gases were expensive and initial applications appeared to be limited.

For many years companies have used resistance spot welding for fabricating parts into subassemblies. Today, however, the size and complexity of these subassemblies have increased to the point where the limitations of the resistance spot welding equipment in both throat dimensions and the need for accessibility to both sides of the joint are restricting design freedom.

The first gas-tungsten-arc spot welding electrode holder or gun was introduced approximately twenty years ago for the one-side welding of thin gage materials. Since then companies have expressed more interest in arc spot welding, for the equipment does not have the limitations of a throat,

224 Fusion Spot Welding (Gas Metal and Tungsten-Arc)

can be portable, and needs access to only one side of the joint. Today the use of the metal and gas-tungsten-arc spot welding processes are widespread in the industry.

Recognizing the advantages of automation, the welding industry has made significant refinements in fusion spot welding so that welding machines are able to meet the high weld production rate required for low cost construction.

THEORY-PRINCIPLES-KEY MECHANISMS OF THE PROCESS

The two types of arc spot welding are gas-tungsten-arc spot welding and gas-metal-arc spot welding. Generally, an arc spot weld is made by applying a welding arc of sufficient heat to penetrate through the top sheet and cause local fusion of two or more sheets of material. This process is also known as fusion spot, heliarc spot, and TIG (tungsten inert gas) spot welding. Specifically in fusion spot the welding action is controlled by the current input to the arc and the time the arc is allowed to dwell.

Figure 5.1 Boron nitride nozzle in operation.

Theory-Principles-Key Mechanisms of the Process

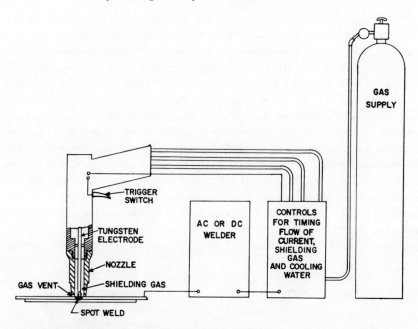

Figure 5.2 Schematic diagram of setup for gas-tungsten-arc spot welding [1].

The arc, tungsten electrode, and fluid puddle are shielded with an inert atmosphere in a manner similar to that employed in the conventional gas-shielded arc-welding process. (See Figure 5.1.) To make a gas-metal-arc spot weld the conditions suitable for the base-metal thicknesses being welded are set on the machine and then the nozzle of the electrode holder (or "gun") is placed on the work with sufficient pressure to bring the members into intimate contact. The weld trigger is depressed and the arc is established through a series of relays and timers for a period long enough to fuse the two pieces.

Illustrated in Figure 5.2 is the manual operation of fusion spot welding with a pistol-like torch having a vented, water-cooled gas nozzle, a concentrically located tungsten electrode, and a trigger switch for controlling the operation. Other types of welding torches that are available for fully automatic application are discussed later in the chapter. Regardless of the type of welding torch used, however, automatic controls for turning on and shutting off the welding power, and cooling water and gas are usually employed. These controls are usually activated by a trigger switch in the torch handle. A fusion-spot welding torch with a trigger switch is seen in Figure 5.1.

226 Fusion Spot Welding (Gas Metal and Tungsten-Arc)

Whether gas-tungsten-arc or gas-metal-arc spot welds are made, the operation is essentially identical. The sequence of welding is simple. The arc is struck to the top surface of the sheet. When a molten pool has been formed and the surface oxide has been disrupted and floated aside, the electrode is moved forward to displace the pool so that the arc force can be exerted on the interfacial oxide surfaces between the two sheets. The electrode is held in this position until the oxide layers have been disrupted and the area of coalescence is of the desired size. Then the electrode is withdrawn to its original position. The arc is maintained until the molten metal flows back into and fills the crater and any gas pores have floated to and escaped through the molten surface. Then the arc is extinguished. As in shielded arc processes the main difference is that in metal arc the electrode is a filler metal and in tungsten arc the electrode is usually tungsten. Filler metal may be added. A portable electrode holder (or "gun") as shown in Figure 5.3 operates from a 50-ft cable. It contains

Figure 5.3 Portable gas-tungsten-arc spot hand electrode holder showing filler metal approach system.

Theory-Principles-Key Mechanisms of the Process

a 1-lb spool of filler metal to allow for uninterrupted production welding. The feed system introduces filler metal to the weld area through an opening at the lower end of the shielding cup. The spot weld is produced from the heat generated by an electric arc from the fixed position of tungsten to the work surface. More on filler metal additions and feed mechanisms is discussed later in this section.

Let us examine the process itself. There is a basic difference between the processes for making continuous fusion welds and fusion spot welds. In a continuous weld the surfaces to be fused are directly exposed to the arc; in a spot weld a solid sheet of metal is between the arc and the surfaces to be fused. An arc spot weld is therefore made in the following five-step operation:

1. The arc is initiated.
2. The metal sheet is penetrated by the arc.
3. The surfaces to be joined are fused.
4. The arc is withdrawn through the metal sheet without leaving a defect in it.
5. The arc is terminated.

These five steps of arc initiation, penetration, welding, withdrawal, and arc termination are more or less one continuous action and are interdependent. Each must be performed correctly and repeated exactly in order to form a series of welds of substantially the same strength. From a more basic point of view this means the proper and reproducible application of heat from the arc and the proper and reproducible transmission of the heat by the work. The reproducibility of heat transmission is strongly affected by the interface between the workpieces. A discussion of key variables follows:

Alternating and Direct Current

Fusion spot welding may be done with alternating or direct current, for suitable equipment is available for both types of power. With alternating current the tungsten electrode is held in position above the workpiece and the arc is initiated by a spark from a high frequency arc-stabilizing unit. When direct current is used, the torch is designed for one of three methods of starting: (a) a spark from a high frequency unit, (b) mechanical retract starting, or (c) pilot arc starting. The touch start system advances and retracts the tungsten electrode, whereas the pilot arc start system is initiated by a low current arc that is maintained between the tungsten electrode and nozzle. With any of the systems the weld cycle timing should begin with the actual flow of welding current. Operation with direct current provides greater penetration than ac operation. Accordingly, direct current is usually preferred if the material is over $\frac{1}{16}$ in. thick.

Arc Initiation and Stabilization

The arc initiating step is defined as that period during which the surface of the work is heated before sheet penetration. It consists of two stages: (a) a stable arc is established and (b) the arc length is shortened to that value at which the arc just begins to penetrate the work. An arc is initiated by fusion of the electrode. When the electrode is brought into contact with the work, a flow of current is started. Resistance heating caused by this short-circuit current increases the temperature of three zones: the interface between the electrode and the contact tube, the interface between the electrode and the work, and the length of electrode bridging these interfaces. When a zone reaches a high enough temperature, fusion acts to break the path of the short circuit current. The inductive reactance of the welding circuit establishes an arc in attempting to maintain the current at its original value.

The rate of temperature rise determines the location of the first zone to reach fusion temperature. The three possibilities are the two interfaces or some randomly located zone on the length of electrode between them. The interfaces are cooled by heat flow into the contact tube and the work. The rate of heating the interface between the electrode and contact tube is the lowest. Electrode and tube are possibly in side-to-side contact at several places along their mutual length. The area for both current and heat transfer is relatively large in comparison with the other zones; consequently, the heating effect is relatively smaller and the cooling effect is relatively larger. This zone can therefore be ignored because one of the other two zones always reaches fusion temperature first.

Which one of these two zones fuses first depends on circumstance. If the interface between the electrode and work is small, the rate of heating is high and the rate of cooling is low. Under this circumstance fusion could occur first at the electrode tip. Conversely, if the area of contact is large, the first fusion will quite probably occur in a zone in the electrode somewhere between the two interfaces.

The location of the zone of fusion determines the degree of heating applied to the surface of the work during the arc-initiating step. Its location may be discussed in terms of arc length because the initial arc length is a measure of the distance from the work to the end of the fused zone that is farthest from the work. The initial arc length may be short as a result of fusion at the tip of the electrode and it then invariably stablizes rapidly and quickly shortens to penetrate the work surface. Conversely, the initial arc length may be long and somewhat unstable; then the time from arc initiation to the start of penetration is greater than with fusion at the electrode tip. As a result more heating of the work surface occurs.

The arc may be so long at initiation that it extinguishes. If this happens, the electrode must be fed forward again to begin a second arc-initiation cycle using the touch start system. If several long arcs extinguish before a stable arc is established, the surface of the work is very strongly heated before penetration begins. In general, all possible variations and combinations among the extremes of the arc initiation step may occur during the welding process.

Heat Transfer

The transfer of heat by the work is affected by the amount and distribution of the heat put into the work surface; it is also affected by the thermal conductivity of both the work and the interface between the two sheets being welded together. For a given material the conductivity of the work is constant; therefore the ability to produce a series of welds of the same strength depends on the ability to maintain constant conditions of heat input and constant thermal conductivity at the interface.

Any heat applied to the top surface of the top sheet during arc initiation causes the top sheet to buckle upward. This doming separates the interface surfaces of the two sheets and thus reduces the thermal conductivity of the interface. A variation in the arc-initiation sequence produces a variation in surface heating, which in turn affects the degree of doming and the interface conductivity. The maximum possible conductivity across the interface, which can only be approached and not achieved, is the conductivity of the metal being welded.

The intimacy of the contact between the surfaces at the interface affects the depth of weld penetration into the bottom sheet. Penetration is minimized with good contact because the two sheets are capable of transferring heat from the arc zone as rapidly as a single sheet equal in thickness to the two together. Conversely, with a gap the two sheets transfer heat away from the arc zone more or less as single sheets. An equipment setting that produces less than complete weld penetration in two sheets having good interface conductivity produces more than complete penetration when a gap exists between the two sheets. Thus the depth of penetration into the bottom sheet is related inversely to the degree of thermal conductivity at the interface.

Furthermore, the area of interfacial fusion is related directly to the degree of interface conductivity since the weld shape approximates that of a cone. When the penetration of the weld beyond the interface increases, the area of the intersection of the weld and the interfacial plane increases. Weld strength therefore also increases since it is related directly to the area of interfacial fusion.

An occasional gap between the sheets can cause a series of welds to have a wide variation of strengths. The welds with the gaps have strengths that are usually greatly in excess of the required average strength of the series. If the interface causes a complete block to heat flow, the bottom surface of the top sheet can fuse beyond the arc zone and fall onto the top surface of the bottom sheet. The strength of the weld then varies in accordance with the amount of bonding that takes place between this additional fused metal and the bottom sheet.

The decrease in the thermal conductivity of the interface may be because of a decrease in the amount of surface in contact, the presence of foreign material to low heat conductivity on the surfaces that are in contact, or both. Foreign materials may include the oxide of the metal concerned. Also, the foreign material can prevent coalition between the surfaces of the sheets even when both surfaces are molten. In summary, control of weld strength depends to a major degree on controlling and minimizing the amount of heat applied to the top surface of the top sheet during arc initiation. Arc initiation at the electrode tip produces the least heating of the top surfaces of the top sheet. Arc initiation at the electrode tip is therefore the most desirable procedure for fusion arc spot welding when using a consumable electrode.

Electrodes—Consumable and Tungsten

PREHEATING

The type and diameter of consumable as well as tungsten electrode and the shielding medium have to be selected to suit the material, metal thickness, and technique of the application. As an aid to arc initiation and stabilization preheating of electrodes to approximately 1450°F temperature has been successfully accomplished. This preheating has achieved improvement in the control of the power input to a weld and much more rapid arc ignition and stabilization. In addition, preheating the electrode assures adequate thoria levels at the tip because first, hot bodies are good electron emitters and less depletion of thoria occurs during ignition and second, the thoria is replenished both during and between welds since the electrode is always hot.

Preheating the electrode also reduces the amount of pitting and therefore the need for frequent regrinding. Preheating reduces the thermal shock during ignition, and more rapid ignition reduces the erosion during high frequency sparking.

Movement

The rate of electrode movement is important. The forward movement must be slow enough to displace the molten pool ahead of the tip or it will stub into the work. Since thicker sheets melt more slowly, the maximum rate varies inversely with sheet thickness. The rate of electrode retraction is also related to the thickness of the top sheet. The electrode must be withdrawn slowly enough to keep the metal molten while it refills the cavity. If withdrawn too rapidly, a large pore is left in the center of the weld.

Shape

The shape, selection, and configuration of the electrode tip require careful consideration. The satisfactory behavior of tungsten inert-gas welding arcs is controlled by electrode composition, and configuration is such that a sharp electrostatic voltage gradient is established at the tip of the electrode. This gradient can be obtained by pointing the electrode, for it confines resistance heating to the tip and sharply limits the possibility of arc oscillation associated with blunt tungsten electrodes. Blunt electrode tips stub into the work when an attempt is made to move them. However, sharp electrode tips, with an included angle of 45° or less, can be advanced into the work without stubbing.

Electrode configuration is important because as occurs in resistance welding and in gas-tungsten-arc spot welding, the electrode tip configuration affects the shape of the welding arc and therefore determines the arc penetration. For example, at high current density and with a needle-pointed electrode tip, splashing or expulsion of the molten weld puddle is easily demonstrated. Furthermore, the electrode life is also affected by the geometry of the tip.

Because of high current densities, the magnetic forces existing in the arc produces a focusing effect of the welding arc. This focusing effect is improved with a tapered tip configuration and weakened by blunt electrode tips. It was found in a recent evaluation of electrode tip diameters for welding several superalloy steels that the ohmic resistance of a 0.010-in. diameter tungsten electrode is many times greater than the resistance of a 0.125-in. diameter electrode. When the same welding current is passed through each electrode, the smaller diameter electrode is resistance heated to emission temperatures for a considerable part of its length, whereas heating of the larger electrode is confined to its tip. With fine diameter electrodes the arc is found to climb up the electrode intermittently because of availability of spots that momentarily have lower emission energy requirements. Although arc starting is not greatly affected as long as a sharp tip is maintained, arc stability is not assured when fine diameter electrodes are used. Therefore the 0.125-in. diameter tungsten electrode containing a nominal 2% thoria is best. In addition, it is found that the degree of taper produced on an electrode governs to some extent the amount of erosion that takes place. The three different electrode preparations evaluated in a series of tests designed to indicate the most suitable shape were those (a) tapered to a point over a length equal to two diameters, (b) to three diameters, and (c) to five diameters.

Accurate symmetry and point angles have been obtained using precision grinding equipment. Results of the tests indicated that after 100 starts and stops there was less erosion on the two-diameter tapered electrode. A $\frac{1}{8}$-in. diameter tungsten electrode was tried with 0.050 in. to 0.080 in. aluminum alloy (2219). Results indicated that this size of tungsten electrode disturbed the nugget formation on thinner base gages and increased the interface discontinuity factor. Interface discontinuity is the designation given to an oxide void that may occur at the faying surfaces. A $\frac{3}{32}$-in. diameter electrode was most suitable for this aluminum alloy and gage. It is therefore obvious from the previous discussion that shape is a critical variable in tungsten electrodes.

REPRODUCIBILITY AND MAINTAINABILITY

The appearance and strength of the gas-tungsten-arc spots produced with several welding electrodes depend on the reproducibility of electrode tip configuration. It is therefore recommended that all the electrode tips be machined automatically.

The tungsten welding electrodes are subjected to constant heating and cooling cycles as well as to impurities in the surrounding atmosphere caused by oil and other residual substances. Therefore it is a practical preventive maintenance requirement to establish the desired tungsten electrode life and change the tips before detrimental effects on weld quality are noticed. On some automotive parts processed without special care for cleanliness, the electrode life of 2000 spots has been recorded.

Arc Length

In discussing arc length the term arc voltage may be used more or less interchangeably. In normal fusion welding the arc length is considered as the distance from the electrode tip to the plane of the workpiece surface. All arc lengths having the electrode tip below the workpiece surface simply described are as being negative. This fictitious and expeditious definition must be replaced by the true definition when discussing arc length in spot welding. The arc length is simply the distance between the terminal ends of the arc on the electrode and the workpiece.

The electrode or the welding torch locating device is responsible for physical arc length reproducibility. Arc length determines the weld heat input into the part. For example, in welding a hem joint with a 0.040-in. thick edge, the optimum arc length was 0.040 in. An increase in arc length may cause undercutting of the edge near the weld and in extreme situations overheating of the joint being welded. A decreased arc length may result in electrode contamination by the material being welded due to electrode expansion as well as to base metal expansion caused by locally concentrated heating of the arc.

The effect of arc length on the shear strength of arc spot welds has been described by Hackman, Copleston and Steer [2,3,4]. As indicated by these authors, shear strength is increased with increasing arc length until a maximum is reached where further increases in arc length causes a decrease in shear strength. Obviously, then, selection of the proper arc length for a given application is extremely important.

Gas Shielding

The choice of shielding gas can affect the quality and shape of the weld and the arc stability. Regarldess of the gas used, the fusion shape of the weld may be considered as being two shapes superimposed. The upper part is semicircular. A triangularly shaped papilla protrudes through the bottom of the semicircle. The size of the semicircle varies little with the choice of shielding gas, but the size of the papilla does. Argon produces a deep, narrow papilla and therefore weld penetration must be very deep if a large weld area is required. Helium produces a wider, shallower papilla that argon and, consequently, a large weld area is achieved with less penetration.

Plate cleaning by cathodic action of the arc and arc stability are better with argon than helium. With good cathodic cleaning the amount of entrapped oxide can be minimized and the likelihood of unfused areas is decreased. Good arc stability is essential for control of the heat that is produced during arc initiation and thus somewhat better stability than that obtained with helium is desirable.

A study of helium-argon mixtures has indicated that a mixture of 75% helium and 25% argon is the best choice for some metal combinations. With this mixture adequate arc stability and cathodic cleaning are achieved and the ratio of the interfacial fused area to the depth of weld penetration approaches that obtained with pure helium. In the gas-metal-arc spot welding of magnesium either argon or helium is satisfactory. Argon is usually the preferred shield gas because excessive weld smoke is generated with helium. In terms of operating conditions helium produces a hotter arc. Consequently, the weld time required to produce the same strength on equivalent thicknesses of base metal is reduced by about 50%.

Shielding gas flow rates as low as 5 cfh are practical in gas-metal-arc spot welding because the nozzle by contacting the workpiece confines the gas to a small area and prevents its loss to the atmosphere.

In fusion spot welding various steels and superalloys, various shielding gases and mixtures have been used. It has been found that 100% argon produced sound welds in the 19-9DL and Type 316 stainless steel materials (two thicknesses of metal) the top sheet did not exceed 0.060-in. thickness. When thickness combinations greater than this were tried, it necessitated the use of the 75% helium-25% argon mixture. This gas mixture was also required on both two-ply and multiple thickness joint combinations on Inconel, Inconel X, Hastelloy X and René 41 materials.

Although argon shielding is normally used for the gas-metal-arc spot welding of aluminum, helium shielding is preferred when welding very thin sheet because it produces a flatter penetration cone. With helium shielding of all gas-metal-arc welds in aluminum the penetration cone is less pointed at the bottom than with argon. Helium shielding has also proved advantageous in making nonpenetrating welds between two sheets of 0.125-in. thick aluminum. Drawbacks to the general use of helium are less smooth surface contour and more spatter.

A precautionary note that applies to fusion spot welding of all metals is that care must be taken to allow about 5 sec gas purge time before making the first weld in a series. Longer times may be necessary to purge hoses at the beginning of a day's work. Contaminating atmosphere present at the beginning of a weld noticeably affects the weld porosity, surface roughness, and wetting. Because spot welding time is short, operators are likely to forget the good gas protection procedures they would follow with conventional gas-tungsten and metal-arc welding. Subsequent welds in a series can be made as quickly as the electrode holder can be repositioned because the gas cup remains purged.

Nozzle Design

Depending on the joint design, electrode holder nozzles vary in their design. Illustrated in Figure 5.4 are several preferred nozzles for magnesium gas-metal-arc welding. Conventional nozzles that contain Vee-notches cut at intervals along the bottom edge of the cup are satisfactory at short weld times; however, they are not desirable at long weld times because the shielding gas tends to force part of the weld puddle out these holes, causing an irregularly shaped spot.

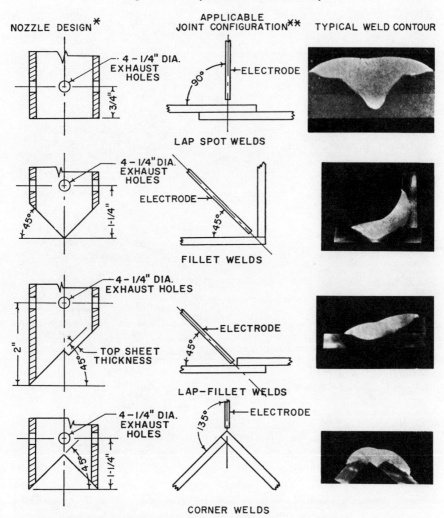

* — Constructed of pure copper: 1.660" O.D. by 0.149" wall.
** — Electrode should strike work in approximate position shown. To accomplish this; move contact tube as close to the work as possible.

Figure 5.4 Nozzle designs for typical gas-metal-arc spot welding applications.

The nozzles shown in Figure 5.4 are designed to produce, as nearly as possible, proper alignment between the electrode and the workpiece when the gun is positioned for welding. It is necessary that the electrode strike the work in the approximate position shown to obtain good bead contour and joint penetration.

In operation in Figure 5.1 is a gas-tungsten-arc nozzle utilized in welding

0.004 to 0.006 in. titanium alloy. Again note the nozzle configuration is made to suit the requirements of the joint geometry.

Weld Backing

Although backing the weld may appear to detract from the "one side" feature of the process, it enables higher currents and longer times to be used, thus leading to an increase in shear strength. Backing also facilitates control of penetration and of surface appearance on both sides. Without a gas backup the under bead would have the typical oxide film or "coke" appearance. A backing atmosphere of 100% argon has been used. In addition, a small amount of development work has been done with ceramic backup coatings in place of the inert gas atmosphere. Initial results appear promising, but additional development work is required to determine the metallurgical effects, if any, of the ceramic on the weld nugget.

Filler Metal Addition

Although usually inconvenient, it is possible to add filler metal in making gas-tungsten-arc spot welds, which is occasionally done. Spot welds made without filler metal normally develop slight sinks or hollows. Filler metal is sometimes added to fill in these hollows for appearance or finish. Filler metal may occasionally be added for metallurgical reasons where materials are used that may be sensitive and subject to cracking in the spot welds.

The filler metal (or wire) approach falls into three categories. First, the approach could be too low, dragging the filler metal across the work surface, and causing the filler metal to build up on the opposite side of filler metal feed-in. When the nugget is built up and filler metal continues to feed into a solidified area, the filler metal sticks. Second, raising the approach of the filler metal to where it centers the arc gap improves the nugget geometry. At this level the filler metal does not stick. The possibility then exists that during filler metal retract a portion of the weld nugget may become attached to the filler metal and, if not above the shave line, could cause cracking. Third, to avoid these undesirable effects the filler metal should be directed toward the tungsten tip. This will aid in burning off all oxide coatings on the filler metal. The highest possible approach finds the center of the filler metal at the point of the tungsten. If the filler metal is raised further and is fed into the side of the tungsten tip, an aborted weld results. The tungsten having adhered to the filler metal requires replacing.

In all instances, the favored side has a high deposit of filler metal, whereas the opposite side is undercut and subject to crater cracking.

EQUIPMENT–ADVANTAGES–ECONOMICS

Equipment and Controls

In gas-tungsten-arc spot welding the usual equipment consists of a power supply for the control of current and time control of the preweld, welding, and postweld operations. Filler material additions may be accomplished by special wire feeders employing rate of feed control and sequence timing so that the wire may be injected into the weld puddle required for any given weld setting.

Power supplies are classified on the basis of their static volt-ampere characteristics. If the output voltage droops, stays flat, or rises with increasing output current, the supply is, respectively, termed a drooping, constant, or rising voltage power supply. Unfortunately, the drooping-voltage power supply is not suitable for gas-metal-arc spot welding thick sheet. At a low current the arc is long enough to distribute the heat input over a surface area that would result in the required size of weld if the arc could play on the surfaces at the interface. The current, however, is too low to produce the necessary penetration of the arc into the work. A rising voltage supply is also not suitable for spot welding thick sheet. Although a large area at the interface can be fused, the fused metal is expelled from the hole and the end result is metal expulsion. The rising voltage supply provides very poor filling and finishing action. The constant voltage supply is the most suitable type for the spot welding of thick sheet. It is not ideal for the purpose, but is a good compromise between the drooping and rising voltage types.

At low voltage settings the current capacity is low and penetration is similar to that of the drooping voltage supply. At high voltage settings the current capacity is high, but the arc is long as it is with the rising voltage supply. In the intermediate settings of voltage the constant voltage supply is intermediate compared with the other two types of supply and produces intermediate results in the penetration, welding, and crater-filling actions.

There is no dictionary in which a engineer, welding manager, or operator can refer in order to correctly set his machine for the correct flow of gas, tungsten electrode diameter, current, amperage, and wire filler additions, if necessary. He must set up schedules such as shown in Table 5.1 depending on the material. When a point is reached where the operator feels satisfied with his test results as well as his quality assurance department, a form is set up and recordings made for future use. Unfortunately, this type of information is not disseminated throughout industry; therefore the expensive duplication of initial development takes place numerous times.

Table 5.1 Gas Tungsten-Arc Spot-Welding Schedules.

Material	Thickness, in. Upper	Thickness, in. Lower	Inert gas Weld gun	Inert gas Back up	Current, amp	Weld time in Cycles — Final slope	Weld time in Cycles — Wire start	Weld time in Cycles — Wire stop	Weld time in Cycles — Spot time	Tensile-shear Avg. Arc spot, lb	Tensile-shear Avg. MIL-W-6858, lb	Weld Diameter Arc spot	Weld Diameter in., MIL-W-6858
19-9DL	0.030	0.030	100%A	100%A	75	120	30	60	300	1190	785	0.16	0.16
19-9DL	0.030	0.040	100%A	100%A	75	120	30	60	300	1150	785	0.16	0.16
19-9DL	0.040	0.030	100%A	100%A	75	120	30	60	300	1120	785	0.16	0.16
19-9DL	0.030	0.050	100%A	100%A	90	120	30	60	300	1265	785	0.18	0.16
19-9DL	0.050	0.050	100%A	100%A	120	210	150	240	600	2540	1855	0.23	0.21
19-9DL	0.040	0.030–0.050	75% HC, 25% A	100%A	75	120	60	120	300	1430	785	0.20[b]	0.16
19-9DL	0.040	0.050–0.050	75% HC, 25% A	100%A	150	120	90	150	390	2705	1310	0.20[b]	0.20
19-9DL	0.080	0.030–0.050	75% HC, 25% A	100%A	150	180	150	240	540	3500[a]	785	0.25[b]	0.16
19-9DL	0.080	0.030	75% HC, 25% A	100%A	200	180	150	240	540	3650[a]	785	0.25[b]	0.16
19-9DL	0.040	0.030–0.080	75% HC, 25% A	100%A	225	180	150	240	240	3640	1310	0.30	0.20
Inconel	0.043	0.043	75% HC, 25% A	100%A	75	210	150	240	600	1858	1150	0.19	0.19
Inconel X	0.062	0.026	75% HC, 25% A	100%A	105	180	90	240	480	4637[c]	2595	0.24	0.24
Hastelloy X	0.062	0.062	75% HC, 25% A	100%A	105	180	90	240	480	2945	2595	0.24	0.24

[a] Tensile shear specimens in double shear.
[b] Smallest diameter of thinnest sheet.
[c] Annealed at 1950° F, double aged at 1525° F for 8 hr and 1325° F for 20 hr.

Of course, this may be required in some instances because companies have new and old equipment. In addition, there may be commercial equipment that is inadequate for a particular assembly and modifications are necessary.

As mentioned previously the preheated electrode gas-tungsten-arc spot welding equipment differs from standard equipment mainly in the provisions for preheating and supporting the electrode. The electrode is held and preheat current is introduced by two tungsten "preheat" jaws. Heat is generated by the contact resistances at the faying surfaces of the jaws and electrode and the resistance of the electrode itself. The preheat current may be either ac or dc.

With the trend toward mass production and welding process automation requirements, equipment has been improved with new power and process control units for simultaneous firing of eight gas-tungsten-arc spot welding electrodes. An automatic multiple-electrode gas-tungsten-arc spot welding machine now has 24 electrodes and is used successfully to spot weld an edge joint of an automobile compartment lid.

Advantages–Disadvantages–Economics

What features make gas tungsten and gas-metal-arc spot welding attractive as a manufacturing tool? Several features include the following:

1. Simple jigging. Access to a lap weld is required on one side only. Multiple welds can be placed close together.
2. Portability. The equipment is mobile, thus facilitating welding on jigged or complicated structures difficult to handle.
3. Semiskilled operation. The process is semiautomatic and requires only occasional changing and resetting of the tungsten electrode.
4. Strength. The weld shear values obtained compare favorably with those made to resistance welding.
5. Shunting. The center-to-center distance of spot welds is not limited as in resistance welding because shunting is not encountered.
6. Multiple thicknesses. It is possible to make welds on multiple thickness joints.
7. In tungsten-arc spot welds it is feasible to have fitup gaps exceeding 0.005 in. between sheets and produce successful welds.
8. Welds are made with only a light force applied to the part. No weld mark appear on the "show" surface.
9. The gas-tungsten-arc spot process requires less line power demand than the resistance welding process. Conservatively, the kVA per tungsten-arc spot is half of the resistance spot.
10. With low gun forces required the need for a massive welding press and expensive backup dies or fixtures is eliminated.

11. The gas-metal-arc spot welding process requires less critical surface cleaning than is required for resistance spot welding. There is no electrode pickup problem.

On the negative side gas-metal-arc spot welds lack the smooth appearance of resistance spot welds because a nugget is deposited on the top weld surface. However, the nugget buildup can be removed, if necessary, by machining, shaving, or other means. The internal quality of welds, particularly in aluminum base alloys, is not as good as that of resistance spot welds because defects such as porosity and peripheral cracking are commonly encountered [5,6,7]. The orientation of these defects, however, do not affect the shear strength of the joint.

Thus arc spot welding, in addition to being very economical, has the advantage of being reliable from the standpoint of producing repetitive high quality welds. The gains made in arc spot welding have been chiefly at the expense of the resistance spot and projection welding processes as well as manual shielded metal-arc welding.

MATERIALS, JOINT DESIGN, AND FUTURE POTENTIAL

Materials

The fusion-spot welding process has become widespread and is now applicable to aluminums, magnesium, all types of steels, refractory metals, and dissimilar metal combinations and over a wide range of metal thicknesses varying from 0.002-in. foil to plate thicknesses of 0.900 in.

ALUMINUM AND ITS ALLOYS

Gas-metal-arc spot welding has gained much favor as a quick and reliable method to join sheet, extrusions, and tubing in aluminum and its alloys. A strong filler metal such as 5556 aluminum produces stronger welds, even in low strength alloys, than base metal filler or unalloyed aluminum. Tensile shear breaking loads for 0.064-in. sheet with "nonpenetrating" and "full penetration" spot welds in a variety of aluminum alloys indicate that strengths are highest with the aluminum-magnesium alloys of the 5000 series. The combination of 5456 aluminum alloy sheet of temper -H343 and 5556 aluminum alloy filler metal produces the highest strength per spot at a given thickness. Nonpenetrating or full penetrating spot welds are dependent on the application as well as compatible alloy combinations shown in Table 5.2. In this respect full penetration welds are stronger, more consistent, and easier to make than nonpenetration welds when sheets of nearly equal thickness are joined. Furthermore, nonpenetration welds,

Table 5.2 Weld Cracking Resistance for Gas-Metal-Arc Spot Welding.

Aluminum base metal alloys	Good filler metal alloys	Usable filler metal alloys
1100, 1060, EC	4043, 1100, 5356, 5556[a]	......
2014, 2024, 2219	2319, 4043	2014
3003	4043, 1100, 5356, 5556[a]	......
5005, 5050	4043, 5356, 5556[a]	......
5052, 5086, 5154	4043, 5356, 5556[a]	5254,[b] 5554
5083, 5456	4043, 5356, 5556[a]	5554
5454	4043, 5356, 5554, 5556[a]	5254[b]
6061, 6062, 6063, 2EC	4043, 5356, 5556[a]	5254,[b] 5554
7075, 7178	4043	2319, 5554, 5556[a]

[a] 5183 may be used in place of 5556.
[b] 5154 may be used in place of 5254.

which allow one surface to be left unmarked, are strongest and most consistent when the bottom sheet is about two and one-half times as thick as the top sheet. In addition. good mating of surfaces is essential for consistent nonpenetration welds. (See Figure 5.5.)

COLUMBIUM AND ITS ALLOYS

Successful tungsten gas-arc spot welding of several columbium alloys has been accomplished.

D-36 Columbium alloy has been successfully fusion spot welded into experimental heat shields shown in a front view in Figure 5.6.

MAGNESIUM AND ITS ALLOYS

At the present time the gas-metal-arc spot welding process has been applied only to AZ31B alloy, but there is no reason to suppose that the process would not also be suitable for welding other wrought Mg-Al-Zn

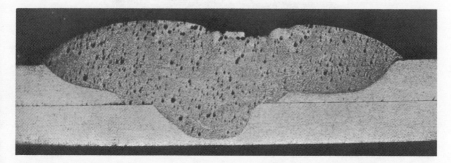

Figure 5.5 Gas-metal-arc spot weld, nonpenetrating type. 0.064-in. thick 1100 aluminum sheet and 4043 aluminum alloy filler metal. ($\times 10$).

242 Fusion Spot Welding (Gas Metal and Tungsten-Arc)

alloys such as AZ10A or AZ80A or the thorium-containing alloys such as HK31A, HM21A, or HM31A. Based on gas-tungsten-arc and gas-metal-arc weld current requirements, it is expected that an increase in weld time and/or current will probably be required when gas-metal-arc spot welding the thorium-containing alloys.

AZ61A electrode is recommended for gas-metal-arc spot welding applications involving AZ31B and other wrought Mg-Al-Zn alloys. EZ33A electrode is recommended for use on the thorium-containing alloys because its elevated temperature strength is more compatible with that of the base metal.

In dissimilar thickness lap joints the weld should always be made from the thinner into the thicker member unless it is permissible to penetrate completely the bottom joint member. Welding in the reverse manner requires extremely close control over welding conditions and good part fitup that usually is not possible or practical to obtain in field use.

For welding alike gages the minimum material thickness is considered to be 0.063 in. For dissimilar base metal thicknesses considerations as thin as 0.040 to 0.090 in. can be joined by this process. Welding through a top member thinner than 0.040 in. is not recommended since there is a tendency to wash the base metal away from the edge of the weld button. Increasing the thickness of the bottom member of the joint has no effect on the operating current and voltage conditions, but the weld time range is extended upward.

The use of complete penetrating welds allows the minimum thickness recommendations previously mentioned to be reduced and welding conditions become less critical. Complete penetrating spot welds also offer a

Figure 5.6 Front view of a D36 columbium heat shield.

Table 5.3 Comparison of Strength of Complete and Partially Penetrating Gas-Metal-Arc Spot Welds in 0.090 in. AZ31B Sheet Welded with AZ61A Electrode.

Types of weld penetration	Arc amp	Arc volt	Weld time, cycles[b]	Button interface diam, in.	Shear strength, lb/spot		
					Min	Max	Avg
Partial	275	26	90[a]	0.30	948	1220	1078
Complete	275	26	120	0.40	1370	1950	1643

[a] Maximum weld cycles before start of complete penetration.
[b] 60 cycle/second base

significant strength advantage over partially penetrating welds as shown in Table 5.3. The biggest disadvantage of complete penetrating welds is that they cannot be used when surface appearance of the bottom member is an important factor because some marking of the surface does occur. This can be minimized, however, by using a smooth steel or copper backing plate against the underside member of the weld joint.

Gas-metal-arc spot welding of magnesium alloys is adaptable to fillet, lap-fillet, lap, and corner joint designs. With these four basic joint designs it is possible to fasten sheet and framing members in a wide variety of configurations and this would be either impossible or impractical to produce by conventional resistance welding techniques.

The strength of gas-metal-arc spot welds is dependent on the diameter of the weld nugget at the joint interface. Because of this relationship, nugget diameter can be used as a quality control tool to assure that adequate nugget strength is being developed.

By proper selection of welding conditions the strength of gas-metal-arc spot welds meets or surpasses the minimum strength requirements of MIL-W-6858B for magnesium resistance spot welds.

STEELS AND SUPERALLOYS

Satisfactory welds have been made in 0.004 in.-thick sheets of Type AM350 stainless steel and 0.002-in. thick foils of Type PH 15-7 Mo stainless steel and 0.004-in. thick foils of René 41. Other superalloys such as 19-9DL, Inconel 600, and Hastelloy X have also been successfully arc spot welded.

TITANIUM AND ITS ALLOYS

Arc spot welds in 6AL-4V titanium alloy have been produced in multiple ply layers of 0.002 in., 0.004 in., 0.008 in., and 0.016 in. foils and sheets. These multiple ply layers are shown in the panels in Figure 5.7. Subsequent

Figure 5.7 Fusion spot welded aerospace structural component.

operations including the inner shell, outer shell assembly, and the complete assembly (Figure 5.7) produced a structural model for research on structures for hypersonic aircraft. The structural assembly shown in Figure 5.7 contains over 400,000 tungsten-gas-arc spot welds in titanium alloy (6AL-4V), René 41 superalloy, and D-36 Columbium.

DISSIMILAR METAL COMBINATIONS

Galvanized aluminum-coated or base steel, copper, and titanium are some of the metals that have been successfully joined to aluminum by gas-metal-arc spot welding. The strong ductile welds have been produced in sheet from 0.005 to 0.250 in. thick. Usually aluminum filler metal is used which penetrates the other metal and fuses into the underlying aluminum sheet. Brittle intermetallic layers, which may be formed at the fusion line between the other metal and the aluminum filler, do not extend across the shear plain. When the metal to be joined to aluminum is less than 0.030 in. thick, the weld can be made directly through the other metal into the aluminum.

Gas-metal-arc spot welds between aluminum and other metals are as strong as typical arc spot welds in the weaker of the two metals being joined. Failure occurs in a ductile manner by shearing the weld button or tearing the member whose product of tensile strength and cross section area is the lowest. Some typical strengths are $\frac{1}{8}$ in. copper to $\frac{1}{8}$ in. 1100-H18 aluminum, 1180 lb/spot; 0.022-in. aluminized steel to 0.064 in. 3003-H 18,600 lb/spot. Average strength and variability appear to meet or exceed any military specifications for resistance spot welds in aluminum.

For simple joint configurations it is preferable to weld through the other metal using aluminum filler metal. In certain applications, however, it is often necessary to weld through aluminum to the other metal using compatible filler metal. Specific joint designs such as aluminum sandwiched between two sheets of copper welded with copper filler metal have been devised to reduce excessive melting of aluminum when a higher melting point filler metal is used. It has been shown that aluminum may be welded to titanium with aluminum filler metal, regardless of whether aluminum or titanium is on the face side of the weld because the intermetallic layers that form are very thin and ductile.

Gas-metal-arc spot welds between aluminum and aluminized steel have exhibited excellent resistance to corrosion and retained a pleasing appearance with no rust staining. Aluminum-to-galvanized steel joints have more noticeably corroded but retained excellent strength. Aluminum-to-bare steel joints have shown severe rusting of the steel, as expected, in intermittent salt spray but retained a high proportion of their original strength after six weeks exposure.

Gas-metal-arc spot welds between aluminum and copper have shown excellent strength and electrical conductivity and have performed well in heat cycle tests.

Joint Design

Gas-metal-arc spot welding has been used in joining aluminum electrical connections. To keep joint resistance no higher than the conductors being joined, the fused area of the weld should equal the cross-sectional area of the conductors. When a small aluminum bus is being joined to a larger bus, this is easily done. When both conductors are of equal thickness, however, full penetration spot welds should be used. Lap fillet spots provide more fused area and better conductivity than lap burnthrough spots and should be used for electrical connections whenever possible. On a thin aluminum sheet conductor the edge of the sheet may tend to melt back from a lap fillet gas-metal-arc spot weld. In this instance full penetration burnthrough spots made against a backing is preferable.

In general, there are four types of joints that are easily welded by either single spot or with a multiple-electrode gas-tungsten-arc spot welding pro-

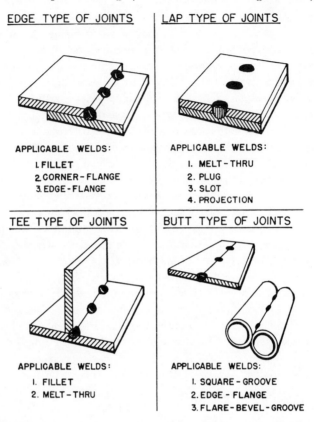

Figure 5.8 Types of joints for multiple electrode gas-tungsten-arc spot welding.

cess: edge, lap, tee, and butt. Figure 5.8 illustrates the four typical joint types.

Applications

Processes for gas tungsten and gas-metal-arc spot welding have been developed for experimental applications as well as high production assemblies. The spot welds meet the requirements of welding specifications (MIL-W-6858B); thus fusion spot welds are as equally reliable as resistance spot welds and permit the greater flexibility needed for welding larger and more complex airframe, missile, nuclear, automotive, and household appliance parts.

The designer and manufacturing welding engineer today must consider many factors in the possible applications of gas-metal and gas-tungsten-arc spot welding: (a) the type and range of material thickness that can be handled with the process and how many sheets can be spot welded together,

(b) the relationship in thickness between two or more sheets which are to be joined, and (c) the permissible center-to-center spacing of spot welds and also permissible spacing from the edge of the sheet.

When considering multiple thickness joint limitations, the manufacturing welding engineer has two alternatives. He can attempt to weld the joint in one operation or revert to a sequence buildup. Recently an intermittent arc spot welding operation produced a multiple pileup joint consisting of 13 0.070-in. thick René 41 sheets. The joint was made by sandwiching two sheets together and making the arc spot welds at the two locations. The face of the weld was flushed using a weld shaver, and the next sheet was placed into position and welded to the first two sheets. This method of manufacturing stackup continued until all 13 sheets were joined together.

Using gas-tungsten-arc spot welding as a repair technique for defects in other types of fusion or resistance welding has great promise. Defects in lineal welding require elaborate repair. Savings can be realized if the defect can be drilled out and filled with a gas-tungsten-arc spot weld. Two possible methods have been tried. In one the defect is drilled halfway into the base metal and then plugged with a spot weld. With deeper defects a hole is drilled through the entire base gage and then filled completely with a spot weld. In both the repair area and the heat-affected zone are held to a minimum.

Aircraft, space boosters, and future aerospace vehicles are utilizing arc spot welding. The aft sections of the B-70 airplane engine shrouds and dividers (see Figure 5.9) are made of René 41. T-angles for spars and webs have been attached to core panels by arc spot welding.

Gas-tungsten-arc spot welding has replaced bonding of instrumentation bracketry on the Saturn S-IC Booster. More than 1500 spot welds per vehicle have been made on base gages of 2219-T87 aluminum from 0.224 in. up to 0.900 in. (See Figure 5.10.)

The outer wall and inner frames of a fuselage for a future aerospace vehicle have been gas-tungsten-arc spot welded as shown in Figure 5.11. This structure has been environmentally and dynamically tested to 1200°F temperature with no failures in the welded joints.

Joining of Zircaloy-2 clad side plates to the Zircaloy-2 clad fuel plates of an experimental boiling water reactor was done by gas-tungsten-arc spot welding when heat generated by the arc produced a molten pool through a side plate into the side clad of the fuel plate. All indexing and welding were controlled automatically to assure reproducibility.

In other fields gas-tungsten and gas-metal-arc spot welding are now used extensively in the assembly of body and frame components for automobiles, trucks, and trailers, transmissions, and other parts as well as in the manufacture of refrigerators, household appliances, sheet metal doors, jacketed kettles, and innumerable other sheet metal items.

Figure 5.9 Shrouds for the jet engines of the B-70 that fit inside aft section (left) are made of René 41. They contain both TIG and resistance spot welds.

Figure 5.10 Instrumentation bracketry gas-tungsten-arc spot welded on an upper LOX bulkhead for the S-IC booster.

Figure 5.11 Closeup view of fusion spot welds in Haynes 25.

Therefore arc spot welding has come of age with its recognized potential as a high production, low cost assembly fabrication process.

REFERENCES

[1] *Welding Handbook,* Section 2, Fifth Edition, p. 27.17 (1963).
[2] R. L. Hackman, "Inert Gas Spot Welding in Aircraft and Missile Industry," *Welding J.,* 25–31 (1961).
[3] F. W. Copleston, and L. M. Gourd, "Developments in the Inert Gas Tungsten Arc Fusion Spot-Welding Process," *Brit. Welding J.* **5,** 394–399 (1958).
[4] J. L. B. Steer, Met. and F. W. Copleston, "Developments in the Argon Arc Spot Welding Process," *Welding and Metal Fabrication,* 282–288 (August 1957).
[5] Eric S. McFall, "Spot Welding of Aluminum," *Welding J.,* 1230–1236 (1960).
[6] R. A. Stoehr, "Gas Metal-Arc Spot Welding of Aluminum," *Ibid.,* 815–821 (1962).
[7] I. A. MacArthur, and Dolomont, A. A., "Aluminum Can Now Be MIG Spot Welded," *Mod. Metals,* XVI (9), pp. 52, 56, 58 (October 1960).
[8] D. J. Maykuth, "Summary of Contractor Results in Support of the Refractory Metals Sheet Rolling Program," DMIC Report #231, p. 55 (December 1, 1966).

Chapter 6

RADIANT HEAT BRAZING

The word "breakthrough" is greatly overworked today, but what appears to be a major breakthrough in space-age brazing technology is the new automatic process to produce honeycomb panels suitable for advanced aircraft and space vehicles.

Nortobraze,* a new process developed by Northrop's Norair Division, combines radiant lamps and electronic controls to produce unusual properties in finished panels. The large amount of time consumed in brazing these components when the operation is conducted in conventional furnaces led to the widespread investigation of brazing processes including research and experimentation and the eventual development of the Nortobraze system which uses banks of electronically controlled radiant quartz lamps to braze honeycomb sandwich and other hollow structures of high temperature, high strength materials. The lamps used in the process are shown in Figure 6.1.

THEORY, PRINCIPLES, KEY VARIABLES

Like most significant developments the process is based essentially on simplification of handling and other mechanical aspects in combination with an improved method of controlling key factors. Radiant quartz lamps supply brazing heat in conjunction with electronic controls (including amplified feedback) to provide continuous, precise control of temperature and time to fractions of degrees and seconds.

* Nortobraze, Copyright of Northrop Corporation Norair Division.

Figure 6.1 The radiant quartz lamp in operation.

In the Nortobraze process the part is suspended from an overhead conveyor and is encased in plain, automotive-body steel that is used as an envelope for atmosphere control. The atmosphere envelope has only a fraction of the strength of the material being brazed, thus creating an inconsequential amount of restraint on part expansion caused by any thermal expansion factors. In addition, the steel envelope serves the same purpose as huge graphite inserts or heavy backup plates ordinarily used with furnaces to insure even heating and support during brazing periods. The envelope is shown in Figure 6.2.

The suspended part is automatically conveyed between banks of radiant lamps with each bank of lamps mounted on a push-bar, permitting the lamps to be adjusted to fit the contour of the part.

In less than a second the radiant lamps reach full brazing heat—up to 5000°F lamp filament temperature—controllable by zones according to part requirements. In Figure 6.3, the banks of opposing quartz lamps are

Figure 6.2 A stainless steel honeycomb panel in its brazing envelope in the load-unload station.

located at the right-hand end of the photograph. Each bank has 19 rows of lamps, as shown in Figure 6.1. The rows are about 4 ft in length and alternately consist of two and three lamps of two standard lengths, which give a staggered effect. Temperatures of 3500°F have been derived from these lamps and there are claims of 5000°F for very short periods of time. The banks of radiant quartz lamps, mounted in specially designed reflectors, are located close to the external surfaces of the envelope. The closeness of the lamps to the envelope permits a high heat flux density to flow into the brazement, resulting in an extremely short heat-up cycle. Control of the heat to local areas in the panel being brazed can be achieved by (a) control of banks of lamps or each separate lamp, (b) coating the braze envelope with a pattern of low or high emissivity coatings, or (c) interposing fixed or changing mask stencils between the radiant source and the panel.

This heating and brazing system provides remarkable flexibility. The

Theory, Principles, Key Variables   253

entire cycle can be programmed to provide precise control, through thermocouples, of temperature, and time-at-temperature, at any point on the part as shown in Figure 6.2. This control can be aided, if needed, by black or gold coatings for heat absorption or reflection.

The system uses true electronic controls, with both feedback and observation or recorder, and loops with each one having its own set of thermocouples. Two thermocouples, or pairs, are used at each desired point rather than a single thermocouple as a precaution against the possibility of one dropping out.

The Nortobraze process also incorporates a simple solution to the problem of rapid chilling inherent in conventional furnace brazing. On completion of the brazing cycle the overhead conveyor moves the package a few feet into the adjacent rapid-chill setup—all in $1\frac{1}{4}$ sec, as shown in Figure 6.3. The resilient-faced chill forms, which utilize a cushion of woven fiberglass, exert a light, even pressure of about 5 psi on the part and are indexed to the envelope by pneumatic pressure. The temperature in the chill platens may also be controlled.

Figure 6.3 The part has just been moved from the banks of radiant lamps at right and is being cooled.

EQUIPMENT AND TOOLING

The Nortobraze equipment consists of two major elements: opposing banks of radiant quartz lamps for brazing and opposing dies that clamp the panel for the chill cycle. (See Figure 6.3.) The hydraulically operated carriage riding single top and bottom rails holds the part in an upright position and automatically positions it between the lamp banks or the chill forms. The lamp banks have a total output of 200,000 W/sq ft of honeycomb panel, although ordinarily only about one-fourth of that amount is used for panels 3 ft by 4 ft. The chill dies have channels for circulating cooling water, although normal heat radiation from the die mass is sufficient for most cooling operations.

There are two groups of control equipment. The machine control console automatically purges air from the brazing envelope and introduces argon gas and controls the positioning of the part. Automatic recording charts in this console provide a complete history of purge cycle negative pressures.

As many banks of lamps can be utilized as desired, depending on part complexity. Time and temperature sequencing of the braze cycle is controlled by a six-channel power controller. Each bank has its own control unit with strip-chart recorders programmed by graphs drawn on ordinary squared chart paper. Scanning devices follow the graph lines and actuate heat control responses. A different time-and-temperature program can be made for each of the six banks, or as many banks as needed, to accommodate thickness variations or other complexities of the panel sections. Closer control or not, as desired, can be achieved by closing or widening the pencil lines forming the path in which the program follower stays.

With the instantaneous reaction of electronic controls, engineers have been unable to measure temperature-time deviation as recorded on the best recording instruments. As a result, they state, conservatively, that deviation is probably controlled within one degree.

With the advent of brazing refractory metals and their alloys at 2500°F temperature and above, new requirements arose. The outer cover of the brazing envelopes could no longer be steel but metals like titanium, tantalum, and columbium. However, all required protective coatings for oxidation protection. Coatings such as alumina-titanium dioxide, titanium-chromium braze alloys, tin-aluminum, and chrome-titanium-silicon have been used. These coatings had to meet the following criteria:

(1) impermeability to air,
(2) high emissivity in the infrared region,
(3) compatible coefficient of thermal expansion with the base materials,
(4) adherence to the base materials,

Equipment and Tooling

(5) compatibility with the base materials with regard to interface reactions.

Impermeability of the coating to air was particularly important in order to minimize oxidation, loss of strength, and embrittlement of the base materials. The envelope had to possess some ductility after exposure to the braze temperature so that it would not be fractured by the stresses caused by the clamping action in the chill dies. The emissivity of the coating also had to be as high as possible to provide efficient thermal radiation transfer from the quartz lamps to the brazing assembly.

In addition to the coatings on the outer envelopes special reflectors and lamp end seals have been developed to permit use of quartz lamps at the high watt densities required to reach and maintain the high brazing temperatures. This has involved the incorporation of water-cooled reflectors and additional lamp cooling facilities for use with the special high wattdensity lamps.

ADVANTAGES

One of the most important advantages is that Nortobraze completes the brazing cycle in 5 to 15 min, depending on the configuration and complexity of the panel being brazed, whereas furnace brazing methods require from 3 to 12 hr. Another advantage is its adaptability to all sorts of configurations. This includes flat, curved, and wedge-shaped tapered panels. (See Figure 6.4.)

The Nortobraze system used for brazing modern-day materials permits the application of varying temperatures and time to different sections of the part to accommodate varying thicknesses or density of the panel section.

Figure 6.4 17-7PH stainless steel wedge panel.

This precise, localized temperature control is extremely critical in that thermocouple feedback from the honeycomb panel and electronically controlled temperature selectivity in the lamp banks provides uniform heating and cool-down throughout the panel.

Stiffbacks, which have been used to support parts in furnace brazing, cause another problem. Parts expand a given distance per length and usually when attached to a stiffback that does not undergo this change, the parts are subject to warping or buildup of stresses. Since parts are suspended during the brazing operations using radiant lamps, this problem is eliminated.

The other unique feature of the Nortobraze system is the device utilizing chilled platens. This has removed the major problem involved in devising handling methods in which parts can be removed from the furnace and transferred to a chill chamber quickly enough to achieve the rapid chilling required for heat-treatable material.

MATERIALS, PROPERTIES, AND FUTURE

The Nortobraze process has been successfully applied to titanium alloys and is adaptable to many of the refractory alloys in the columbium and molybdenum groups as well as stainless, superalloy, and precipitation-hardened steels. It has been able to obtain up to 245,000 psi in 17-7PH and PH 15-7 Mo stainless steel with 6% or better elongation using the RH950 subsequent heat treat.

René 41 honeycomb panels have been successfully brazed by the quartz lamp process. Note in Figure 6.5 the complete filleting of braze alloy on 0.002-in. foil with no erosive effects.

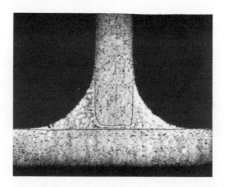

Figure 6.5 Microstructure of René 41 Tee joint brazed with Ni-Cr-Si-Mn alloy [1].

Figure 6.6 A TZM molybdenum honeycomb panel locked inside an airtight metal envelope begins a brazing cycle.

A honeycomb sandwich panel of a molybdenum alloy, TZM, is shown at the onset of a brazing cycle in Figure 6.6. The panel is encased in a coated tantalum envelope. Currently, stationary lamps are used and the work is moved into position. In the future it is conceivable that the work will remain stationary and the lamps will be moved, thus improving the process capabilities. With its conveyor system and in-line brazing and chilling units, the process permits high rate production, with the parts conveyed through brazing, chilling, and continuing to other operations as desired.

In addition to brazing the Nortobraze process has been applied to diffusion welding techniques and the system offers metallurgists a potential new way to study heat treatments of metals and their resultant structures.

The extraordinarily quick response of thin sections to heating and cooling appears to be an excellent tool in which metallurgists can examine unsuspected changes that take place in precipitation hardening steels. Careful redrawing of phase-cooling diagrams should reveal many physical changes not detected in heavier bars that are usually studied.

REFERENCES

[1] L. H. Stone, A. H. Freedman, and E. B. Mikus, "Brazing of Thin Gage René 41 Honeycomb," *Welding J.*, 397-S–403-S (September 1963).

[2] D. B. Hugill, B. Gaiennié, "Quartz Lamp Radiant Heat Brazing of Large Refractory Metal Honeycomb Sandwich Panels," IR-7-937A(VII), Northrop Norair under AF33(657)-8910 (January 1964 to April 1964).

Chapter 7

EXOTHERMIC BRAZING AND EXO-FLUX BONDING

The use of exothermic reactions for industrial brazing applications has been limited because little information has been made available concerning exothermic brazing materials and the economics of their use.

In order to understand exothermic reactions more fully, this section elaborates on several different processing methods, currently in practice, which use controllable exothermic reactions. One method is exo-flux welding, is also referred to as "exo-flux bonding." A second technique is "exo-adhesive bonding." This is a process in which the exo-reactant materials are not only used as a heat source but the metallic product of the reaction is the bonding adhesive. When compared with the first two methods, exothermic brazing, which usually requires a preplaced braze filler alloy, is a simpler approach. The first two require considerably more exacting formulations of the exo-reactant mixtures to allow the products of the reaction to form an acceptable metal alloy, which, when molten, will wet and bond the adjoining metal surfaces.

PRINCIPLES, KEY VARIABLES, AND MECHANISMS

Exothermic Brazing

Exothermic brazing refers to a process in which the heat required to melt and flow a commercial filler metal alloy is generated by a solid-state

Table 7.1 Metal-Metal Oxide Reactions [2].

Formula no.	Exo-reactants	Products	Theoretical heat evolved,[a] K cal/g
(a)	2Fe + Bi_2O_3	Fe_2O_3 + Bi	0.11
(b)	2Fe + 3CuO	Fe_2O_3 + 3Cu	0.24
(c)	Mn + CuO	MnO + Cu	0.40
(d)	10B + $3V_2O_5$	$5B_2O_3$ + 6V	0.73
(e)	2B + 3CuO	B_2O_3 + 3Cu	0.73
(f)	2Al + MoO_3	Al_2O_3 + Mo	1.14
(g)	2Mg + MnO_2	2MgO + Mn	1.20
(h)	4Li + MnO_2	$2Li_2O$ + Mn	1.40

[a] K cal per gram of reactant mixture based on $-H\ 298°$ K.

exothermic chemical reaction. An exothermic chemical reaction is defined as any reaction between two or more reactants in which heat is given off due to the free energy of the system. Nature has provided us with countless numbers of these reactions; only the solid-state or nearly solid-state metal-metal oxide reactions, however, are suitable in exothermic brazing units [1]. In Table 7.1 are some typical metal-metal oxide reduction reactions, their products, and the heat evolved. As shown in the last column of Table 7.1 the calculated quantity of heat released by the reaction is dependent on the choice of the reacting materials and the weight of the reacting mixture. A reaction typical of those utilized is:

$$2\ Mg + MnO_2 \rightarrow 2\ MgO + Mn + 1200\ cal/g \quad \text{of reactants}$$

In order to understand further the basis for the development of exothermic brazing, examine Tables 7.2 and 7.3. The first column in Table 7.2 lists the more common oxides, the next three columns show the heat of formation for these oxides, and the third column indicates heat of formation per gram of oxide formed. Table 7.3 shows a sample calculation for estimating the heat evolved by a typical exo-reactant system at 25°C. This value, of course, changes slightly when the temperature increases and considerable

Table 7.2 "Heat of Formation" Data for Oxides (Kcal) [2].

Oxides	−H 298° K per mole	−H 298° K per oxygen atom	−H 298° K per gram oxides
Ag_2O	7.3	7.3	0.03
Al_2O_3	400	143.8	3.93
B_2O_3	302	101	4.34
Bi_2O_3	138	46	0.3
Cr_2O_3	270	90	1.78
Fe_2O_3	197	65.5	1.21
Li_2O	143	142.6	4.75
MgO	144	143.8	3.56
MnO_2	125	62.3	1.42
MoO_3	180	60.1	1.25
SiO_2	211	105.2	2.33
V_2O_5	373	74.5	2.06
ZrO_2	262	130.8	2.13

calculation is required to determine exactly the theoretical heat evolved at, say, the ignition temperature of the mixture [2].

From these tables, it can be seen readily that numerous formulations have been developed to exothermically braze a variety of base metals. Most of the formulations currently in use contain a number of oxide-metal components blended to provide the proper thermal pulse for brazing, compatibility with the base metal, and easy removal from the surface of the base metal following brazing.

The use of these mixtures to supply heat involves a knowledge of some specific exothermic parameters: ignition temperature, reaction duration, and net available energy (useful heat).

Table 7.3 Calculation of Heat Evolved in an Exothermic Reaction at 298°K [2].

(1) $2Mg + MnO_2 = 2MgO + Mn$
From the heat of formation at 298° K of the oxides, the heat evolved in eq 1 can be calculated.
(2) $2Mg + O_2 = 2MgO + 287.6$ Kcal
(3) $Mn + O_2 = MnO_2 + 124.5$ Kcal
Subtracting (3) from (2) gives a value of H at 298° K of 163.1 Kcal. Also since the gram molecular weight of the reactants is equal 135.6 g then *1.2 Kcal of heat will be released for each gram of exo-reactant mixture.

Ignition Temperature

All the mixtures in exothermic brazing that produce heat are capable of propagating from the point of ignition throughout the rest of the reaction mass at some specific rate. Local ignition of the exotherm mixture is accomplished by a variety of methods. The easiest and simplest method is to ignite it with a flame, pass current from a 6-V battery through a "hot wire" bridge or hot spark, and allow the reaction to propagate like a fuse. Simultaneous ignition of the entire exothermic mixture can be obtained only by heating the mixture evenly to the ambient ignition temperature. Ignition temperature thus signifies the temperature at which the system initiates spontaneous propagative reaction. The requirements for propagative reaction are based on the rate theory [2]. The essential requirement is that the net heat gained from the reaction, plus any heat supplied externally, must be greater than the heat energy absorbed when activating additional adjacent reacting molecules.

Therefore the rate of the ignition reaction is dependent primarily on the effective concentration of the reacting materials and in solid-solid reactions; this is determined by the amount of particle surface areas in contact or in near contact.

Table 7.4 gives the ignition temperatures (°F) obtained with various

Table 7.4 Average Ignition Temperatures, °F, and Approximate Duration Times of Stoichiometric Mixtures [2].

Oxide	Avg. particle size, Micron	Mg +40 Mesh	Mg −100 Mesh	Al 5	Ti −325	Si 5	B 1	ZrH_2 4	Zn 15	Fe 3
CuO	4	(12)[a] 1045	1090	(1)[a] 1640	(2)[a] 970	1745	(5)[a] 930	(6)[a] 770	(6)[a] 1095	(13)[a] 860
CuO	1	(2)[a] 1020	(1)[a] 1030	...	...	...	(11)[a] 930	...	...	(5)[a] 700
Bi_2O_3	1	940	...	1260	...	...	...	...	1110	510
MoO_3	4	1220	...	1220	...	...	...	...	...	...
MnO_2	3	(25)[a] 1020	(32)[a] 1140	1650	...	(1)[a] 1880	(5)[a] 1080	...	(25)[a] 1040	b
Fe_2O_3	0.4	(18)[a] 1140	...	(8)[a] 1700	(20)[a] 1480	1650	(8)[a] 1070	(2)[a] 880	(8)[a] 1390	b
SnO_2	0.6	...	1070	(8)[a] 1702	...	...	1090	...	...	b
V_2O_5	1	1550	1520	(4)[a] 890	(5)[a] 1010	...	(19)[a] 860	(9)[a] 790	...	b
Mn_3O_4	3	...	1100	(7)[a] 1750	...	...	(13)[a] 1350	...	...	b

[a] Reaction duration time in seconds (shown in parentheses).
[b] Instantaneous reaction.

metal-metal oxide mixtures. The particle sizes are listed because of their basic influence. This table shows the greatest heat release reactions in the upper left-hand corner, with the amount of heat release decreasing as one proceeds downward and/or to the right.

Ignition temperature appears to be a more or less intrinsic value. However, a number of interesting points are quickly observed. All aluminum containing mixtures have high ignition temperatures, except when aluminum is mixed with vanadium pentoxide. Magnesium mixtures are just the opposite. No definite explanation can be made although many theories can be advanced. The lowest ignition temperature in a recent investigation was 450°F for the combination silver oxide and very fine iron powder. The highest, 1880°F, was for the reaction silicon and manganese dioxide.

Reaction Duration

For a given quantity of reactants the duration is limited by the rate at which the reactants are consumed and products formed. Thus reaction durations can be varied by

(1) the addition of inert materials that absorb heat.
(2) adding a third substance that reacts at a different rate.
(3) using nonstoichiometric proportions.
(4) varying powder particle sizes.

Reaction duration is seemingly not related to the position of the reactants in Table 7.4. In general, however, heats of reaction of less intensity seem to give longer reaction durations, provided particle sizes are similar.

The influence of reactant mass on ignition is of most importance, particularly for those combinations that are less exothermic (those located downward and to the right in Table 7.4). In many instances, even with some of the more exothermic reactants, 1-g disc pellets would not ignite when sandwiched between two 50 mil stainless steel sheets but would ignite between 25 mil sheets. A thin adjacent metal does not prevent propagation; a thicker metal does. A metal of higher thermal conductivity prevents propagation, whereas one of lower conductivity allows propagation. These are mainly factors only when working with a small mass of mixture or working with mixtures that have low heat releases.

In order to determine experimentally the effect of adding inert materials, additions of stable oxides have been made. The composite system AL-(V_2O_5 and Fe_2O_3) was used with increasing percent additions of about 44 to 60 μ alumina (aluminum oxide). The reaction duration can be almost tripled with a 20 wt.% alumina addition without affecting the ignition temperature; greater weight additions had no further effect on duration but eventually stopped the ignition. Other data show similarly that silica additions can be made without detriment to the reaction characteristics. Another investigation evaluated the ignition characteristics of various exo-reactant mixtures in argon atmosphere and vacuum. Results showed an increase in ignition temperature up to 100°F over that of the same mixtures ignited in air. Other characteristics were the same.

Net Available Energy

This defines the "useful heat" of any particular reaction and can be theoretically calculated by use of the previously mentioned procedures [2]. Experimentally, this heat can be determined by measuring the area beneath the time-temperature curve for any particular reaction. It is more advisable, from an applied research approach, to establish a standard procedure by using one standard reaction mixture and comparing all others with it. The heat losses caused by this technique can be considered near equal and therefore the results can be compared.

Exo-Flux Bonding

Two distinctly different phases to the bonding mechanism in this concept have been developed by the Narmco Research and Development Division [3]. The prefix, "exo," refers to the novel, externally applied heating tape that is employed to provide localized high heat energy inputs to the parts to be joined. The flux term refers to the elemental metal donating compound (e.g., boron oxide, B_2O_3) which supplies elemental boron.

The concept relies on the fact that it is well known in the metallurgical field that certain metal combinations result in an alloy having a lower melting point than either of the original two metals. For example, a mixture of metal A plus metal B melts lower than either metal A or metal B. This lower melting alloy is often used to join metal parts together. The exo-flux concept [3] utilizes the metal elements already in the surface of the metal part and a metal element from heat reducible metal oxide (like B_2O_3) to form this low-melting alloy. For example, if metal element A was part of the metal part being joined and metal element B came from a reducible oxide compound, a low melting alloy, consisting of A and B, would result. The bond chemical structure is therefore composed of a metal alloy in which the greatest percentage of metal (approximately 98%) is taken from the metal surfaces being joined together, and the remainder 2% is boron from the B_2O_3 compound or similar-type compound.

In order to verify this concept a number of exacting conditions must occur simultaneously.

1. The donor supplier (B_2O_3, etc.) must give up its boron element at a temperature below the damaging temperature of the metal part to be joined.
2. The metal part surface must not oxidize appreciably before the boron is taken up by the metal surface to form the eutectic alloy.
3. The oxygen of the donor material (B_2O_3, etc.) must be removable from the joint area.

Why only through the use of an exothermic heating method, in which the metal surface reaches a temperature of over 2000°F in less than 2 sec, will these conditions happen? Why not a furnace, quartz lamps and/or induction heating methods? It appears that there is no chemical breakdown of the B_2O_3. It acts as a fluxing agent and dissolves metal oxides from the surface until it is saturated. This saturation tends to prevent the B_2O_3 from releasing boron when in contact with a metal. Therefore exothermic heating serves two purposes:

1. It minimizes the oxidation of the surface of the steel because of the exceptional fast heat-up time.
2. The rapid heating rate seems to catalyze the breakdown of the boron oxide. The compound is important, not only because of its supplying of a donor metal but because it acts initially as a flux and protects the metal surface so that a eutectic alloy can be formed when the correct temperature is reached.

The combination therefore of the proper choice of the donor compound and a rapid heat source makes the concept workable. In addition, the use of a small percentage of a glass-forming cryolite assists in flux wetting and removal of the oxide from the joint interface.

Exo-Adhesive Bonding

This is a lower temperature (below 2000°F) method of exothermic joining with controllable exothermic reactions which was described in the introduction. The processing method involves oxide-metal systems and metal-glass forming systems. Table 7.5 lists some of the combinations that have been investigated. The first three are examples of the oxide-metal system, the next two of the metal-glass forming system, and the last is an example of a combination system. The results of these early developments revealed that a combination system was necessary to assure an all-metallic braze bond. The oxide-metal systems would not allow for separation of the oxide byproducts from the metallic byproducts. The metal-glass forming systems produced bonds with an excess glass phase, brittle, and of low strength.

Table 7.5 Exo-Adhesive Reactant Mixtures and Their Products [2].

System Formula No.	Exo-Reactants	Adhesive Products	Theoretical Heat Evolved* Kcal/g
B-1	5% Ti + 5.6% Mg + 85% Cu_2O + 9.4% SnO_2	26.3% CuSn + 72.7% oxide	0.6
I	12.1% Al + 78.1% Cu_2O + 9.8% SnO_2	80% CuSn + 20% Al_2O_3	0.64
E	25% Mg + 52.3% NiO + 22.8% MnO_2	59% NiMn + 41% MgO	0.96
K	7.1% Borax + 16.4% SiO_2 + 6.4% B + 70.1% CuO	55.3% Cu + 44.7% glass	0.73
P	7.4% Borax + 17.6% SiO_2 + 6.6% B + 68.4% NiO	54% Ni + 46% glass	0.51
R	System K + System B-1	...	...

* Based on $-H\ 298°\ K$.

Combination systems in which the mixture components were not optimumly balanced showed metallic bonding with included glass and joint lap shear strengths were over 2000 psi.

The reaction rate of the bonding process is so rapid (1 to 5 sec) that compositional balances in the exo-reactant mixture are important factors. For example, it is necessary to control the flow characteristics of the glass component produced during the reaction and remove this glass from the joint area while it is still fluid. Therefore the composition of this formed glass is important in that composition affects fluidity, and since the fluidity of the glass is a function of temperature, the useful heat produced by the reaction becomes of basic importance. The formed glass composition can be controlled by additions to the original basic mixture of inerts like silica, lithia, alumina, and borax. Boron additions can also be made to the original mixture when it is felt that the boric oxide content of the glass should be increased.

The use of pressure to assist in squeezing out the glass phase is also important. Pressures from 50 to 200 psi have been used; however, too great a pressure causes not only extrusion of the glassy phase but most of the metal phase. The applied pressure must be adjusted for each exothermic mixture.

A most important factor in braze bonding is the wetting of the metal surface by the metallic byproduct of the exothermic reaction. All work to date has been done with stainless, precipitation-hardened and high strength, superiron base steels. Since the key to productive bonding is to perform the operation in an air atmosphere, extensive oxide formation on the steel surface during any heat-up period before ignition and/or during ignition must be prevented; otherwise metallic wetting will not occur. To resolve this problem, since the glass component is already present, it is fairly simple to increase the wetting and oxide-absorbing efficiency of this glass. By adding up to 6% of various glass soluble halides, metallic wetting improves.

Another improvement is the addition of approximately 2 wt% of either copper or silver metal powder. The theory behind this is that, with this small metal addition, metallic wetting can be initiated during the early portion of the reaction time period and subsequent wetting by the metallic products, when formed during the reaction, is easier to accomplish. With these results the small metal additions is now part of all basic exo-reactant formulations. Variations in these compositions can be made, but a balance in "useful heat" produced and in "reaction products formed" must be maintained.

EQUIPMENT, TOOLING, ECONOMICS

Exothermic Brazing

This type of brazing employs simplified tooling and equipment. The process utilizes the reaction heat in bringing an adjoining or a nearby metal interface to a temperature at which an already preplaced braze alloy melts and wets the metal interface surfaces. The braze alloy can be any commercially available brazing material having suitable flow temperatures. The only limitations may be due to the thickness of the metal that one must heat through and the effects of this heat, or any previous heat treatment, on the metal properties. The first two steps are the selection of a heat source and braze alloy. Second, the parts are cleaned chemically and mechanically if necessary, the exo-reactant mixture suspended in alcohol is painted on the outside of the parts to be brazed, and the complete assembly with preplaced braze filler alloy is then wrapped in a Fiberfrax blanket, then heat lamp dried.

The assembly is inserted into a furnace containing an inert atmosphere and the furnace temperature increased to the point at which ignition of the heat source can occur. For example, if the boron-vanadium pentoxide system were selected, the ignition temperature would be approximately 900 to 950°F. Finally, the external reaction products are washed away with hot water and the assembly is complete and ready for use.

A recent program was conducted to develop and evaluate exothermic brazing units for making permanent connections in AM-350 tubing. An

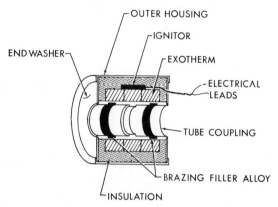

Figure 7.1 Cutaway view of exothermic tube brazing unit showing components [1].

integrated design approach was taken in developing exothermic tube brazing units, as illustrated in Figure 7.1. The unit was complete with couplings, filler metal alloy, and heat source in one package. Each unit was therefore designed for a specific tube size, wall thickness, material, filler alloy, as well as an outside package limit size.

An insulative package is used to conserve the generated heat and to contain the molten reaction products. Of the various forms of insulation that have been used, a package containing alternating layers of a porous insulation and metallic heat shields has been most efficient. Blankets of bulk ceramic fiber were used for insulation because of excellent heat resistance and permeability that permitted capture and slow escape of gases generated as a result of the exothermic reaction. Tin-plated carbon-steel (0.010 in. thick) was interposed between the ceramic blanket layers to act as reflective heat shields and to block a direct path from the exothermic reaction to the outer housing. This multilayered construction permitted the use of a minimum diameter package.

Exo-Flux Bonding

Figure 7.2 illustrates the components necessary to bond metals together exothermically. In this instance overlap test specimens are to be tested; therefore the fixturing is suitable for an overlap configuration. Fixturing is made to suit the particular parts to be joined. In operation the cleaned metal surfaces (most work to date has been on stainless steel) to be joined are coated with a thin, powdered, inorganic fluxing agent such as the low-melting B_2O_3. Experience has shown that the metal surfaces can be cleaned by techniques no more sophisticated than scrubbing with a mild alkali commercial cleanser. The metal surfaces are placed in contact with one another in proper alignment and the exo-reactant heating tape is positioned to the outside of the assembly. This assembly, including any necessary holding fixtures, is then placed in an oven or furnace and brought as rapidly as possible to an ignition temperature of approximately 1000°F. At this temperature ignition of the exo-reactant occurs and, within a matter of seconds, the reaction supplies sufficient heat to bring the joint surfaces to the bonding temperature. This instant heat normally takes less than 10 sec. During this period the high strength bonds are achieved and the bonded component can be withdrawn immediately from the ovens and the holding fixtures removed.

The process can be considered analogous to welding, except that the

Figure 7.2 Overlap bonding layup. *A* (*top*). Pictured, from the bottom, are fiberfrax, specimen parts and exothermic wafer, aligning fixture, fiberfrax and pressure block. *B* (*bottom*). Components are assembled and ready to be placed in a furnace, have pressure applied and temperature increased until ignition and bonding occur [3].

metals being joined do not have to be heated to their melting points, but to some lower temperature where the joining takes place. High temperature brazing is similar, except no filler metal is used.

Several donor materials such as B_2O_3 have been examined. The following have shown the most promise:

1. Soda-lime-lithia-silicate glass.
2. Sodium pyrophosphate.
3. Halides.
4. Borax glass.
5. Boron oxide.
6. Sodium-borosilicate glass.
7. Boron oxide-cryolite.

Depending on the base metal, the donor with the highest quality bond is selected. The most practical method of applying the donor to the base metal is to dip the metal into a thin very fluid suspension of the donor.

The other important ingredient for exo-flux bonding is the exothermic heat source, which can be applied by two methods. It is either painted on as a pasty suspension in isopropyl alcohol or applied as a pressed wafer approximately 0.125 in. thick and with lateral dimensions corresponding to those of the area to be bonded. Several exothermic systems utilized in exo-flux bonding are shown in Table 7.6. These systems differ from those of Table 7.4 in that the heat of formation will generate 2000°F bond-line temperatures which are necessary to bond structures that must operate close to 2000°F. Figure 7.3 illustrates the preparation of system E of Table 7.6.

A recent study was made comparing the costs of exothermic brazing versus induction as applied to the brazing of tubing illustrated in Figure 7.1. The comparison of average costs indicated that exothermic brazing saves approximately $16.00 per joint since capital equipment is unnecessary

Table 7.6 Exothermic Systems [3].

System	Composition	Heat of formation, cal/g
AF	24.1% Mg + 47.3% Cu_2O + 28.6% MnO_2 = 39.9% MgO + 42.0% Cu + 18.1% Mn	880
E	24.9% Mg + 52.3% NiO + 22.8% MnO = 41.1% MgO + 58.9% NiMn	757
BA	67.6% Cr_2O_3 + 32.4% Mg = 53.8% MgO + 46.2% Cr	712
BC	73.8% Cr_2O_3 + 26.2% Al = 49.5% Al_2O_3 + 50.5% Cr	633
A	15.6% Mg + 75.5% Cu_2O + 8.9% Sn_2O = 26.0% MgO + 74.0% CuSn	630
AN	20.5% Mg + 79.5% Ag_2O = 13.8% MgO + 74.2% Ag + 12.0% Mg	468

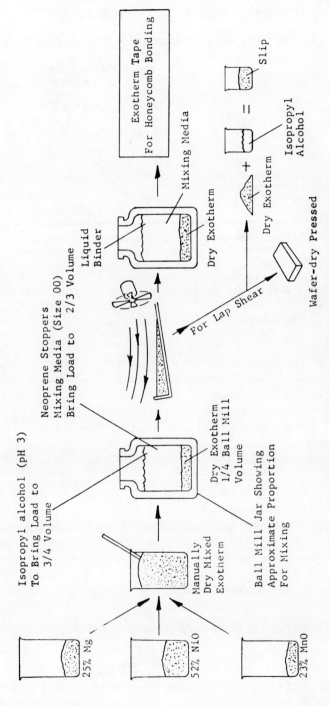

Figure 7.3 Flow sheet showing typical exotherm preparation [3].

Table 7.7 Comparative Costs of Induction and Exothermic Brazing [1].

Item	Average cost per joint, $	
	Induction brazing	Exothermic brazing
Hardware	12.87[a]	25.70[b]
Manufacturing time	8.40	5.02
Purge (argon)	2.54	1.27
Amortized capital Equipment	13.07	...
Generator and tool Maintenance	11.27	...
	48.15	31.99

[a] Fitting—filler metal alloy ring assembly.
[b] Exothermic unit—fitting—filler alloy rings—exothermic package.

(see Table 7.7) and satisfactory brazes can be made consistently and with relative ease.

For this application the exothermic brazing process offers the following advantages:

1. No power requirements. Only a 6V battery is needed to ignite the exothermic reaction. This means that no capital equipment investment is required and it permits brazing to be accomplished in place where it is difficult or impossible to use more bulky equipment.
2. Short heating times. There are minimum thermal effects on the materials.
3. Localized heat. Only the area immediately adjacent to the braze is affected.
4. Ease of handling. A minimum of hand manipulation is required to assemble a joint and make a braze.
5. Pre-engineered units. Each unit is designed to accomplish a particular braze. Control of temperature is a function of unit design and therefore is not dependent on equipment settings and operator control.
6. Reduced costs. Because capital equipment is not required and the operation itself is easily accomplished, cost savings are realized.

MATERIALS—JOINT DESIGN AND TESTS

Refractory Metals

Three different concepts for brazing the refractory alloys of tungsten, tantalum, columbium, and molybdenum, utilizing exothermic reactions have

Materials—Joint Design and Tests

been evaluated. These included (a) an exotherm heat source external to a joint in conjunction with a standard braze alloy; (b) internal exotherm whose reaction products included a metal phase that served as the braze alloy; and (c) an external exotherm heat source used to produce a joint by means of a fluxing ceramic phase. The first technique produced high-strength brazed joints with all the refractory metals, with minimum damage to the base metal properties. Limited studies on the second concept yielded a satisfactory nickel-manganese braze alloy. The feasibility of using a boron "donor" compound to produce an "in-situ" braze metal (concept 3) was also satisfactory, especially with molybdenum.

Superalloys—René 41

Several superalloys including nickel and cobalt base alloys were joined using a "nickel exo-reactant" technique. The René 41 bonds made by this nickel exo-adhesive technique were tested in shear from room temperature to 1700°F. Failures were in the base metal adjacent to the bond zone, with no evidence of either intergranular attack or erosion on the René 41. Metallographic studies revealed that the bonds had good continuity, no bondline segregation or intermetallic formation, and little interaction between the base metal and braze metal.

Steels–Stainless and Precipitation–Hardened

By using exothermic brazing techniques and standard heating methods several lap shear specimens of 17-7 PH steel were evaluated. Shear strengths were equivalent. It is of interest to note that an exo-brazed specimen produced the highest shear strength, which was about 15% higher than that obtained by standard furnace procedure. This is not unexpected, for better wetting and flow of a noneutectic silver braze alloy should be obtained under fast heating conditions. In order to evaluate fully the systems that are exo-reactant adhesives and exothermic brazed joints, a set of lap shear specimens was tested and data are shown in Table 7.8. It is interesting to note that all lap joints produced failed in the parent metal adjacent to the lap shear joint, even though the overlap was decreased to bare minimum.

With the development of exo-flux bonding techniques stainless steel types PH15-7Mo, AM 350, AM 355, A-286, and 17-7 PH have all been successfully bonded. Overlap brazed specimens of the 17-7 PH displayed a tensile strength within 16% of that of the base metal, with most failures occurring

Table 7.8 Tensile Lap Shear Strengths—Age Hardenable Stainless Steels [2].

Type of steel, braze and heat treatment	Test temperature, °F	Avg. joint shear[a] strength, psi	Avg. base metal tensile strength at failure, psi
17-7 (TH 950)			
(Ag Formulation 1)	80	>9,000	180,500
(TH 1050)	600	>7,300	88,000
(TH 1050)	800	>8,350	96,000
15-7 Mo (TH 1050)	80	>13,750	150,000
(Ag Formulation 1)	600	>11,100	140,000
	800	>10,700	120,000
AM 350 (hardened)	80	>8,600	110,000
(Ag Formulation 1)	600	>8,900	110,000
	800	>7,700	95,000
17-7 (RH 950)[b]	80	>5,900	122,000
(Cu Formulation)	1000	>2,800	61,500

[a] All specimens failed outside lap joint—data avg. 3 test specimens.
Aged only at 950° F after exo-brazing and cold treatment.

outside the bond area. Exo-flux bonded honeycomb sandwich compared favorably with those brazed by conventional techniques.

Light Metals—Aluminum, Magnesium

Through the use of an exo-adhesive system consisting primarily of boron and aluminum metal and vanadium pentoxide the soldering of aluminum alloy 6061-T6 has been successfully accomplished. In addition, the same general principles involved in exothermic brazing can be applied successfully to the bonding of lower temperature metals with organic adhesives. Structural adhesives are available that will provide excellent joint strengths when exothermically bonded in a vacuum. These systems provide most of the attributes associated with adhesive bonding, such as joining dissimilar materials, good distribution of applied stresses, and permanency of bond.

In addition, this approach, which allows utilization of quick-acting thermoplasticfilms for bonding, requires a very short bonding time—only long enough for the adhesive to melt and solidify. Furthermore, the thermoplastic adhesives can utilize fully cured polymers of sufficient molecular weight to preclude problems of volatilization. In addition, molten thermosetting bonding adhesives require only contact pressures to provide densification of the glueline.

Structural thermoplastic adhesives are available in the form of readily handlable films based on nylon, modified nylon, polyesters, and phenoxy resins. These systems, in combination with controlled heat from a suitably designed exothermic package, offer the potential of structural bonding of all applicable metals.

APPLICATIONS AND FUTURE POTENTIAL

Exothermic joining has proved applicable in numerous potential areas. The method has been used to braze honeycomb sandwich panels, to braze inserts into panels locally, to attach small parts to larger structures, and is being considered as a field repair mechanism for aircraft structural honeycomb panels. This type of heat source has also been used as a preheating method before sheet metal forming and/or explosive forming.

Figure 7.4 illustrates several hundred exothermically brazed tube connections that were used with the hydraulic lines on the XB-70 airplane. The connections were made in place on the aircraft. Second, the photograph shows the interior of a fuel tank in which exothermic brazing proved useful because of limited access and cramped quarters. Several hydraulic and fuel lines can be seen along the walls of the tank.

In the reproduction of permanent tubing joints the following criteria were met:

1. Brazements made exothermically surpassed the requirements for flow with little void content.

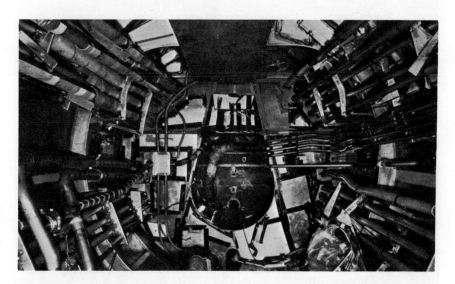

Figure 7.4 Interior of XB-70 fuel tank in which exothermic tube brazing was used [1].

276 Exothermic Brazing and Exo-Flux Bonding

2. Tubing joints made by exothermic brazing passed all proof and burst test requirements. Burst failures were always in the tubing away from the joint.
3. Exothermically brazed tubing joints greatly surpassed the design requirements for flexure-impulse loading.

Tee-type tube fittings can be placed in service in less than 5 min after ignition. Ignition can be accomplished by applying a match to the ignition paste or by using a welding or cutting tool to ignite directly the sleeve. This simple way to join tubing involves no special tools, no special skills, and high integrity connections, and it removes the human element from heating by providing a uniform heat source at every point on the brazing ring.

A new environment to be encountered by man is the hard vacuum of space. The joining of structural materials in spacecraft or space environments is a complex process involving human manipulation in restricted areas and/or under severe environmental conditions. Another consideration is the limited amount of boost force that can be delegated to place a unit amount of the joining material and equipment into the mission path.

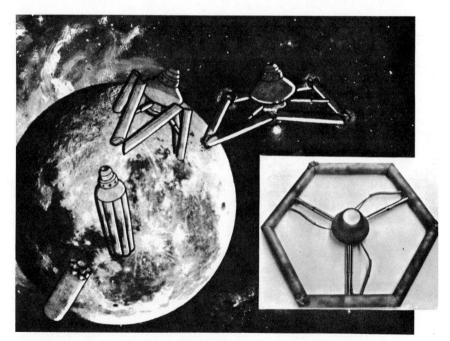

Figure 7.5 Exothermic joining in a vacuum environment.

The ideal minimum weight joining process of in-space applications is one that incorporates the following characteristics:

1. Requires little auxiliary equipment other than the joining material and associated structure.
2. Operates from lightweight, stored-energy sources such as solar batteries and storage batteries.
3. Produces reliable and reproducible joints.

Exothermic joining appears to be capable of fulfilling these requirements and its techniques are extremely economical and simple when compared to heat-source systems that depend on solar-energy conversion. These systems, except when the sun's rays are focused directly on the work area, require the conversion of solar energy to electricity and the associated storage and conversion of the electrical energy to heat. To demonstrate the effectiveness of exothermic joining in vacuum the module shown in Figure 7.5 was assembled remotely and rigidized in a vacuum chamber by an electrical impulse from a conventional dry cell battery. As illustrated by the sequential drawing, the finalized exothermic joint designs in this structural model include hinged, telescopic, butt, and swivel joints typical of those visualized for future space stations.

REFERENCES

[1] N. E. Weare, and R. A. Long, "Exothermically Brazed Hydraulic Fittings for Aircraft," *Welding J.*, 29–38 (January, 1967).
[2] R. A. Long, "Exothermic Brazing," *Welding J.*, 346-S–350-S (August, 1961).
[3] R. A. Long, R. A. Caughey, and W. Bassett, "Exo-Flux Welding of Stainless Steel Type Alloys," *Welding J.*, 259-S-264-S (June, 1964).

Chapter 8

VACUUM BRAZING

Brazing is one of the oldest joining processes. An estimated 2800 years ago and probably more, brazing was first used to join ornamental gold fabrications. Gold-silver and gold-copper-silver alloys were used as filler metals. These alloys, still called gold solders, are true brazing alloys melting above 800°F but well below the melting point of pure gold. With the advent of the iron age and ferrous technology, copper-zinc alloys, called spelter, were developed for joining iron and steel. These alloys were strong and easily melted.

The early craftsmen were not employing technique in the modern sense of the word: this was "art." The uneducated craftsman did not know why he was doing these things, only that they "worked."

Paralleling the development of low-melting-point gold alloys for joining gold, silversmiths discovered silver-base alloys of a low melting point for similar use in the craft. These silver "solders," now carefully formulated on the basis of sound metallurgical principles, survive as silver-base brazing alloys of ever-increasing importance to the metals engineering industry. Brazing has embarked on a vastly successful "second childhood." Sparking this progress in recent years had been the continuing development of more and better alloys and the introduction of machines that have brought brazing processes to the fore in high-production situations.

With these metals and alloys coming into use during the last fifty years, new brazing compositions were developed to join them. Aluminum is brazed with low-melting-point aluminum-base alloys. Silver has been found effective for joining titanium and some of its alloys. A number of brazing alloys of high hot-strength and good resistance to oxidation has been developed for the heat-resistant alloys so important to jet flight and the exploration of space. The development of new brazing techniques like vacuum has kept pace with filler metal development. The brazing of ceramics to metals, for example, is now an accepted technique.

The use of vacuum in metal processing has grown steadily since the first announcement in 1948 of vacuum-melted metals and alloys. Vacuum brazing, a logical evolution in vacuum metallurgy arising from vacuum sintering and vacuum melting, is currently being investigated and used and shows promise of becoming an increasingly important production process.

Vacuum brazing is the most modern brazing process and is at present far from being completely understood. It was first used, possibly accidentally, for assembling evacuated electronic valves. Added impetus was given to the development of the process by the demand for a flux-free method of brazing for those heat-resisting alloys that cannot be processed in controlled atmospheres. By brazing under high vacuum, in an atmosphere free of oxidizing gases, a superior product, with greater strength, ductility, and uniformity, may be obtained more economically than by brazing methods previously used.

THEORY AND PRINCIPLES OF PROCESS

Vacuum brazing has not only been found to be successful in the production of joints of unequaled quality in difficult metals but also to impart a standard of cleanliness to the work that could not be achieved by any other method. This second aspect is so important in some instances that components that could be brazed by easier methods are being vacuum brazed. Early investigators believed that the success of vacuum brazing relied on the metal oxides present being dissociated when heated under low gas pressures. This now appears to be unlikely, for the dissociation pressures of all of the more stable oxides are considerably lower than those required for successful brazing. There are several other possible explanations, of which the most logical is that a vacuum is simply an effective means of screening the work from oxidizing gases and other impurities.

Let us examine the combination of the two technologies, vacuum and brazing.

Vacuum

The pursuit of low pressure phenomena probably began with Otto von Guericke and his Magdeburg spheres. About 200 years ago a key discovery leading to present developments was made when Bernoulli found that pressure dropped when gas velocity increased past an orifice. Invention of the rotary mechanical vacuum pump (which operates in the viscous flow range) and the diffusion pump were big steps toward ultra-high vacuum. These came within the last 50 years.

Vacuum is correctly defined as a state which exists in a completely sealed space from which all gases and vapors have been removed. Since no method of producing an absolute vacuum has yet been devised, progress toward that goal must be described in terms of "degree" of vacuum attained. Pressure attained is limited by the materials chosen to enclose such a space, nature of the gases and vapors to be removed, and methods of pumping used. Degree of vacuum bears a definite relation to atmospheric pressure. (At sea-level this pressure is about 760 mm of mercury.) The list below shows the relationship existing between the degree of vacuum and the pressure.

Condition	Pressure Range
Low vacuum	760 to 25 torr
Medium vacuum	25 to 1.0 torr
Fine vacuum	1.0 to 1.0×10^{-3} torr
High vacuum	1.0×10^{-3} to 1.0×10^{-6} torr
Very high vacuum	1.0×10^{-6} to 1.0×10^{-9} torr
Ultra high vacuum	1.0×10^{-9} and below torr

Behavior of gases and vapors becomes more involved as conditions of pressure and temperature change. Ideal gases obey Boyle's law as well as Charles' law, but noncondensable gases and vapors do not. Generally speaking, in any range of vacuum being used, one is confronted with a complex of gases and vapors. In 1 cc of gas, at normal atmospheric pressure (760 torr), there are about $2.7 \times 10^{+19}$ molecules. If we reduce the pressure to 1.0×10^{-7} torr, we would have about $2.7 \times 10^{+9}$ molecules.

Molecules of air, under standard conditions, collide with each other after traveling an average distance of 9.5×16^{-6} cm. The average distance a particle travels between successive collisions is called the mean free path. At a pressure of 1.0×10^{-7} torr this increases to 14.0 m.

Vacuum is created when molecules of gas are removed from a chamber until as few as possible remain. Present methods of eliminating gas from a chamber involve pumping. When pumping is applied to a system at atmospheric pressure, gas flows from the direction of the volume being evacuated to the pump. The gas expands into a chamber defined by a fixed stator and a rotary element. The gas expands into the chamber and is compressed by vanes until it is finally ejected at the outlet through a valve. The entire mechanism is usually submerged in a bath of special low-vapor-pressure vacuum oil to prevent air from leaking back into the chamber.

A point is reached, however, where pressure here is so low that the random molecular motion is no longer directed. There is no longer enough gas to constitute a flow. Thereafter pressure can be reduced further only by catching a molecule wandering into this area and dragging it out. A mechanical pump cannot do this, but the diffusion pump can. The diffusion pump is an early development, dating back to Gaede, who in 1915 did the fundamental work that resulted in a basic diffusion pump. It was capable of pumping large volumes of gas at relatively low pressures. The diffusion pump operates on the principle that when a liquid, oil, or mercury is boiled, the vapor rises. Heat drives the vapor up the chimney where it hits jets that force it against the walls of the chamber. The walls are kept cool by a coil arrangement so that this working fluid condenses, flows back into the boiler, is reheated, and refluxes all over again. Any molecules of gas in the chamber to be evacuated, which wander into the oil stream, are captured and directed toward the base of the pump. Here the pressure is built up to a point at which a mechanical pump can remove the gas. From the above you can see that the diffusion pump acts as a compressor.

Brazing

Brazing is defined as a group of welding processes in which coalescence is produced by heating to suitable temperatures above 800°F and by using a ferrous and/or nonferrous filler metal that must have a liquidus temperature above 800°F but below that of the base metals. The filler metal is distributed between the closely fitted surfaces of the joint by capillary attraction.

This definition brings out four distinct parts:

1. The coalescence, joining or uniting of an assembly of two or more parts into one structure is achieved by heating the assembly or the region of the parts to be joined to a temperature of 800°F or above.
2. Assembled parts and filler metal alloy are heated to a temperature high enough to melt the alloy but not the parts.
3. The molten alloy spreads into the joint.
4. The parts are cooled to freeze the alloy and anchor the parts together.

A comprehensive theory of the wetting or spreading of liquids on solid surfaces has been presented by Hawkins [1]. The complete and detailed derivation of the quantitative relationships is not repeated here; only the necessary definitions and equations relating to the discussion are presented.

---

* In recent years, the term "torr" has been generally accepted as the international standard term equivalent to a column of mercury 1.0 mm high. ("Torr" is in honor of Evangelista Torricelli, who invented the barometer in 1643.)

The free surface energy (or surface tension) of a substance at constant temperature, T, pressure, p, and concentration, N is defined as

$$\frac{\partial F}{\partial A} p, T, N \equiv \gamma \equiv \text{free surface energy (usually dynes) per square centimeter} \tag{1}$$

where F = the free energy of the substance,
A = its surface (or interfacial) area.

The thermodynamic condition for spreading to occur is that, for an incremental increase in area, the entire system will have

$$dF < 0 \tag{2}$$

Conversely, for nonspreading

$$dF > 0 \tag{3}$$

If it is assumed that in the course of spreading of a liquid, L, in equilibrium with its own vapor, V, on a surface, S, in equilibrium with its own vapor, the following area relation exists:

$$dA_{LV} = dA_{SL} = -dA_{SV} \tag{4}$$

then

$$\frac{\partial F}{\partial A} p, T = \gamma LV + \gamma SL - \gamma SV \tag{5}$$

where γLV = liquid-vapor surface energy
γSL = solid-liquid surface energy
γSV = the solid-vapor surface energy

Let $-dF/dA$ be designated as the final spreading coefficient, S_{LS}; then at constant temperature and pressure

$$S_{LS} = \gamma SV - (\gamma LV + \gamma SL) \tag{6}$$

For a measure of the attraction between different materials the work of adhesion can be obtained as

$$W_{ad} = (\gamma LV + \gamma SV) - \gamma LS \tag{7}$$

The work of adhesion represents the decrease in energy in bringing together a unit area of liquid surface and a unit area of solid surface to form a unit area of interface.

Experimentally, it is observed that liquids placed on solid surfaces usually do not completely wet but rather remain as a drop having a definite contact angle between the liquid and solid phases. This condition is illustrated

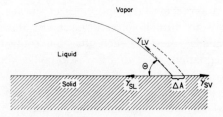

Figure 8.1 Sessile drop showing the vectors of the surface energies for the system.

in Figure 8.1. The Young and Dupré equation [2] permits the determination of change in surface free energy, ΔF^s, accompanying a small change in solid surface covered, ΔA. Thus

$$\Delta F^s = \Delta A(\gamma SL - \gamma SV) + \Delta A \gamma LV \cos(\sigma - \Delta\sigma)$$

At equilibrium,

$$\lim_{\Delta A \to 0} \frac{\Delta F^s}{\Delta A} = 0$$

and

$$\gamma SL - \gamma SV + \gamma LV \cos \sigma = 0 \qquad (9)$$

or

$$\gamma SL = \gamma SV - \gamma LV \cos \sigma \qquad (10)$$

In Equations 9 and 10 it can be seen that σ will be greater than 90° when γSL is larger than γSV, as shown in Figure 8.2a and the liquid drop tends to spheroidize.

σ is less than 90° when the reverse is true as shown in Figure 8.2b and the liquid drop flattens out and wets the solid. If the balance is such that σ is zero and greater wetting is desired, σ should be as small as possible so that $\cos \sigma$ approaches unity and the liquid spreads over the solid surface.

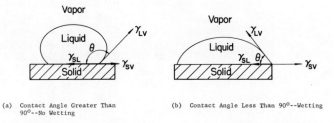

(a) Contact Angle Greater Than 90°--No Wetting

(b) Contact Angle Less Than 90°--Wetting

Figure 8.2 Sessile drops and interfacial energies.

These considerations show the importance of surface energies in brazing. If a braze alloy is to form a joint, it must wet the solid material. The surface energy balance must be such that the contact angle is less than 90°. The energy equation shows that if σ is to be less than 90° ($\cos \sigma > 0$), γSV must be greater than γSL. For example, a series of contact angle measurements of several liquid metals on beryllium at various test temperatures in argon and vacuum are presented in Table 8.1.

Table 8.1 Contact Angles for Various Times, Temperatures, and Atmospheres.

System	Temperature (°C)	Atmosphere	Contact Angle for Elapsed Time (Deg)				
			0 sec	15 sec	45 sec	90 sec	150 sec
Ag	1,010	Vacuum	142.3	141.1	140.6	141.2	140.2
	1,060	Vacuum	135.4	135.6	131.4	133.2	127.9
	1,010	Argon	148.1	148.9	150.7	147.3	143.4
	1,060	Argon	61.0	59.9	57.2	54.4	52.6
Au	1,070	Vacuum	58.6	58.0	57.2	55.7	54.6
	1,070	Argon	57.3	56.6	57.9	56.9	55.8
Cu	1,108	Vacuum	129.0	128.5	124.2	124.1	123.6
	1,133	Vacuum	126.6	122.5	121.8	120.0	120.0
	1,108	Argon	Extreme Base Metal Solution				
	1,133	Argon	Extreme Base Metal Solution				
Ge	1,000	Vacuum	145.5	145.3	144.3	143.7	143.6
	1,050	Vacuum	102.9	100.0	100.9	96.1	95.4
	1,000	Argon	99.3	99.6	98.9	98.2	97.1
	1,050	Argon	89.2	89.2	88.9	87.1	87.1
Pd-2.1 wt% Be	980	Vacuum	77.0	74.0	73.2	70.8	69.9
	1,030	Vacuum	74.2	73.2	68.7	66.5	65.6
	980	Argon	77.2	76.5	76.0	75.0	72.7
	1,030	Argon	70.2	63.1	61.7	57.2	57.2
Al	710	Vacuum	155.1	154.1	151.3	146.6	142.0
	760	Vacuum	140.0	138.8	136.6	134.1	132.0
	710	Argon	146.2	139.6	130.5	120.1	116.6
	760	Argon	108.2	106.6	102.3	100.0	99.9
Zr-5 wt% Be	1,030	Vacuum	88.0	85.6	83.9	81.8	80.9
	1,080	Vacuum	69.8	67.7	67.3	66.5	64.7
	1,030	Argon	101.4	98.8	99.2	98.3	96.2
	1,080	Argon	83.6	84.2	84.9	82.4	79.7
Ti-6 wt% Be	1,075	Vacuum	15.8	14.8	14.5	14.0	13.2
	1,125	Vacuum	14.2	13.4	12.1	11.2	10.7
	1,075	Argon	13.4	12.0	11.9	9.7	8.6
	1,125	Argon	10.0	7.9	7.8	6.8	5.6

Table 8.2 Surface Tension Calculations.

System	Temperature °C	Atmosphere	Contact* Angle (deg)	γ_{LV}	γ_{LS} (dynes/cm)	W_{ad}	S_{LS}
Al	710	Vacuum	142.0	831	2,520	176	−1,490
Al	760	Vacuum	132.0	726	2,350	239	−1,210
Al	710	Argon	116.7	787	2,220	434	−1,140
Al	760	Argon	99.9	561	1,960	464	−658
Zr-5 wt % Be	1,030	Vacuum	79.8	625	1,940	549	−702
Zr-5 wt % Be	1,080	Vacuum	64.6	520	1,770	614	−427
Zr-5 wt % Be	1,030	Argon	97.0	488	1,780	576	−405
Zr-5 wt % Be	1,080	Argon	79.8	172	1,690	569	−225
Ti-6 wt % Be	1,075	Vacuum	13.2	492†	1,380	970	−9.00
Ti-6 wt % Be	1,125	Vacuum	10.8	277	1,590	548	−4.00
Ti-6 wt % Be	1,075	Argon	8.6	424	1,440	844	−1.00
Ti-6 wt % Be	1,125	Argon	5.6	329†	1,530	660	+3.00

* Average of two tests.
† One test only.

Although during the testing period minor fluctuations in angle were noted, the overall tendency was for the contact angles to decrease with increasing time at test temperature. This behavior is reflected in the data of Table 8.2. The fluctuations may be attributed to alloying effects because, as alloying progresses, composition and temperature variations change the interfacial energies.

These data provide a means of comparison of the wettability on beryllium of the various liquid metals. If one assumes wetting for contact angles less than 90° and nonwetting for angles greater than 90°, then these data permit qualitative separation of the systems into these two classes. Gold, Pd-2.1 wt% Be, and Ti-6 wt% Be were found to wet beryllium at both temperatures and in both argon and vacuum atmospheres. The Zr-5 wt% Be alloy wets beryllium in vacuum at both temperatures. This alloy, as well as silver and gold, did not wet when tested in argon atmospheres at the lower temperature. It was found, however, that an increase of 50°C resulted in wetting. Aluminum exhibited a nonwetting behavior at both temperatures and in both atmospheres. Copper was nonwetting at both temperatures in vacuum, but underwent extensive alloying with the solid beryllium at these temperatures in argon. From these data it appears that the binary alloys have the greatest tendency to wet beryllium [3].

Surface energy calculations have been performed on systems of Be with

Al, Zr-5 wt% Be and Ti-6 wt% Be. The calculated spreading coefficients were negative for all system test conditions except with Ti-6 wt% Be. Negative spreading coefficients indicated a very definite trend. The greater the system wettability, as indicated by lower contact angles and liquid-vapor surface tensions, the greater the numerical magnitude of the spreading coefficient.

Of the three systems used, the Ti-6 wt% Be was observed to provide the best wetting on beryllium; the Zr-5 wt% Be was next in order. Although no basic significance can be given to the absolute values of the surface tension properties calculated, they are useful in indicating the relative tendencies for these liquids to wet beryllium.

From the above, some liquid metal systems will wet beryllium, whereas some systems which should wet do not in actual practice. The problem is more complex than it seems at first.

Consider what effect the changes in the surface or interfacial energies have on the contact angle. If the initial contact angle is less than 90°, the following will decrease the contact angle: (a) increase γSV, (b) decrease γSL, and (c) decrease γLV.

As mentioned previously, the presence of adsorbed molecules on a metal surface markedly reduces the surface energy γSV. This increases the contact angle and retards wetting. The importance of clean surfaces and their maintenance by brazing in vacuum is apparent. The presence of small amounts of impurities can affect the liquid/vapor surface energy.

It can be concluded that wetting is the ability of the molten brazing alloy to adhere to the surface of a metal in the solid state and, when cooled below its solidus temperature (see discussion later in chapter), to make a strong bond to that metal. Wetting is a function not only of the brazing alloy but of the nature of the metal or metals to be joined. There is considerable evidence that in order to wet well a molten metal must be capable of dissolving, or alloying with, some of the metal on which it flows.

Flow

Wetting is only one important facet of the brazing process. If the molten alloy does not flow into the joint, the use of the alloy is greatly restricted. Flow is facilitated by capillary attraction, which in turn results from surface energy effects. As pointed out in the previous section, a liquid metal meets the solid at a characteristic angle σ. If the liquid is confined in a narrow joint, such as between two flat plates, the surface of the liquid must curve in order for the two contact angles to form [4]. This curvature results in a pressure difference across the surface. The liquid flows into the gap until the pressure caused by the surface energy is equal to the hydrostatic pressure. It can be shown that the pressure difference, ΔP, across the curved interface is given by

$$\Delta P = \frac{2\gamma LV \cos \sigma}{D}.$$

For vertical plates the maximum height h that the liquid rises is given by

$$h = \frac{2\gamma LV \cos \sigma}{\rho g D},$$

where g = gravitational constant,
ρ = density of the liquid.

This equation gives the maximum height, but does not furnish any information on the rate of flow. By applying the principles of fluid flow, it can be shown that the distance d covered in time t is given by

$$d = \frac{\sqrt{Dt_\gamma LV \cos \sigma}}{3\eta},$$

where η = viscosity of the liquid,
D = width of gap.

It is apparent that a high liquid surface tension, a low contact angle, and low viscosity are desirable. Thus a low contact angle, which implies wetting, is a necessary but not a sufficient condition for flow. Viscosity is also important. Alloys with narrow freezing ranges that are close to the eutectic composition (see discussion later in chapter) generally have lower viscosity than those with wide freezing ranges. For most metals and brazing alloys, the maximum distance d is reached in a very short time, usually less than 1 sec.

Capillary attraction does some unusual but interesting tricks. For instance, if two clean glass plates are immersed in water and held closely together, the water wets the glass and rises in a column between the plates. If glass plates are immersed in mercury, however, no wetting takes place and the column of liquid is actually depressed below the surface of the mercury.

Water and mercury do not behave the same, however, with all materials. For example, if paraffin plates are immersed in water, no wetting results and we get a depressed column, the same as with glass in mercury, and if amalgamated zinc plates are immersed in mercury, good wetting results and the column rises. The smaller the gap between the plates, the higher the column will rise. Obviously, then, if we wish to have a liquid drawn between two surfaces, the liquid must wet the surfaces and the surfaces should be very close to one another.

What does all this have to do with vacuum furnace brazing? It has been found that if assemblies with clean surfaces are put together so that their joints are snug and tight, brazing metal applied near the joints melts and wets the parent metal and creeps into the joints by capillary attraction, providing that a suitable vacuum atmosphere is maintained within the furnace to assure that the surfaces stay clean and free from oxides.

To illustrate further the principles of capillary attraction and its relationship to vacuum brazing, glass plates spaced apart with different gaps using India ink and held in position with sealing wax show very clearly the relationship between the wetting action and capillary attraction and the desirability of cleanliness of metal surfaces within joints to be vacuum brazed. Two pieces of glass were fastened to a second set of glass plates and strips of glass were held in intimate contact throughout their length and immersed in ink. The surfaces of glass in one case were chemically cleaned to remove oil films and assure good wetting. The surfaces of glass in the second set were fairly clean, but fingerprints were purposely impressed on both strips of glass to assure the presence of oil films. The result is obvious. In the first set the ink rose as high as possible by capillary attraction, uniformly wetting the glass. In the second set the ink rose until it reached the fingerprints, where wetting became difficult. However, passages around the fingerprints were soon established and the ink continued to creep upward, going around small oil spots on its way. If the ink were brazing metal instead, the bond would not be expected to have maximum strength or tightness where the surfaces were unclean. In another test all pieces of glass were chemically cleaned to assure good wetting, but various gaps were used between the strips to show the effect of the width of the gap on the height of the ink column and to illustrate the desirability of having uniform fits in assemblies to be vacuum brazed. The spacings of the glass strips were as follows:

(1) uniform contact (no gap) throughout.
(2) uniform gap of 0.001 in. throughout.
(3) tapered gap from bottom to top: intimate contact (no gap) at bottom, 0.020 in gap at top.
(4) tapered gap from left to right: intimate contact (no gap) at left, 0.020 in. gap at right.

In (1) the ink rose in a very thin film to the top of the small glass strip. In (2) the ink rose as high as its surface tension could draw it through the uniform gap of 0.001 in. In (3) the ink started to rise from the bottom where there was no gap, but it soon reached a level where the gap was so wide that the surface tension could pull it no higher. The width of the gap at this point was about 0.006 in. In (4) the ink rose a short distance above its surface between the glass strips that were touching at the left and were 0.020 in. open at the right, but soon stopped rising along the right edge. It continued to rise along the left edge, however, until it reached top, and the height of the column between the left and right edges adjusted itself to the varying width of the gap.

The test demonstrates that the distance of flow of brazing metal by capillary attraction is directly dependent on the width of the gap; the travel becomes greater as the gap diminishes; a uniform clean gap gives a uniform bond; a gap can become so wide that it completely stops the flow; and line contact or nonuniform contact, rather than uniform contact throughout a joint, results in a nonuniform bond.

A fact not illustrated here, but worthy of consideration when studying the behavior of brazing metals, is that viscous or pasty alloys bridges wider gaps than will those of low viscosity. The temperature, nature, and specific gravity of the liquid are also factors.

Flow is the property of a brazing alloy that determines the distance it will travel away from its original position because of the action of capillary forces. To flow well an alloy must not suffer an appreciable increase in its liquidus temperature even though its composition is altered by the addition of the metal it has dissolved. This is important because the brazing operation is carried out at temperatures just above the liquidus.

These discussions are based on noninteracting systems. The compositions and thus the surface energies of the liquid and solid are assumed to remain constant. In real systems, however, the following interactions can occur [5]:

(1) alloy formation between liquid and base metal,
(2) diffusion of base metal into filler metal,
(3) diffusion of filler metal into grains of base metal,
(4) penetration of filler metal along grain boundaries, and
(5) formation of intermetallic compounds.

These interactions can alter surface energies and viscosities, thus leading to changes in wetting and flow. The rate at which these interactions take place depends on the temperature, the time at temperature, and the materials involved. Generally, the extent and type of interaction, for simple systems, can be predicted from the appropriate phase diagrams and from diffusion data. However, if either the filler or base metal, or both, are alloys, these predictions become very difficult and experimental data are required.

In understanding these base metal/alloy interactions, for they can affect the wetting and flowing of the braze alloy, an analysis of the theory of an equilibrium diagram makes the preceding discussion clearer. Figure 8.3

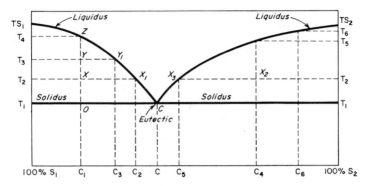

Figure 8.3 Equilibrium diagram of two metals.

is a hypothetical diagram of a plot of the liquid and solid states of all possible compositions consisting of metals S_1 and S_2. The following terms must be defined.

Liquidus. The two heavy curves, TS_1–C–TS_2, represent the liquidus; above this temperature any composition is liquid. (The liquidus temperature of composition C_1 is T_4, as indicated by point Z.)

Solidus. The heavy horizontal line T_1–C–T_1 is the solidus; below this line all compositions are entirely solid. The solidus temperature is T for all compositions (e.g., points C and O).

Eutectic. Composition C is the eutectic. When an alloy of this composition is heated, it remains entirely in the solid state until temperature T_1 is reached, at which point it becomes entirely liquid. T_1 is both the solidus and liquidus temperature of composition C. The eutectic is entirely liquid at a lower temperature than any other composition in the diagram.

Liquid-solid state. Unlike composition C, any other composition will not convert directly from solid to liquid at one temperature, although all compositions begin this conversion at T_1. Composition C_1, for example, will be partly solid and partly liquid at any temperature between T_1 and T_4 (such as T_2 or T_3, indicated by points X and Y).

Composition of each state. At temperature T_2 (point X) composition C_1 is separated into a solid of composition S_1 (the pure metal) and a liquid of composition C_2 (point X_1). The liquid must be of composition C_2 because this is the only composition that can be liquid at temperature T_2. (It cannot be C_5 because the original composition contained more S_1 than S_2.)

Brazing alloys that are not eutectic compositions present a danger if parts being brazed are disturbed during the time the alloy passes through the part solid-part liquid condition. Any movement during this period may produce cracks in the braze metal.

A second type of difficulty can be illustrated by assuming that S_1 is silver, S_2 is copper, and the parts to be brazed are copper. If the brazing alloy has composition C_4 at T_1, then at T_2 the liquid portion changes to composition C_5 having a higher copper content. But the copper in the base metal is also soluble in this liquid phase. If the rise in temperature is not rapid enough between T_1 and T_5 (the liquidus temperature of C_4), the alloy may dissolve enough copper to change its composition to C_6 with a resultant higher liquidus T_6. If the furnace temperature has been set for T_5, the braze metal will never become completely liquid and will not flow adequately.

Another aspect of the same phenomenon occurs if the brazing alloy is allowed to remain at a temperature above the liquidus for so long a time that another copper is dissolved to alter the composition to one with a higher liquidus. In this case a solid phase slowly crystallizes from the melt and weakens the brazed joint.

In practice, these interactions are usually minimized by (a) selecting the proper filler alloy, (b) keeping the brazing temperature as low as possible, but high enough to get flow, and (c) keeping the time at temperature short and cooling down as quickly as possible without cracking or distortion.

MECHANISMS AND KEY VARIABLES

A careful and intelligent appraisal of the following variables will produce satisfactory brazed joints.

- Base metal characteristics
- Filler metal characteristics
- Surface preparation
- Joint design
- Temperature
- Time
- Rate of heating

Base-Metal Characteristics

The base metal has a prime effect on joint strength. A high-strength base metal produces joints of greater strength than those made with softer base metals (other factors being equal). When hardenable metals are brazed, the joint strengths become less predictable. This is because there are more complex metallurgical reactions involved between hardenable base metals and the brazing filler metals. These reactions can cause changes in the base metal hardenability and can create residual stresses. Table 8.3 clearly shows that base metal strength has a greater effect on brazed joint strength than does the brazing filler metal or furnace atmosphere used.

Table 8.3 Strength of Lap Joint Specimens of Various Base Metals Brazed with AMS-4777 Filler Metal (Nickel-Chromium-Silicon) [7].[a]

Furnace atmosphere	Joint overlap, T	Test temp.	Base metal stress, psi	Fracture location
Base metal A-286[b]				
Vacuum	5.0	Room	126,000	B.M.
Vacuum	5.0	Room	123,500	B.M.
Vacuum	5.0	1200° F	82,037	Braze
Hydrogen	4.9	Room	88,500	B.M.
Hydrogen	4.9	Room	131,000	Braze
Hydrogen	4.9	1200° F	89,310	Braze
Base metal AM-350[c]				
Vacuum	4.7	Room	85,500	Braze
Vacuum	4.7	Room	87,800	Braze
Vacuum	4.7	1000° F	68,548	Braze
Hydrogen	3.8	Room	76,000	Braze
Hydrogen	3.8	Room	82,500	Braze
Hydrogen	3.8	1000° F	72,222	B.M.

[a] Results of shear tests on brazed joints in hardenable base metals.
[b] All specimens extremely strong. Test results show considerable variations.
[c] All joints very strong. Stronger than brazed soft metals but not as strong as hardened base metal.

Some of these metallurgical phenomena that influence the behavior of brazed joints and in some instances necessitate special procedures include these base-metal effects: (a) carbide precipitation, (b) hydrogen and sulfur embrittlement, and (c) oxide stability.

There are also base-metal-filler metal interaction effects: (a) alloying, (b) stress cracking, and (c) phosphorus embrittlement. The extent of interaction varies greatly depending on compositions (base metal and filler metal) and thermal cycles. There is always some interaction, except when mutual insolubility permits practically no metallurgical interaction.

The term "alloying" is a general term covering almost all aspects of interaction. Let us, however, be more specific about the components of the term "alloying." First, the molten filler metal can dissolve base metals. Second, constituents of the filler metal can diffuse into the base metal either through the bulk of the grains or along the grain boundaries or can penetrate the grain boundaries as a liquid. The results of these base metal dissolution or filler metal diffusion may be to raise or lower the liquidus or solidus temperatures of the filler metal layer, depending on composition and thermal cycle. Whenever excessive base-metal dissolution and diffusion are liable to occur, the brazing should be done in as short a time and at as low a temperature as possible. Sufficient filler metal should be present to fill the joint completely, but excess filler metal is both undesirable and uneconomical.

Another factor included under the general term "alloying" is the formation of intermetallic compounds resulting from interaction between constituents of the base and filler metals. These compounds are usually brittle. Whether or not the compounds form depends on base and filler metal compositions, time and temperature. Because intermetallic compounds do form, it does not necessarily follow that the joint is so embrittled that it loses engineering utility. This depends on the nature of the specific compound, its quantity, and distribution.

Many high-strength materials have a tendency to crack during brazing when in a highly stressed condition and in contact with molten brazing filler metal. Materials that have high annealing temperatures are subject to this stress-cracking phenomenon. The process may be described as stress-corrosion cracking in which the molten filler metal is considered the corrosive medium. Stresses set up by changes in the volume of parts made from some special alloys when they pass through a phase change on cooling may also cause cracking. This cracking is not usually catastrophic, but may weaken the brazed joint. When stress cracking is encountered, its cause usually can be determined from a critical analysis of the brazing procedure. The common remedy is to remove the source of stress.

Filler Metal Characteristics

The second material involved in joint structures is the brazing filler metal. The term "brazing filler metal" is essentially synonymous with the commonly employed "brazing alloy." Its selection is important, but not for the reasons many engineers think. A specific filler metal cannot be chosen to produce a specific joint strength, which is unfortunate, but true. Actually, strong joints can be brazed with almost any good commercial brazing alloy if brazing methods and joint design are done correctly.

Composition of the brazing filler metal does become important when the brazement is to be subjected to chemically corrosive or high temperature service conditions. The brazing filler metal must be able to resist oxidation or chemical corrosion and must not remelt in service. Another factor is that among the heat and corrosion-resistant brazing filler metals some will diffuse quite readily into the base metal. It is incorrect, however, to say that to be "good" brazing alloys must be either diffusible or nondiffusible. The degree of diffusibility must be matched to the application. A highly diffusible brazing filler metal might seriously erode the base metal adjacent to the joint in an assembly employing extremely thin sections (such as a honeycomb structure). In another application the same diffusing type of filler metal might be desirable to attain a high enough remelt temperature. Table 8.4 lists several nickel base brazing filler metals, giving a few criteria to consider when making a selection.

Several other characteristics that braze filler metals must possess are:

(1) proper fluidity at brazing temperatures to assure flow by capillary action and provide full alloy distribution;
(2) stability to avoid premature release of low-melting point elements in the alloy;
(3) ability to wet the base metal joint surfaces,
(4) the alloying elements of the braze filler metal must have low volitalization characteristics at brazing temperatures;
(5) it is desirable that it alloy or combine with the parent metal to form an alloy with a higher melting temperature;
(6) washing or erosion between the brazing filler metal and the parent metal must be controllable within the limits required for the brazing operation;
(7) depending on the requirements, the ability to produce or avoid base metal-filler metal interactions.

One of the most broadly misunderstood facts relating to brazing filler metals is that brazed joint strength is in no way related directly to the melting method used. This fact is hard to accept because it seems to contradict a long-established metallurgical truth with regard to the manufacture of steels or other constructional metals. The effect of melting practice on brazing alloys, however, is not the same as that of melting practice on steels.

If constructional metals are produced by vacuum melting, for example, there is a definite relationship between the vacuum melting practice and the final strength of the ingot, bar, or rolled sheet. That is not true with a brazing alloy, since joint strength is dependent on such factors as joint design, brazing temperature, amount of alloy applied, location and method of application, heating rate, and many, many other considerations that make up what is termed "brazing technique."

Melting practice only affects the purity of the metal fillers. Purity in turn only determines whether or not the filler will braze. "Purity" in a brazing alloy means primarily freedom from refractory oxides (titanium, aluminum, or silicon oxides), although any oxidized metal or certain dissolved gases, such as oxygen or nitrogen, are undesirable in brazing alloys. Metal oxides and oxygen or nitrogen inclusions may affect brazing alloy flow and thereby impair brazeability. Dissolved hydrogen, on the other

Table 8.4 Comparison of Various Characteristics of Twelve Nickel-Base Brazing Filler Metals [7].[a]

	Boron-containing alloys								Silicon-containing alloys			Phosphorus-containing alloys
	AMS 4775	AMS 4776	AMS 4777	AMS 4778	WG	E	F	G	B	D	A	C
Property												
Joint strength[b]	1	1	1	2	3	3	3	1	1	3	4	2
Diffusion with base metal	1	1	2	2	1	2	3	3	4	4	4	5
Fluidity	3	4	2	2	5	2	4	4	2	1	1	1
Application												
High-temperature, high-stress, moving components	1	1	2	2	3	1	3	1	1	3	3	3
Heavy, stationary structures (variable gaps)	1	1	1	2	1	3	1	3	2	3	3	3
Honeycomb and other thin materials	3	3	2	2	3	2	3	3	1	1	1	1
Atomic reactor core assemblies[c]	X	X	X	X	X	X	X	X	1	1	2	1
Large, machinable, or soft fillets	2	2	3	3	1	3	2	2	3	3	3	3
Contact with liquid metals (NaK, Hg)	1	1	1	1	2	1	1	1	1	1	1	1
Tight or deep joints	3	3	2	2	3	2	3	3	2	2	2	1
Torch brazing (with special flux)	1	1	1	1	1	1	1	1	2	2	2	2

[a] Rating of 1 is best choice, 2, next best, etc.
[b] Joint strength also depends on brazing cycle, joint design and joint clearance.
[c] Boron-containing alloys are not suitable for atomic reactor core assemblies. These metals (rated X for this application) have high neutron-absorption ratios.

hand, is not harmful in a brazing alloy because it is released during the brazing operation [8].

The degree to which braze filler metal penetrates and alloys with the base metal during brazing is referred to as diffusion. In applications requiring strong joints for high temperature, high stress service conditions (such as turbine rotor assemblies and jet engine components), it is generally good practice to specify a filler metal that has high diffusion and solution properties with the base metal. When the assembly is constructed of extremely thin base metals (as in honeycomb structures and some heat exchangers), good practice generally calls for a filler metal with a low diffusion characteristic relative to the base metal being used. Diffusion is a normal part of the metallurgical process that can contribute to good brazed joints when brazing, for example, high temperature metals with nickel base filler alloys.

Solution exists to some degree in every brazed joint, but is harmful only when the amount is sufficient to cause erosion. The amount of solution of the base metal is affected by such factors as solubility of the brazing alloy in the base metal, the quantity of brazing alloy applied to the joint, and the brazing cycle (time and temperature). Erosion (caused by excessive solution) is not affected directly by the manner in which the brazing alloy is manufactured. The filler metal physical properties determining solution, erosion, strength, corrosion resistance, etc., are inherent chemical characteristics which are not changed by melting practice or even by the purity of the brazing alloy.

The amount of impurities in the brazing alloy does not change its basic solubility in a given base metal. Should a very impure filler metal be used, however, there probably would be no erosion because there would be no braze. The refractory oxides in the brazing alloy would prevent satisfactory flow. When there is no braze, there can be no erosion.

Probably the most important factor affecting degree of erosion for a given filler-base metal combination is the brazing technique, which encompasses assembly and fixturing practices, alloy application, and the brazing cycle itself [8].

The melting behavior and nature of brazing filler metals were discussed previously (See Figure 8.3).

Surface Preparation

In vacuum furnace brazing the wetting action is a little different from water on glass because we are dealing with metals at elevated temperatures and in most situations the attraction between the brazing metal and the parent metal is sufficiently great to cause alloying between the two. The condition of the surface of the parent metal, however, greatly influences the behavior of the brazing metal from the standpoint of wetting and the tendency to creep or to ball up. These tendencies depend on the relative surface tension—whether there is sufficient attraction by the surface of the parent metal to draw the brazing metal out in a thin film or whether the surface tension of the brazing metal is sufficient to draw it up into balls or lumps on the surface of the parent metal when compared with a lack of attraction by the surface to draw it out in a film.

The first thought might be that the molten brazing metal would creep best on a highly polished surface, but it can be said that exactly the opposite is true. The brazing metal creeps best and farthest on a clean, roughened surface. For instance, molten copper will wet clean polished steel, but it will not have much of a tendency to creep out on its surfaces and distribute itself evenly in the presence of the protective atmosphere in the furnace. However, if this surface should be roughened by shot blasting, rough grinding, rough machining, pickling, or some other process, the copper will spread out in a thin film and tend to distribute itself evenly over the surface. This is because the liquid is drawn through the tiny scratches and pores on the surface by capillary attraction. The reader can easily demonstrate this effect and other phases of the wetting action for himself with some ink and several grades of paper.

This test, illustrates the wetting action of ink on five different grades of paper, which, although not identical in performance with respect to brazing metals on parent metals, can be somewhat likened to certain conditions encountered in vacuum furnace brazing. Studying the wetting action in this manner at least gives one a better understanding of some of the results obtained with the furnace-brazing process. Suppose a drop of ink is placed on each of ten pieces of paper while all are in the horizontal plane, and then the back panel is raised 90° into the vertical plane. This allows the ink to creep out or draw together on a horizontal surface or to cling or flow downward by force of gravity on a vertical surface, if it so desired.

Take one piece of the blotting paper. The ink soaks into it almost instantaneously on striking its surface and is so completely absorbed by the body of the paper that it is unable to spread. This can be likened to some extent to the action of copper on nickel, relatively high-melting silver solder on copper, or similar combinations in which the metals are mutually soluble to a high degree. Because of this great mutual solubility, it is usually desirable to supply a number of reservoirs of brazing metal to a joint when the parent metal is likely to absorb the brazing metal readily, rather than to expect the brazing metal to creep appreciable distances by capillary attraction, say $\frac{1}{4}$ in. or more. This can be accomplished by machining one or more grooves in the joint to retain the wire rings; by drilling feeder holes to the joint, filled with slugs; or by including foil within the joint if practicable.

This test does not illustrate, however, another effect of high mutual solubility of the brazing metal and the parent metal, the "freezing" of the brazing metal due to a pickup of the higher-melting parent metal. This also retards the flow of the brazing metal and sometimes stops its penetration into a joint or leaves collections of it on the surfaces of the parts instead of allowing the excess to creep out freely into a thin film.

Suppose a piece of facial tissue paper, noted for its highly absorbent properties, is used in a second test. Because it is extremely thin, very little ink is required to soak through it, and the remainder has spread widely because of the capillary attraction. This can be compared with the action of copper on clean low-carbon steel with a matte finish, where the copper is only slightly soluble in the steel and penetrates only "skin deep," but has great ability to creep on the surface by capillary attraction. Similarly, copper plating on such a matte surface tends to retain its even distribution in the furnace-brazing cycle.

In a third test take a piece of poster board (a grade of cardboard) which is readily wetted by the ink but neither absorbed by it nor is it drawn out over its surface. The surface tension of the ink has held the drop in position even in the vertical plane. Copper or silver solder, for instance, sometimes behaves in this manner on steel if the wetting action is not quite right, which results from unclean surfaces.

Fourth, take a piece of bristol board, another grade of cardboard. Although it is wetted by the ink, the ink has drawn itself together instead of creeping out in the horizontal plane, but the surface tension of the ink is not sufficiently great to hold the drop intact in the vertical plane which has resulted in downward flow. This action can be likened to some extent to copper on highly polished low-carbon steel, where it is sometimes found that the brazing metal gets beyond control and runs downward, forming collections that solidify on projections or on low portions of assemblies. Sometimes this property can be used to great advantage and at other times it is better avoided.

Finally, a piece of cellophane, which very effectively resists the wetting action of the ink, allows the surface tension of the ink to draw the ink together into a small drop when resting in the horizontal plane and to run freely down the vertical surface. This can be likened to the very poor wetting action of copper on steels that have protective surface films of oxides of chromium, manganese, vanadium, aluminum, or silicon, formed by impurities in the controlled atmosphere, and explains the necessity of preventing or removing such oxide films.

A clean, oxide-free surface is imperative to ensure uniform quality and sound brazed joints. A sound joint may be obtained more readily if all grease, oil, dirt, and oxides have been carefully removed from the base and filler metals before brazing because only then can uniform capillary attraction be obtained. It is recommended that brazing be done as soon as possible after the material has been cleaned. The length of time the cleaning remains effective depends on the metals involved, atmospheric conditions, storage, handling, etc. Cleaning is commonly divided into two major categories: chemical and mechanical. Chemical cleaning is the most effective means of removing all traces of oil or grease. Trichlorethylene and trisodium phosphate are the usual cleaning agents employed. Various types of oxides and scale that cannot be eliminated by these cleaners are removed by other chemical means. Many commercial products are available for this purpose.

The selection of the chemical cleaning agent depends on the nature of the contaminant, the base metal, the surface condition, and the joint design. Regardless of the cleaning agent or the method used, it is important that all residue or surface film be removed from the cleaned parts by adequate rinsing to prevent the formation of other equally undesirable films on the faying surfaces. Objectionable surface conditions may be removed by mechanical means such as grinding, filing, wire brushing, or any form of machining, provided that joint clearances are not disturbed. When grinding surfaces of parts to be brazed, care also should be taken to see that the coolant is clean and free from impurities so that the finished surfaces does not have these impurities ground into them.

When faying surfaces of parts to be brazed are prepared by blasting techniques, several factors should be understood and considered. The purpose of blasting parts to be brazed is to remove any oxide film and to roughen the mating surfaces so that capillary attraction of the brazing filler metal is increased. The blasting material must be clean and of a substance that does not leave a deposit on the surfaces to be joined that restricts filler metal flow or impair brazing. They should be fragmented materials rather than spherical so that the blasted parts are lightly roughened rather than peened. The operation should be done so that delicate parts are not distorted or otherwise harmed. Vapor blasting and similar wet blasting methods require care because of possible surface contamination.

Joint Design

It is assumed that the designer has a basic knowledge of the part or assembly to be designed and its intended functions. From this information the base metal and filler metal that best suit the application are determined.

A further and more detailed discussion of joint designs follows later in the chapter.

Time and Temperature

The temperature of the furnace naturally has an important effect on the wetting action because the wetting and alloying action improves as the temperature increases. Of course, the temperature must be above the melting point of the brazing metal and below the melting point of the parent metal. Within this range a temperature is generally selected that will give the best satisfaction from an overall standpoint.

Usually, the lowest brazing temperatures are preferred to (a) economize on heat energy required, (b) minimize heat affect on base metal (annealing, grain growth or warpage, for example), (c) minimize base metal-filler metal interactions, and (d) increase the life of fixtures, jigs, or other tools.

Higher brazing temperatures may be desirable to (a) use a higher melting but more economical brazing filler metal, (b) combine annealing, stress relief, or heat treatment of the base metal with brazing, (c) permit subsequent processing at elevated temperatures, (d) promote base metal-filler metal interactions in order to modify the brazing filler metal (this technique is usually used to increase the remelt temperature of the joint), (e) effectively remove surface contaminants and oxides with vacuum brazing, and (f) to avoid stress cracking [9].

The time in the furnace also affects the wetting action, particularly with respect to the distance of creep of the brazing metal. If there is a tendency to creep, the distance generally increases with time. The alloying action between brazing metal and parent metal is, of course, a function of both temperature and time. In general, for production work, both temperature and time are kept at a minimum consistent with good quality.

In conclusion, the alloy and process for brazing must be selected with a true understanding of both the physical metallurgy of the base material and the interactions of the base material with brazing alloys. The effect of brazing thermal cycles on properties of the parent material must also be assessed to see whether compromises will be required in design assumptions. If these factors are properly considered, sound, reliable brazed parts will result.

TYPES OF EQUIPMENT AND TOOLING

Equipment

Modern machines ensure accurate and repetitive brazing results through a system of automatic controls regulating temperature, time and, when applicable, atmosphere. These three factors incidentally are the bases for selecting a proper brazing furnace.

Each brazing application deserves individual study concerning the best means of heating, cooling, protective atmosphere, etc. Size, weight, configuration, number of joints, type of base metal, type of filler metal, and its melting point are all important factors in the selection. If the part diameter is small and the area to be brazed is only a small portion of the total, such as a ferrule on the end of a tube, or if annealing of components by overall heating is objectionable, then localized heating, such as induction brazing, might well be the best choice. If, on the other hand, the size of the assembly is large, the shape is complex, or there are multiple joints to be brazed, possibly furnace heating would be a better selection. With furnaces the heat source is usually electric; with induction heating the retorts for heating and cooling are usually glass bell jars or quartz tubes.

In the search for higher temperatures and faster heating rates, the use of induction heating combined with a vacuum has exhibited some interesting applications. Leads from the induction generator are brought through the walls of the vacuum chamber and surround a graphite cylinder, called a susceptor which is heated by the induction field. The theoretical top temperature attainable is the sublimation temperature of graphite. For furnace brazing metal retorts made of heat-resisting alloy or mild steel are utilized in batch-type furnaces, and muffles of heat-resisting alloy or ceramics are employed in continuous or semicontinuous types. Electric furnaces are available with heating units of various types such as nickel-chromium ribbon or rod for furnace temperatures up to about 2200°F, silicon carbide bars up to about 2500°F, and molybdenum rod or wire up to about 3100°F.

The furnace shown in Figure 8.4 illustrates a cold-wall design with heating elements of nickel-chrome resistance ribbon stock attached to revolving reflector panels. These enclose the heated work area during the braze cycle and turn 90° during the cool-down cycle. This releases the heat from the work rapidly by radiation to the water-jacketed walls. The panels are rotated from the outside the furnace by shafts and cranks. This method of panel rotation control also permits observation of the part and its tooling through several ports periodically during the brazing operation.

Batch-type furnaces for brazing include the box-type with in-and-out retorts or built-in muffles, and the pit, bell, bell-car, elevator, and car types with retorts. The elevator-type has its own hoist that can be operated by the furnace attendant to raise and lower the assemblies in and out of the bottom opening of the furnace. The bell-type has stationary bases with permanent piping for atmosphere gas and cooling water, if used. There

Figure 8.4 High temperature vacuum furnace. Open door contains one of the panels of nickel-chromium heating elements that surround the work.

is no moving of the work when changing the furnace from one position to the other.

A number of vacuum brazing furnaces, now operating successfully in the field, are in effect modifications of conventional pit-type or bell-type atmosphere furnaces. They employ leak-tight vacuum muffles with suitable closures and are exhausted by high-vacuum diffusion and mechanical pumping systems. These units provide good service at temperatures from 1600 to 1850°F. They are not distinguished by their flexibility because their massive alloy retorts heat and cool slowly. For temperatures above 1850°F the single vacuum-tight alloy muffle cannot resist air in-leakage and maintain its shape against the external pressure of the atmosphere. High-temperature creep causes the muffle to sag out of shape. The end result of this effect over the life of the retort may be collapse with the loss of a costly product load.

This shortcoming of the hot-retort furnace can be overcome. Placing a secondary vacuum chamber around the hot retort, the heating elements, and the refractory thermal insulation permits an increase in operating temperature to 2150°F. Above 2150°F even the double-retort type of hot-wall furnace becomes marginal because of (a) in-leakages from the guard vac-

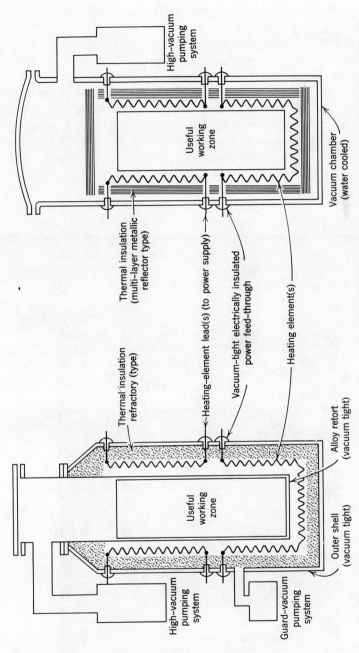

Figure 8.5 Schematics of hot retort and cold wall furnace. The former (left) operates well from 1000 to 2150°F. The cold wall (right) serves the 1700 to 5000°F Temperature range.

uum into the high vacuum interior through the retort and (b) the tendency of the retort to creep-sag of its own weight.

The cold-wall vacuum furnace represents an approach that avoids many of the disadvantages of the hot-retort type. It is useful at the lower temperatures because of its simplicity, fast heating, and cooling characteristics. These are ideal for batch cycling, rapid response to temperature control, unlimited retort life, and accessibility of heater elements for inspection or replacement. At higher temperatures the cold-wall vacuum furnace provieds the only method to achieve temperatures up to 4500°F and even for short periods to 5000°F. It is particularly useful when high vacuums are needed during most of the cycle. A schematic comparison of the two types of furnaces is illustrated in Figure 8.5.

Design of the cold-wall furnace utilizes the insulating characteristics of an evacuated space in which convection and gaseous conduction effects are eliminated. Radiant heating elements that shine directly on the workload are insulated from the water-cooled external retort by radiation shields that reflect the heat toward the work zone. In this design there is no refractory thermal insulation employed. (See Figure 8.6.)

Figure 8.6 Serpentine-design heating elements.

Elements

The center of any high temperature vacuum furnace is the refractory metal-heating elements and radiation shields. Requirements for the heating elements are low vapor pressure, high elevated temperature strength, and (since the heating is totally by radiation) a large heating surface which is necessary for rapid heat transfer.

One of the first tantalum or tungsten heating elements was introduced in 1950. The maximum operating temperature of this type of element was approximately 1900°C. Above this temperature the quartz or other insulating material used between the two tabs disintegrated. In addition, since the tantalum or tungsten was clamped over its entire length, distortion occurred because of expansion stresses. It also had a tendency to sag because it was supported only on one side, further reducing the useful operating life. An improvement on this design was made in 1956. This was a dual tungsten element design which eliminated the use of insulators, and the operating temperature increased to 2200°C. Above this operating temperature distortion due to thermal stresses was still a problem even though sagging was greatly minimized.

In 1959 the free hanging element design was introduced. This design, permitting both lateral and vertical thermal expansion, resulted in an element capable of enduring higher (up to 3000°C) temperatures and increased the operating life three to four times more than previous types. The three vertical segments, each with its own connector, also permitted the use of three-phase ac current.

This latest development in heating element design has lengthened operating life of the element and increased temperature uniformity by 60%. Rod and strip heating element designs have also been used. The latter is shown in Figure 8.6. Heating elements, depending on temperature range, may be nichrome, molybdenum, tantalum, or tungsten. Graphite is sometimes used, but in general it has been found satisfactory only when a carbonaceous atmosphere in the furnace can be tolerated. For the temperature range to about 1300°C the materials used almost exclusively have been metallic alloys with high electrical resistance and excellent scale stability such as Ni-Cr, Fe-Ni-Cr, Fe-Cr-Al, and Fe-Co-Cr-Al.

In the working temperature range higher than 1300°C suitable metallic and nonmetallic materials have been used for heating elements. The nonmetallic types are based on silicon carbide (Silit, Globar rods, etc.) and carbon (carbon or graphite tubes and rods) which permit working temperatures up to 1450°C. and some even over 2000°C. Platinum and platinum alloys have been used in small and medium-sized laboratory furnaces working at temperatures up to 1700°C. Owing to high cost, relatively high vapor pressure at high temperature and reactivity toward certain impurities such as silicon, the metals of the platinum group are practically out of the question for larger industrial furnaces. Therefore only the following metallic materials for high temperature furnaces remain—molybdenum, tungsten, and tantalum. The main requirements are high electrical resistance and stability toward the furnace atmosphere combined with satisfactory mechanical properties.

MOLYBDENUM

Because of reaction between molybdenum and ceramic furnace parts at very high temperatures, destruction of the element is very common. By allowing the elements to hang freely or by supporting them at points only on suitable intermediary materials, this damage through reaction can be avoided to a large extent.

The high melting point of molybdenum (2360°C) permits continuous operating temperatures of up to 1900°C; only then does the evaporation of molybdenum become at all serious. In high vacuum (10^{-4} torr) account must be taken of the evaporation of molybdenum, which becomes considerable above 1600°C. It is therefore better to use tantalum or tungsten in high vacuum furnaces working above 1700°C.

TUNGSTEN

Tungsten, with a melting point of 3410°C, provides a heating element for the highest working temperatures. In most applications the working temperature of the metal may reach 2000°C and may even reach 2500°C in furnaces with tungsten heating elements, provided that no contact is permitted between the metal and ceramics, that is, in so-called "ceramic-free construction." Because tungsten is much more difficult to work than molybdenum or tantalum, it is used most often in the form of round rods, sheet strips, or even better, roughly swaged tubes as heating elements. Tungsten, like molybdenum, is used in a vacuum of less than 10^{-2} torr. The evaporation rate is very low even at high temperatures, and brittleness can be kept down by additives that prevent recrystallization.

TANTALUM

Tantalum, with its high melting point of 3030°C, can also be heated to a high temperature and the evaporation rate is low (less than that of molybdenum and greater than that of tungsten). Columbium has high temperature properties that equal those of tantalum. At present its low melting point is against its use for heating elements.

Radiation-Heat Shields

Radiant-heat shields have been fabricated of stainless steel, molybdenum, tantalum, tungsten, or combinations of these metals. The metal used depends on the operating temperature range and the presence of partial pressure of certain atmospheres. (See Figure 8.6.) Usually the radiation shields are arranged in packs. Each package assembly consists usually of shields secured together with pins. These assemblies are usually mounted from the walls of the vacuum chamber on brackets.

The entire hot zone, consisting of heaters, radiation shields, and hearth, is surrounded by a water-cooled chamber. This chamber has a stainless steel interior to maintain a clean gas-free surface. Full water jacketing on top, bottom, and sides assures rapid cooling, which is increased when gaseous convection and conduction are restored by backfilling with an inert gas.

Power supplies are conventional low-output voltage transformers. Stepless controls are usually employed to take full advantage of the rapid temperature response of the system. In addition, it is necessary to point up the importance of the availability of well-qualified and thoroughly skilled technicians who must operate these new cold wall furnaces. Accurate temperature measurements are usually taken by thermocouples (platinum-rhodium and tungsten-rhenium) or by optical methods such as Rayotube. The thermocouple method, although it employs ceramic insulators that surround the thermocouple wire, is usually preferred. The drawback when using optical methods such as a Rayotube or any other similar unit for temperature indication is condensation. It is extremely difficult to prevent condensation from occurring on the sight glass that is used for Rayotube viewing.

The thermocouple insulators are either thoria (ThO_2) or beryllia (BeO) when temperatures exceed 2500°F. If thoria is used, a very minor danger is radioactivity. X-ray badges need not be worn; however, it is wise to always wash one's hands after handling. If beryllia insulators are used, care and precaution must be taken because of the toxicity of beryllia powders in very fine form. One way is to avoid chipping or grinding insulators to fine granular form so that the powder becomes fine enough to enter the respiratory tract. This can be harmful, however, a considerable amount of powder must be inhaled to cause any trouble. Special attention must be taken to keep thoria and beryllia insulators in proper containers in order to avoid any mixup with ordinary lower temperature mullite insulators.

A specialized and novel vacuum furnace was recently built and installed and used exclusively for honeycomb panels. This vacuum furnace, largest of the type in use, has a graphite base plate. Basically, it is a bell jar

Figure 8.7 Exposed view of furnace and graphite-base for specialized vacuum brazing.

that opens horizontally. The significant and attractive feature of this furnace is its self-contained inner retort which is sealed by rubber O-rings, in place of welding.

The retort is closed by cold-rolled steel cover sheets, top and bottom of the picture frame as shown in Figure 8.7. Heat is applied by silicon-carbide Globar resistance elements mounted in the top and bottom multiple-electrode carriers. The cover sheets are sealed firmly around the work by the downward pressure of hydraulic cylinders around the perimeter of the working zone. This creates the envelope around the work.

The space between the two cover plates is evacuated during heating to a pressure of less than 0.5 μ. A differential pressure of 3 cm is maintained automatically outside the working zone and provides a uniform bonding pressure against the work while it is being heated. No mechanical pressure is applied directly to the panel during brazing.

Another technique has been the application of electron beam that has been used by various companies for brazing through the use of coating difficult-to-braze metals.

Tooling

When putting assemblies through the furnace to be vacuum brazed, the relationship of the parts must be maintained, of course, from start to finish. There are a number of ways this can be done. When an assembly is complicated, it might be necessary to resort to a combination of several different methods.

The method of holding the assemblies together, the design of the joints, and the means of applying the brazing metal are all very closely related. Because the success of a furnace-brazing job depends largely on these fundamentals, some practical hints or reminders follow. Undoubtedly, some of the discussion may be considered quite elementary and perhaps rather simple. It is this very simplicity, however, that lends a certain fascination to the vacuum furnace-brazing process and makes it so easy to use in production.

It is possible to design the parts so that they may be assembled and brazed without the use of fixtures. In other instances it may be necessary to provide clamps and supports to ensure that the parts are held in correct alignment.

One of the most important things to remember in furnace brazing is the force of gravity. Assemblies have a tendency to fall apart after they become heated and the joints become loosened because of expansion. Also the brazing-metal naturally tends to flow downward more than in any other direction. It creeps horizontally or upward on the surfaces of the metal, but it flows downward quite freely and collects at low spots if applied in excess quantity. Therefore in designing an assembly for vacuum furnace brazing, one must keep in mind how the assembly will be held together within the furnace and how it will be set up in the furnace both to direct the flow of the brazing metal into the joints to the best advantage and to give minimum distortion or movement of the parts. These points are generally easy to determine by cut-and-try methods and when a proper procedure is found, the brazing metal can be made to flow into all joints leaving neat fillets and clean surrounding surfaces; the job can usually be done without distortion.

Some of the various methods of holding assemblies together within the furnace follow.

LAYING PARTS TOGETHER

Perhaps the simplest method of joining two parts is simply to lay one on top of the other with brazing metal either in between the members or wrapped around one of the members near the joint. In this case the weight of the upper member would be required to assure good metal-to-metal contact or a weight could be added to assure it. The scheme sometimes lacks the advantage of having a definite means of indexing or keeping the parts from moving in relationship to one another.

Figure 8.8 Vacuum brazed ordnance projectile.

PRESSING PARTS TOGETHER

The most common method of assembling parts for vacuum furnace brazing is simply to press them together. In general, regardless of the degree of tightness, some scheme is usually employed in order to prevent slippage of the parts when they become heated up in the furnace, particularly if the joint has a vertical axis. Figure 8.8 has a shoulder formed on one member to accomplish this stability.

SNUG FIT DESIRABLE

In pressing parts together the usual tolerances from machining the parts naturally give variations in the amount of press fit which cannot be avoided. An effort should be made, however, to have a snug fit at all times if possible. Sometimes a heavy press fit results in distortion of the parts due to stretching them beyond their elastic limit when hot; resulting effects are opened-up or weak joints. When brazing with lower-melting alloys, particularly on nonferrous assemblies, it is desirable to have a clearance of 0.001 to 0.003 in. within the joint to permit the best flow of the brazing metal.

SPOT WELDING AND TACK WELDING PARTS TOGETHER

Spot welding is frequently employed for holding definite relationships between the parts assembled for vacuum furnace brazing. It is fast and inexpensive and the job generally a neat one.

AUXILIARY FIXTURES

All of the foregoing methods of holding assemblies together are used for vacuum furnace brazing, the choice of any method depending on the characteristics of each individual product. In some instances, however, it is found impracticable to use any of the suggested methods and it is then necessary to resort to auxiliary fixtures to locate properly the members with respect to one another while being furnace brazed. These fixtures sometimes take the form of graphite blocks, heat-resisting superalloy and refractory alloy supports, or clamps.

Auxiliary fixtures have several disadvantages. First, they present additional mass that must be heated up, are subject to warpage that might make them unusable, and they present an extra item of maintenance expense. However, two examples where auxiliary fixtures have been used to advantage are as follows.

Blocks. This type of tooling consisting of weighted blocks has been successfully utilized to produce refractory metal components. The tools are usually of the same material as the parts being brazed, thus minimizing any thermal expansion problems.

Figure 8.9 Tungsten pellets used as tooling material for brazing.

Pellets. This new approach to tooling has resulted in a quantum step to remove the distortion problems in elevated temperature brazing (1600°F to 4500°F) and produce a fluid-type pressure over the assembly to be brazed. Thus a control over the mass of the fixturing, weight of the fixturing, and continuous use of the fixturing without any thermal warpage has been attained. An example of this type of tooling is shown in Figure 8.9.

Therefore when selecting fixtures for assembling parts for vacuum brazing, the following factors should be considered:

1. The mass of the fixture should be kept to the minimum that will adequately accomplish its purpose. The fixture should be designed to provide minimum interference with even heating of the parts by removing heat by conduction from the brazing area. It is also important that the fixture not hamper the flow of the brazing filler metal.
2. Vacuum is a determining factor in the selection of the material to be used in the fixture, these materials must withstand the temperatures involved without being appreciably weakened, distorted, or vaporized.
3. Consideration should be given to the expansion and contraction of the fixture in relation to the parts being brazed to assure a combination that will maintain the proper joint clearance and alignment at the brazing temperature. Therefore the coefficient of expansion of the fixture material as well as of the parts should be considered.
4. The fixture should be made of material that will not readily alloy at elevated temperatures with the material of the assembled parts. The fixture should be so designed that it will not contact the brazing filler metal.

There are certain other requirements for brazing fixtures that are peculiar to the vacuum brazing process. In vacuum brazing fixture materials should be selected that will not expel gas or otherwise contaminate the inner furnace environment. Whereas graphite and various steels and superalloy heat-resisting materials are satisfactory up to 2000°F brazing temperatures, the tooling material for use at 3000°F is limited. Refractory metals and/or their alloys, ceramics, graphite, newly developed carbides or borides, and refractory-coated graphite are the only materials available in the 3000°F and above furnace temperature range.

Of the materials discussed, ceramics such as alumina, zirconia, and beryllia cannot be used in vacuum atmospheres due to contamination, reaction with refractory metals, and thermal expansion and contraction. Graphite is another unsatisfactory material. It has excellent dimensional stability; however, as shown by Table 8.5 it embrittles the refractory metals that are usually the heating element material of the furnace. The newly developed carbides and borides (ZrB, ZrC, TiC) are still in the developmental

Types of Equipment and Tooling

Table 8.5 Refractory Metals and Their Reactions in Vacuum [10^{-4}] Torr, with Refractory Materials.

Material	Molybdenum	Tungsten	Tantalum	Columbium
Graphite	Strong carbide formation beyond 1,2000°C.	Strong carbide formation beyond 1,400°C.	Strong carbide formation beyond 1,000°C.	Strong carbide formation beyond 1,000°C.
Al_2O_3	Up to 1,900°C.	Up to 1,900°C.	Up to 1,900°C.	Up to 1,900°C.
BeO	Up to 1,900°C.	Up to 2,000°C.*	Up to 1,600°C.	Up to 1,600°C.
MgO	Up to 1,800°C.	Up to 2,000°C.* (Strong magnesia evaporation)	Up to 1,800°C.	Up to 1,800°C.
ZrO_2	Up to 1,900°C.* (Strong molybdenum evaporation)	Up to 1,600°C.*	Up to 1,600°C.	Up to 1,600°C.
ThO_2	Up to 1,900°C.*	Up to 2,200°C.*	Up to 1,900°C.	Up to 1,900°C.
Sillimanite	Up to about 1,700°C.	Up to about 1,700°C.	Up to about 1,600°C.	Up to about 1,600°C.
Firebrick	Up to about 1,200°C.	Up to about 1,200°C.	Up to about 1,200°C.	Up to about 1,200°C.
Magnesite Brick	Up to about 1,600°C.	Up to about 1,600°C.	Up to about 1,500°C.	Up to about 1,500°C.

stages and are therefore in limited use. This leaves only the refractory metals and/or their alloys.

Stop Off and Parting Agents

Frequently, it is necessary to prevent the filler metal from wetting portions of assemblies, fixtures, and metallic supports. It is customary to turn to refractory oxides: levigated alumina, magnesium oxide, magnesium hydroxide, and titanium dioxide. They are used as extremely fine powders suspended in alcohol, lacquer, acryloid cement, or water. Slurries usually are brushed on.

Although these coatings can prevent wetting by filler metals on portions of assemblies that contact the coated surfaces, they will not necessarily form barriers for creep of filler metals on the assemblies. Filler metals often creep beneath the coatings. This problem becomes more acute when brazing in vacuum. Because many stop-off materials are oxides, the vacuum does remove the oxide, leaving a slight residue that is ineffective as a stop-off material. Some of the most recent work has been done on parting agent materials, zirconium oxide. When zirconium oxide (ZrO_2) is mixed with a suitable nitrocellulose lacquer, it produces an effective parting agent for brazing in vacuum above 1600°F. Since ZrO_2 is only a parting agent and not a stop-off, usually thin separator sheets are used between the braze tools and the assembly to be brazed. The surfaces in contact with the assembly must be coated.

In certain types of work it is necessary to confine the flow of brazing alloy to definite areas. It may sometimes be accomplished by controlling the amount of brazing filler metal used and its placement in the assembly.

It is quite evident that the current trend in the brazing equipment industry is leaning toward cold-wall furnaces. Tooling is available for all temperature ranges as well as some stop-off materials. Numerous cold-wall vacuum furnaces have been built in sizes ranging from a few cubic inches to several hundred cubic feet. The cold wall permits constructing units of virtually unlimited size. The initial costs are high, but the reliability and quality of the product meet the stringent requirements intended for the end use. More on this subject follows in the next section.

ADVANTAGES–DISADVANTAGES–ECONOMICS

The brazing process itself has distinct advantages over other joining media. When metals can be joined without melting them, the problems caused by high heat are eliminated or reduced to a minimum. Parts do not tend to warp or burn because the application of heat is general. The entire joint area is evenly heated to the flow point of the brazing alloy. In welding the heat is intense and localized, applied directly to the filler metal and the joint.

Furthermore, almost any metal can be formed in a single assembly by simply selecting a brazing filler alloy that melts at a temperature lower than that of the metals to be joined. Assemblies incorporating steel, copper, and brass, for example, are extremely difficult to weld, but they are commonplace in brazing.

This ability to join dissimilar metals permits the use of the metals best suited for their functions in the assembly. Some further advantages of the vacuum brazing process are increased strength of brazed joints, increased wetting and flow of brazing alloy, bright surfaces that dispense with expensive postcleaning operations, removal of unwanted gases, minimum interference with basic properties of material, improved corrosion resistance that is essential in many industries, such as food, chemical, and drug processing as well as in most laboratory, kitchen, and hospital equipment. Many metals are easy prey for harmful gases, such as oxygen and nitrogen, which is found in the cleanest of furnace brazing atmospheres. Oxides and nitrides form on the parts being brazed and these films prevent a good braze. Vacuum eliminates these films.

The way a metal brazes in a vacuum is affected by surface reactions and the pressure of the gaseous reactants. Elements of commercial alloys form compounds that affect vapor pressure as well as the surface oxides. At high temperature a vacuum drives surface reactions by reducing the pressure of the gaseous elements.

The microstructure of an alloy's surface layer also affects the way alloy constituents vaporize. A recent metallurgical study of oxide behavior on pure chromium and Type 430 stainless steel found that oxide films can be removed by vacuum thermal treatment.

Vacuum brazing is economical because it is adaptable. If there are a few parts to be joined, it is economical to use brazing. If there are hundreds of thousands of parts to join, it is economical to use vacuum brazing, with modern mass production methods. This adaptability is the reason vacuum brazing is becoming one of the most widely used of metal-joining processes.

Why is vacuum brazing so adaptable? Because it is simple. To braze metal parts these three elements are needed: a brazing alloy to fill the joint, vacuum to prevent oxidation, and a way of getting heat to the parts.

MATERIALS

The advancements in the field of brazing during the past ten years are impressive. The need to fabricate complex assemblies for service at elevated temperatures along with the need to develop brazing filler metals that are compatible with the base metals being used has changed the nature of the brazing process itself. The introduction of vacuum into the brazing process used with many metals, nonmetals, and alloys previously brazed with other protective atmospheres and heating methods has been a part of this change. In fact, it is generally recognized today that brazing has become a science. This concept is reflected by the growing number of research programs directed toward a fundamental understanding of brazing. Much of the credit for this intense interest belongs to the aircraft and missile manufacturers, to firms engaged in developing improved base metals and filler metals, and to the government agencies that have stimulated, sponsored, and guided research in this area of joining.

Because of the ever-changing requirements in all fields and new data constantly being generated, this section attempts to cover as many of the numerous new metals and alloys as well as the standard ones in table form. Many of the metals listed can be joined by more than one braze filler dependent on the end use (temperature and function) and when applicable, this will be illustrated.

Beryllium and Its Alloys

Beryllium is not considered to be easily brazed since beryllium's chemical activity requires that care be taken to prevent contamination and oxidation of joints. Brazeability is affected by process as well as the alloy used. Problems in brazing beryllium increase with increasing temperatures; those not familiar with beryllium should choose the brazing alloy with the lowest brazing temperature and suitable mechanical properties.

Several commercially available braze alloys have been used to braze beryllium in vacuum as well as several newly developed braze alloys. Vacuum furnace brazing has been used to join relatively simple shapes. Aluminum, silver, the silver-aluminum (28% Al) eutectic, and silver-copper eutectic (28% Cu) have been found to be some of the best filler metals for joining beryllium to itself. In addition, various investigators have concluded that aluminum-12% silicon, titanium-6% beryllium, titanium-copper-indium, silver-15 wt% lead-20 wt% copper, silver-5 wt% lead, and copper-18 wt% lead also successfully vacuum braze beryllium joints.

Normal brazing times cannot be used when silver or silver-copper eutectic alloys are used. The joint must be held at brazing temperature only long enough for the brazing material to melt and flow properly. Satisfactory brazes have been made in the temperature range of 1435 to 1650°F, although the lower temperatures are recommended to lessen the rate of diffusion between the beryllium and the brazing alloy and the formation of weak, brittle intermetallics (beryllides). Even with a vacuum of better than 1 torr of mercury, some oxidation of the beryllium surface occurs [10]. However, the silver-copper eutectic alloy is able to wet the base metal.

Figure 8.10 is a beryllium-to-beryllium joint made with the silver-copper eutectic alloy. Voids can be noted in the beryllium adjacent to the brazing metal. Apparently, these are caused by diffusion of beryllium into the braze. There is a possibility that these voids may be caused by some migration of the braze similar to that which occurs when aluminum is used as the brazing material. Except for applications involving very short term brazing cycles, the eutectic composition is not recommended. The sterling composition containing less than 8% copper and 93% silver is less sensitive to time at the brazing temperature and is sometimes used to limit diffusion of silver and recrystallization of beryllium.

Generally, the sterling composition exhibits more desirable wetting characteristics than pure silver although joint strength is lowered by prolonged brazing cycles.

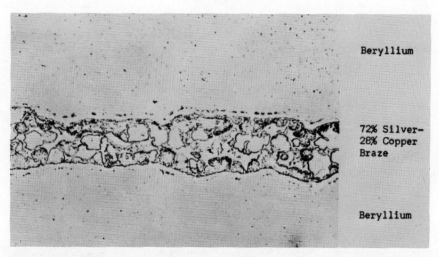

Figure 8.10 Beryllium-to-beryllium braze made with silver-copper eutectic alloy.

A critical factor when brazing beryllium with aluminum is the time at brazing temperature. If the joint is held too long at temperature, the braze material migrates away from the joint. Voids then form both at the interface and in the beryllium-base material. Higher brazing temperatures increase the rate of migration. Holding the joint at temperature for excessive periods of time causes complete failure of the joint due to the formation of voids [11]. However, recently preliminary studies [12] utilizing vacuum evaporation and deposition of titanium vapor were found to be very effective in promoting wetting on beryllium. The titanium evaporation treatment drastically changes and promotes extensive spreading and wetting of the beryllium by the aluminum. The ultimate strength obtained at room temperature for joints butt brazed with commercially pure aluminum compares very favorably with that for the beryllium base metal. Some of these results are discussed later in the chapter. Beryllium has been electroplated with silver and/or copper and successfully brazed with the pure silver and silver-copper alloys.

In addition to the use of vapor deposition and plating beryllium, a technique of applying a continuous strip of titanium in the joint area has also promoted the flow of the braze alloy.

Aluminum and Its Alloys

All previous means (torch, furnace, and dip) of brazing aluminum and its alloys has required the use of fluxes. Within the last two years vacuum brazing of aluminum and several alloys has been developed. The braze alloy primarily used has been 88% aluminum-12% silicon and the following alloys have been joined:

1. 6061
2. 1100
3. 3003
4. 2219
5. X7005
6. 6071
7. 6062
8. 6066
9. 6070
10. 356 Casting

All brazing has been conducted at 1090°F temperature and 1×10^{-6} torr. A typical vacuum brazed joint in 6061 aluminum is shown in Figure 8.8.

Columbium and Its Alloys

The refractory metals family in which columbium is a member has attracted considerable interest because of their high melting points and excellent high temperature strength. Alloys of the refractory metals are continually being evaluated with new braze alloy combinations. In selecting new elevated temperature braze alloys (2000 to 4000°F) consideration must not only be given to atmosphere purity and recrystallization behavior of the refractory metals but also to the effects of the braze alloy on the base metal. An example of a braze alloy that meets these criteria is shown in Figure 8.11; the braze alloy does not cause severe degradation of the base metal through alloying, formation of brittle constituents, and erosion.

D-36 (10 Ti-5Zr-Cb) alloy. Many braze alloys have been evaluated for use with D-36 columbium. In Figure 8.12 D-36 was brazed with palladium at 2900°F. Here, alloying and erosion of the base material are most severe. Figures 8.11 and 8.12 show rather significantly that what may be an acceptable filler alloy for one base metal may not be acceptable for another, although the alloy is in the same base metal system.

The most successful braze alloy used to date on D-36 has been B-120 VCA, a chromium-vanadium-aluminum-titanium alloy that flows at 2950°F. Analysis of the microstructure of brazed joints shows metallurgical bonding along the braze alloy-base metal interface and complete alloying of the braze alloy especially with honeycomb foils. The joint is very ductile and can be bent double without fracture.

Recently consideration has been given to brazing systems in which higher remelt temperatures can be obtained by modifying the brazement chemistry subsequent to brazing. The concepts for higher remelt temperature alloys are not new, but only a limited amount of the work to date has been relevant to columbium alloys. Three higher remelt temperature techniques have been considered.

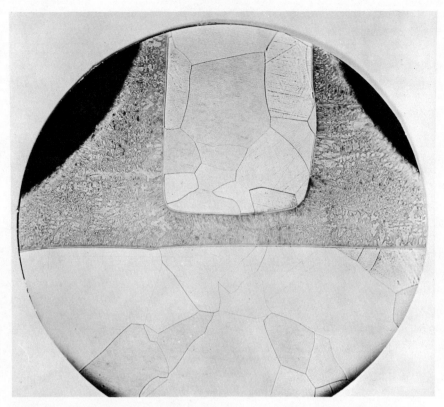

Figure 8.11 Columbium brazed at 2900°F with palladium. No erosion of the base metal is indicated by a very narrow alloy zone.

VOLATILIZATION

Volatilization is a two-step process [14] in which elements with high vapor pressure, such as lithium, phosphorus, and indium, are added to the basic filler alloy to depress the liquidus temperature. The initial braze is made in an inert atmosphere at atmospheric pressure to retain the volatile constituent until melting. The brazement is then subjected to low pressures (vacuum) where removal of the liquidus depressant by volatilization is and the remelt temperature is accordingly increased. Increases in melt temperature on the order of 800°F have been realized.

EXOTHERMIC REACTION BRAZING

This is a process [15] in which both the heat for brazing and the braze alloy are produced by the exothermic reduction of compounds of the alloy

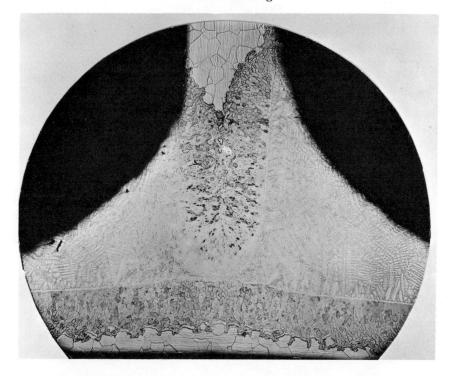

Figure 8.12 Photomicrograph of a D36 columbium alloy brazed with palladium at 2900°F. Severe alloying has occurred between the base and filler material.

elements by a more reactive element. Suitable systems provide brazements with considerably higher melting temperatures than those of the original compounds in the system.

DIFFUSION ALLOYING

In this process elements such as silicon and boron are added to the basic filler alloy to depress the liquidus temperature (e.g., by the formation of a low melting eutectic) and the melting point is then raised by altering the chemistry of the brazement by diffusion alloying [16,17]. For example, platinum-3.5% boron melts at a temperature approximately 1500°F below that of platinum. Braze-remelt temperatures as high as 3800°F have been reported for this system in the brazing of tungsten [16].

As a result of high remelt temperature brazing techniques based on diffusion sink and reactive brazing concepts, two new alloys have been developed for D-36 columbium. One is a Ti-33Cr alloy employed for diffusion sink brazing requiring a brazing temperature of 2650 to 2700°F followed by a diffusion treatment of 16 hr at 2400°F. A comparison of as-brazed and diffusion-treated Ti-33Cr specimens has shown that diffusion treatment raised remelt temperatures from approximately 2900 to 3500°F, essentially doubled lap-shear strength at 2500°F, and provided excellent strength at 3000°F.

The other alloy is a Ti-30 V-4Be reactive braze alloy that required a brazing temperature of 2350 to 2400°F followed by a diffusion treatment of 4 hr at 2050°F plus 16 hr at 2400°F. These processing temperatures were 600 to 900°F below those required for conventional brazing approaches.

F-48 (15 W + 5 Mo + 1 Zr + Cb) alloy. Two brazing alloys having good metallurgical compatibility and high temperature strength were recently evaluated on F-48 columbium alloy. The braze alloys (60 V-30 Cb-10 Ti and 60 V-30 Cb-10 Zr) showed excellent service at 2500°F and F-48 was brazed at 3200°F.

FS-82 (33 Ta + 0.8 Zr + Cb) alloy. Pure titanium used as a braze alloy (3300°F braze temperature) was found to be the best braze alloy for FS-82.

D-43 (10 W + 1 Zr + 0.1 C + Cb) alloy, Cb-752 (10 W + 2.5 Zr + Cb) alloy, C 129Y (10 W + 11 Hf + 0.07 Y + Cb) alloy. Two braze alloys, B 120VCA and Ti-8.5% Si, have been used successfully with these columbium alloys. Ti-8.5% Si is illustrated in Figure 8.13.

Figure 8.13 Cb-752 columbium honeycomb sandwich panel brazed with titanium-8.5% silicon at 2650°F.

Magnesium and Its Alloys

Until 1966 no work had ever been reported on the vacuum brazing of magnesium or its alloys. The author has conducted a preliminary wettability program on a magnesium-zinc-rare earth alloy (ZE 10) and magnesium-lithium alloy (LA 141A) using AZ 125 as the braze alloy. Tests were conducted on bare material as well as silver-plated specimens. The silver-plated specimens showed wetting at brazing temperatures of 1080 to 1100°F. Considerable work is still necessary before magnesium can be considered brazeable by vacuum techniques.

Molybdenum and Its Alloys

There are three basic limits to brazing molybdenum and its alloys for use above 1800°F. The three metallurgical problems are the following:

(1) recrystallization of molybdenum,
(2) formation of intermetallics between refractory metal and brazing filler metals shown in Figure 8.14, and
(3) relative weakness of brazing filler materials at elevated temperatures.

Recrystallization is a process in which the distorted grain structure of cold-worked metals is replaced by a new, strain-free grain structure during annealing above a specific minimum temperature.

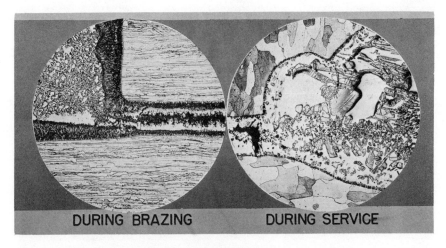

Figure 8.14 The problem of intermetallics is shown very clearly by the erosion during service of the brazed joint on the right [20].

Materials

The deleterious effects of recrystallization on the ductile-brittle transition temperature and the general mechanical properties of molybdenum and its alloys are well known. This condition also occurs with tungsten and its alloys. Recrystallization of molybdenum as a consequence of brazing can occur two ways. If the brazing temperature and time are in a range that will cause molybdenum recrystallization, this will occur grossly and completely during brazing. The ductile-brittle transition temperature is the temperature of transformation from one solid crystalline form of a substance to another, or more broadly, the point at which different phases can exist in equilibrium.

The formation of intermetallics between molybdenum and brazing filler metals is harmful due to the brittleness of the intermetallics that may fracture at relatively low loads when the joint is stressed. In the same way that recrystallization may occur either during brazing or elevated temperature service, intermetallics forms under either condition.

The relative weakness of filler materials at elevated temperatures poses a basic limitation for using brazed molybdenum assemblies. Most of the nickel base elevated temperature service filler materials melt in the range 1800 to 2100°F, where the superior elevated temperature strength of molybdenum begins to manifest itself. Special high melting point filler materials for molybdenum and its alloys are clearly indicated, provided of course, that they do not cause recrystallization or form intermetallics with the metal.

Very few braze alloys possess useful strength at 1800°F. The best filler materials as listed below all contain elements forming intermetallic compounds with molybdenum.

	Composition	Liquidus, °F
Pure nickel	100 Ni	2650
Pure platinum	100 Pt	3300
Monel	70 Ni-30 Cu	2370
Copper	100 Cu	1981
84 Ni-16 Ti	84 Ni-16 Ti	2350
52 Cb-48 Ni	52 Cb-48 Ni	2175
Inconel	80 Ni-14 Cr-6 Fe	2540
L-605	55 Co-20 Cr-15 Mo-10 Ni	2600
Palladium-copper	60 Pd-40 Cu	2200
Palladium-nickel	53 Pd-47 Ni	2200
Palladium-silver	60 Pd-40 Ag	2500

Molybdenum + 0.5 titanium alloy. The same problems discussed above for pure molybdenum exist with Mo-0.5% titanium alloy and are shown in Figure 8.15.

TZM (0.5 Ti + 0.07 Zr + Mo) alloy. A concentrated effort was made in the early 1960s and continues today to solve the problems presented

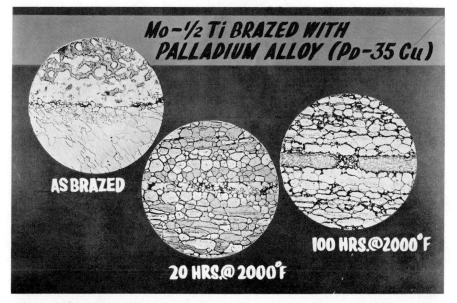

Figure 8.15 The problem of recrystallization is depicted in the photomicrograph of braze joint exposure [20].

and make molybdenum and its alloys useful above 2000°F. The use of a diffusion sink alloy, Ti-8.5% Si, that melts at approximately 2425°F, about 600°F below the melting point of pure titanium, was developed. Brazements exhibited excellent filleting and wetting and were ductile and free from cracks. Other studies were conducted to determine if the remelt temperature of Ti-8.5% Si brazements could be increased to 3000°F or higher by diffusion heat treatment. Specimens with molybdenum powder added to the brazing alloy powder were brazed at 2550°F.

Figure 8.16 shows a TZM T-joint brazed at 2550°F with molybdenum powder added to the Ti-8.5% Si filler alloy. Very little alloying of the filler with the base material and molybdenum powder occurred. The matrix is of eutectic composition with a dispersion of molybdenum powder.

Other work in this same area has shown that other alloys can be used. Currently being developed is a Ti-25 Cr-13 Ni braze alloy (braze temperature 2300°F) that has produced the highest remelt temperatures on TZM. Tee and lap joint remelt temperatures were above 3100°F as shown in Table 8.6.

Steels–Stainless–Precipitation Hardened

Practically every braze alloy that has been manufactured or recently developed for use above 1600°F will wet steel in vacuum. Since the list

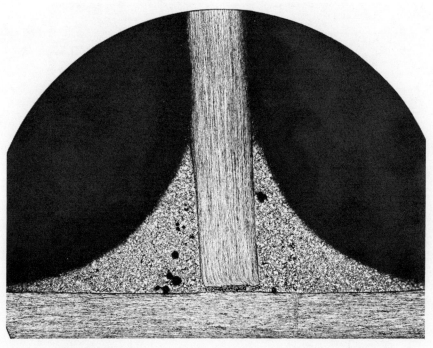

Figure 8.16 TZM T-Joint-Titanium-8.5% silicon [21]. (75×).

of steels and alloys are too numerous, a few examples are briefly discussed. No major problems to date have been encountered in using the high temperature silver, gold, palladium, copper, and nickel-chrome base alloy systems in vacuum brazing steels. Depending on the application and temperature environment, one of these brazing alloy systems has been adequate and suitable for the job requirements.

A 54% Pd, 36% Ni, and 10% Cr braze alloy has produced ductile joints on Type 316 stainless steel after they were exposed to air for 500 hr at 1800°F plus an additional 90 hr at 2000°F. A 52.5% copper, 9.5% nickel alloy has successfully brazed type 304 stainless steel and tests have shown the following:

	Nominal Diameter of Rod Tested	Average (psi) (5 joints)	Max. (psi)	Min. (psi)
Room Temperature	0.250 in.	73,300	85,400	58,400
800°F	0.250 in.	55,600	58,600	47,200

Table 8.6 TZM Lap Joint Remelt Temperatures and Shear Strengths [18].

TZM LAP JOINT REMELT TEMPERATURES AND SHEAR STRENGTHS

REMELT TESTS

SAMPLE NO.	BRAZE ALLOY	*BRAZE ALLOY PLACEMENT	APPROX. JOINT THICKNESS	DIFFUSION TREATMENT	REMELT TEMP.-F	REMELT TEMP. MINUS BRAZE ALLOY LIQUIDUS-F
ML-19	33Zr-34Ti-33V	A	.003 inches	2050F, 20 hrs.	2320	-160
ML-9	Ti-25Cr-10Ni	A	.0016	2050F, 20 hrs.	2440	120
ML-34	Ti-25Cr-13Ni	A	.002	1900F, 16 hrs. + 1950F, 8 hrs. + 2000F,	< 2600	< 500
ML-35	Ti-25Cr-13Ni	A	.002		< 2600	< 400
ML-49	Ti-25Cr-13Ni	A	.0016	40 hrs. + 2050F, 56 hrs.	2680	480
ML-50	Ti-25Cr-13Ni	A	.0016		2770	570
ML-41	Ti-25Cr-13Ni	B	.001		3470	1270
ML-55	Ti-25Cr-13Ni	C	.001		3220	1020
ML-46	Ti-25Cr-13Ni	A	.0016	Same as above + repeat	3780	1580
ML-39	Ti-10Ni-6Si	B	.001	Same as ML-34	3425	1265

SHEAR STRENGTH TESTS

SAMPLE NO.	BRAZE ALLOY	*BRAZE ALLOY PLACEMENT	APPROX. JOINT THICKNESS	DIFFUSION TREATMENT	TEST TEMP.-F	SHEAR STRENGTH	BASE METAL STRESS
ML-57	Ti-25Cr-13Ni	C	.0006 inches	1900F, 16 hr + 1950F, 8 hr + 2000F	2600	830 psi	2500 psi
ML-58	Ti-25Cr-13Ni	C	.0008		3000	310	920
ML-56	Ti-25Cr-13Ni	C	.001	40 hr + 2050F, 56 hrs	3000	270	910

NOTES: All specimens were .040 inches thick with a 3 ± .5 T lap. Remelt tests conducted with 120 psi tension on base metal. Shear strength tests conducted at .05 inches/min. head speed. All specimens failed in braze joint.

* A = powdered braze alloy placed over lap.
  B = particle of braze alloy placed at edge of lap.
  C = powdered braze alloy placed at edge of lap.

Superalloys

The term "superalloy" applies to a large family of alloys that show strengths and oxidation resistance at elevated temperatures that are superior to low alloy steels. The three distinct groups of these alloys based on their major or basic alloying ingredient are the following.

Nickel Base (René 41, Hastelloy R-235, Nimonic Family, Inconel 700 Group, and Others; melting Point 2650°F). Two basic brazing alloy systems encompass most of the alloys with only a few exceptions: precious metal base and nickel base. Only a limited amount of basic new brazing alloy development work has been conducted since 1961. Most of the reported successful results have required some processing gimmick to overcome the inherent characteristics of the available brazing alloys. A review of reported information has shown that the following alloys are the best available: (a) Ni-Mn-Si-Cu, (b) Au-Ni-Cr, (c) Ni-Cr-Mn-Si, (d) Au-Pd-Ni, (e) Mn-Ni-Co-B, (f) Pd-Ni-Cr-Si, (g) Ni-Cr-Si, (h) Cu-Mn-Co-Ni, and (i) Pd-Ni-Cr.

The brazing of age-hardenable nickel-base alloys (René 41, Hastelloy R-235, Inconel 718) presents a different and a much more complex problem than the brazing of other nickel-base alloys (Monel, Inconel, Hastelloy C). In industry where these alloys must be brazed they are called "problem" alloys. The problem arises because they contain aluminum and/or titanium. In general, the magnitude of the problem varies directly with the sum of the amount of these elements present. The difference between lower and higher levels of aluminum and titanium, however, does not warrant concern because it is best to use the brazing procedure that assures good brazing under the worst possible conditions.

Note that the brazing filler metals listed are all high temperature (above 1800°F) and oxidation resistant. This is because they are the only materials compatible with the use to which most age-hardenable nickel-base alloys are subjected. The interactions between base metals and filler metals is applicable also to solid-solution-strengthening and nonaging nickel alloys if these filler metals are used on them [23].

The age-hardening nickel-base alloys have titanium and aluminum oxides on their surfaces. These oxides cause difficulty because they are easily formed on exposure to air or other oxidizing conditions and cannot be reduced by even very dry pure hydrogen atmospheres. The results of great effort in developing brazing techniques for fabricating high-temperature, high-strength, nickel-base alloys into useful parts have shown that the problem alloys can be satisfactorily brazed by vacuum brazing at pressures of 2 μ or lower.

The presence of titanium and/or aluminum in age-hardenable nickel-base alloys is not the only problem encountered when brazing these alloys. The interactions between the base metal and the brazing filler metal at or near brazing temperature must also be recognized. The practical aspects of these interactions are manifest most when it becomes necessary to braze assemblies with relatively thin cross sections or when vibrational loading is anticipated. Penetration of elements such as carbon, boron, and sometimes silicon which reduce the effective thickness of ductile base metal and replace it with nonductile metal compounds is not tolerable. Erosion or dissolution of the base metal is also encountered when brazing nickel-base alloys. This too is intolerable, especially in thin sections.

A great amount of effort has been expended in overcoming the problems arising from the interactions between nickel-base alloys and nickel-base filler metals. Solutions have been found, but none is completely satisfactory in all respects.

Electroplating is one apparent solution, especially since it also aids in overcoming the effects of titanium and aluminum oxides. Electroplating of René 41 with nickel has reduced erosion and grain-boundary penetration [24], but the resultant joints were weaker than joints brazed without plating in a vacuum. Table 8.7 presents the results of one such study. These data might also be used to show that vacuum brazing gives a stronger joint than hydrogen brazing and that the nickel plate is not the cause of the difference. These conclusions have been made and have become the subject of some controversy. There is no fundamental reason why vacuum-brazed joints should be stronger than hydrogen-brazed joints. The differences are caused by an inability to produce the extremely dry hydrogen atmospheres needed for these alloys in ordinary furnaces.

Another means for minimizing the interaction between the base metal and the brazing filler metal involves strict control of the amount of brazing filler metal used and of the brazing temperature cycle. The practice of very carefully controlling these two variables is probably the most widely used method for limiting the undesirable interactions.

Most of the high temperature brazing filler metals are powders, which permit the use of another method to control filler metal base-metal reactions. Pure metal powders, such as nickel, or a powder with the base-metal composition added to the filler metal, inhibits reaction with the base metal. This technique, however, has not been widely used.

Many investigations have been undertaken to develop brazing filler metals that fulfill all other requirements and do not react with the nickel-base metal. A few of these have been successful, but have not been widely accepted because they are high in expensive noble metal (gold and palladium) content or do not always have the desired strength and oxidation

Table 8.7 Brazed Lap Joints in René 41 Sheet. Braze Alloy, AMS4775 (Nickel, Chromium, Boron, Silicon, Iron), Braze Temperature, 2150°F, Braze Time, 10 Min, Overlap, 0.100 In [24].

Test Temperature, F	Cleaned and Vacuum Brazed			Nickel Plated and Hydrogen Brazed		
	Indicated Shear Stress(a), psi	Indicated Tensile Strength, psi	Failure Location	Indicated Shear Stress, psi	Indicated Tensile Strength(b), psi	Failure Location
78	38,000	62,000	PM(c)	31,000	53,000	Braze
78	29,000	45,000	PM	29,000	59,000	Braze
1200	37,000	56,000	PM	26,000	52,000	Braze
1200	37,000	60,000	PM	32,000(a)	58,000	PM
1400	27,000	44,000	PM	24,000	38,000	Braze
1400	28,000	45,000	PM	15,000	28,000	Braze
1600	46,000	74,000	PM	15,000	23,000	Braze
1600	41,000	66,000	PM	12,000	20,000	Braze
1800	22,000	36,000(b)	Braze	5,000	10,000	Braze
1800	22,000	35,000	PM	6,000	9,000	Braze

(a) Shear stress for load at tensile failure.
(b) Tensile stress for load at shear failure.
(c) Parent metal.

resistance. One such filler metal has been used for some time, but was only recently reported in the literature. It may overcome some of the cited disadvantages. Its composition is 33% chromium, 24% palladium, 4% silicon, and the balance nickel.

1. René 41 (20Cr-10Ni-15W-2Fe-Bal Co). The Ni-Cr-Pd-Si alloy has been found to exhibit excellent wetting and flow characteristics on René 41. This alloy has provided the desired brazing temperature in combination with good wetting and flow on René 41 without excessive erosion or reaction with the base metal. The optimum brazing temperature for vacuum brazing René 41 is 2150°F and the alloy has the ability to braze René 41 thickness less than 0.015 in. without erosion or embrittling effects.

2. Inconel 700 group and other Inconels (Ni-Cr-Fe-2Ti-Al). Inconel 718 can be brazed relatively easier than can most nickel-base alloys hardened by the precipitation of aluminum and/or titanium compounds. The lower concentration of these oxide-forming elements permits the practice of brazing procedures similar to those used for the precipitation-hardening stainless steels. Nickel-base and silver-base brazing filler metals have been very successfully used on Inconel 718. Inconel X has been very successfully brazed with an 82% gold-18% nickel alloy that has been proven the strongest braze alloy between room temperature and 800°F, whereas above 800°F, a nickel-silicon-boron alloy (AMS 4778) was equal to or better in tensile strength than the gold-nickel alloy. 33Cr-24Pd-4Si-Rem Ni has also been successful.

3. Hastelloy R-235 (64Ni-16Cr-5.5Mo-8Fe–Ti–1.5Al). This alloy has been successfully brazed in vacuum and results of tests are shown in Table 8.8.

Iron Base (A-286) (Melting Point 2800°F). The Ni-Cr-Si braze alloys are the best for A-286 (26Ni-15Cr-53Fe-2Ti-.17Al). Tables 8.7 and 8.8 reflect comparable results with A-286 for typical room and elevated temperature shear strengths in separate studies and actual production use.

Cobalt Base (L-605 or Haynes 25) (10Ni-19.5 Cr–14.5 Mo–51.5 Co) (Melting point 2720°F). The following braze alloys are the best available for cobalt alloys: (a) Au-Pd-Ni, (b) Ni-Cr-Si, (c) Co-Ni-Cr-Si-W-B, and (d) Pd-Ni-Cr.

This discussion of the brazing of superalloys may seem unduly pessimistic. It is not intended to be. It must be noted that with due cognizance of the characteristics of all the materials involved many complex structures

Table 8.8 "Indicated" Shear Strengths* of Single-Lap Zero-Clearance Joints Brazed in R-235 with Ni-Cr-Si Alloy [38].

Test Temp.	Hydrogen Brazed Nickel Plate psi	Flux psi	Vacuum Brazed psi
Room	38,200	35,100	54,200+
Room	38,900	37,300+	49,400
1200°F	25,800	31,000+	41,100+
1200°F	29,100	30,800	37,800+
1400°F	22,500	34,700	45,000
1400°F	21,100	40,000	45,200+
1600°F	19,400	23,500	26,000
1600°F	18,000	20,000+	27,100
1800°F	9,300	7,700	10,700
1800°F	7,300	10,700+	10,200

All joints were solution treated at 1950°F for 1/2 hour, air cooled and aged 4 hours at 1650°F after brazing

. . . .

* See text for definition
+ "indicated" shear stress at time of base-metal failure

have been built that have withstood very severe service conditions. This discussion has shown that the base metal, the brazing filler metal, and the procedures for brazing must be selected very carefully. A thorough understanding of the physical metallurgy of the base metal-filler metal interactions is necessary. In addition, the effect of brazing thermal cycles on the properties of the superalloy considered must be determined. If this effect is significant, design changes may be necessary.

TD-Nickel and TD-Nickel-Chromium

A new brazing alloy for service with TD-nickel at 2200°F has been developed that shows excellent stability and compatibility with the base metal. The nominal composition is 50Ni-16Cr-25Mo-5W-4Si and is used at 2375°F. Although brazed butt joints are not usually used for brazing, a recent evaluation was made to determine the strength of butt joints made with this alloy, to assess the strength of such joints, and to obtain a comparison of the tensile strengths of three braze alloys used with TD-nickel. The tensile data obtained are given in Table 8.9 [25]. It can be concluded that the data indicate that butt joints made by brazing TD-nickel would have tensile joint strengths approaching 50% of the base-metal strength.

In the vacuum brazing of TD-nickel-chromium two recently developed braze alloys have exhibited excellent wetting characteristics and strength at a 2000°F operational environment. The first alloy is listed in Table 8.9 as TD-6, whereas the second has been designated TD-50 (20Cr-10Si-9Mo-21Fe-2Co and balance Ni). The effect of gap distance on the load-carrying ability of TD-6 brazed joints is shown in Table 8.10.

Table 8.9 Strength of Brazed Butt Joints— 0.050-In. Thick TD-Nickel [25].

Joint Gap - Inches	Alloy	Tensile Strength - Psi		
		TD-5	TD-6	TD-20
.005		7,280	9,620	3,780
.005		7,130	5,500	3,440
.005		-	10,230	-
	Avg.	7,205	8,450	3,610
.015		8,580	4,030	6,760
.015		6,900	3,760	6,470
.015		-	4,930	-
	Avg.	7,720	4,240	6,615
.030		7,390	2,130	7,860
.030		7,970	6,220	8,160
.030		-	1,690	4,950
	Avg.	7,680	3,343	6,990

	Ni	Cr	Mo	W	Si
TD-5	41.5	22	9	-	4
TD-6	58	16	17	5	4
TD-20	50	16	25	5	4

Materials

Table 8.10 Effect of Gap on the Mechanical Properties of TD-6 Brazed Joints—0.060 In. Sheet, 0.120 In. (2T) Overlap at 2000°F in Air [26].

Gap (In.)	Load at Failure (lb)	Indicated Shear Stress (psi)	Mode of Failure	Base-Metal Stress at Failure (psi)
0.005	319	4,960	Base metal	10,600
0.005	283	4,440	Base metal	9,430
0.005	301	4,660	Base metal	10,050
	Average 301	4,680		10,020
0.015	290	4,270	Base metal	9,080
0.015	213.5	3,120	Base metal	6,650
0.015	224	3,100	Base metal	7,250
	242	3,493		7,660
0.030	192	3,010	Base metal	6,030
0.030	199	2,760	Base metal	6,440
0.030	186	2,680	Base metal	5,870
	192	2,817		6,110

Tantalum and Its Alloys

Because most uses of tantalum and its alloys are for elevated temperature applications (3000°F and above), investigations to the present time have examined conventional brazing, reactive brazing, and diffusion sink brazing concepts. The reactive brazing concept uses a braze alloy containing a strong melting temperature depressant. The depressant is selected to react with the base material or powder additions to form a high melting intermetallic compound during a postbraze diffusion treatment. By removing the depressant in this manner, the joint remelt temperature is increased.

Successful application of this concept appears highly dependent on controlling the intermetallic compound reaction to form discrete particles. If continuous intermetallic compound films are present in the grain boundaries or along the base metal-fillet interfaces, joint ductility could be seriously impaired.

Several alloy systems investigated for conventional brazing have been the following:

Cb-vanadium   Ta-boron
Cb-titanium   Cb-boron
Ta-vanadium

Preliminary results of one of these investigations shows that a Cb-1.3 boron braze alloy has excellent potential for conventional brazing of closed cell honeycomb panels in vacuum.

The diffusion sink concept that was discussed under columbium alloys has been applied to tantalum alloys with excellent results. The following metal combinations have been investigated as diffusion sink systems: (a) pure titanium, (b) titanium-vanadium, and (c) zirconium-titanium-vanadium.

Diffusion sink brazing with titanium and Ti-30V (braze temperature 3050 to 3200°F) has produced remelt temperatures exceeding 3800°F on tee and lap joints. Diffusion sink brazing with 33Zr-34Ti-33V (braze temperature 2600°F) has produced remelt temperatures exceeding 3200°F. The remelt temperatures indicate that maximum service temperature for the titanium and Ti-30V alloys could be 3500°F and the 33Zr-34Ti-33V alloy could be 3000°F.

A problem area with most of these alloy systems is that they can cause embrittlement if used to produce brazed core in honeycomb structures. The braze alloy volatilizes within the sealed honeycomb cells, deposits on the foil, and diffuses inward. This lowers the vapor pressure in the cells and additional material volatilizes from the fillets. This embrittlement effect

Figure 8.17 A photomicrograph of a new tantalum alloy (T-222) brazed with a combination of pure titanium and D36 columbium alloy.

seen microscopically with diffusion sink alloys was not found in the Cb-1.3 boron conventional brazing alloy. Shown in Figure 8.17 is T-111 (Ta-8W-2Hf) brazed with a combination of titanium and columbium at 3700°F.

Although the base-metal producers are developing newer alloys to meet varying environments (creep strength at 3000°F, oxidation resistance, etc.) new braze alloy development is also continuing. Currently 90 to 95% of the available alloys are in powder form, which at times is difficult to work with at elevated temperatures. New powder braze alloys such as Hf-7Mo, Hf-40Ta, and Hf-19Ta-2.5Mo are examples of new tantalum braze alloys that are being developed into foil-type braze alloys either by direct rolling of the alloy or pack-diffusion heat treatment of hafnium foil.

Titanium and Its Alloys (Ti-6AL-4V) (Ti-5AL-2.5 Sn) (Ti-8AL-1Mo-1V)

Titanium and its alloys have certain chemical and metallurgical characteristics that are important in brazing operations. An understanding of these characteristics has helped to ensure the selection of suitable brazing filler metals and brazing procedures for use in joining titanium. One characteristic is that titanium readily alloys with most molten brazing filler metals. Because of this characteristic, molten filler metals readily wet and flow on titanium, but there is a strong tendency for excessive alloying to occur in the brazed joint. Excessive alloying, which results in undercutting of the titanium and the formation of brittle intermetallic compounds in the joints, should be avoided because it lowers joint strength and ductility and may embrittle the base metals when thin foils and fine wires are being joined.

Several precautions have been taken to minimize alloying in brazed joints. One is the development of brazing filler metals that alloy only slightly with titanium. Another is to use brazing cycles in which the filler metals which, although they alloy with titanium, do not form brittle intermetallic compounds.

Another characteristic that affects brazing operations for titanium and its alloys is their extreme sensitivity to embrittlement by oxygen, nitrogen and by hydrogen contamination. The presence of oxygen and nitrogen in the brazing atmosphere should be avoided to prevent the formation of brittle-contaminated surfaces on the titanium. The use of vacuum for brazing prevents this embrittlement of the titanium surface.

Vacuum brazing of titanium has been accomplished with silver base alloys as the filler metal. The melting point of pure silver is above 1775°F, which is too high for brazing most of the titanium alloys. Alloys that have been added to silver to lower the melting point of the brazing alloys used on titanium include lithium, cadmium, zinc, copper, aluminum, and tin either as binary or more complex alloys. Most of these elements tend to form brittle phases in titanium alloys or promote other deficiencies such as poor oxidation or corrosion resistance, poor fillet formation, and excessive erosion of the base metal.

The most satisfactory alloy for brazing titanium alloys has been 95Ag-5Al. Others have included 90 Ag-10Sn, 94Ag-5Al-0.2Mn and 56Ag-24Cu-19Ge-1Ti. Depending on the particular end use, problems have been encountered with these braze alloys:

- The formation of continuous TiAg intermetallic films along the joint interface.
- The tendency for severe crevice corrosion on the joint interface (<100 hr) in typical marine and industrial environments.
- The tendency for prolonged braze cycles to result in gradual deterioration of braze fillets because of sweating or flowaway of fillet braze material.

The Ag-Cu-Ge braze alloy had satisfactory flow at 1100°F temperature, but the mechanical properties of joints made with this alloy were poor in comparison with the mechanical properties of similar joints made with Ag-5Al braze alloy that flowed at 1625°F. The poor mechanical properties of the joints made with the Ag-Cu-Ge braze alloy were caused by an intermetallic compound that formed as a thin film at the boundary between the braze alloy and the parent metal. The desired combination of flow characteristics and mechanical properties has not been obtained to date for braze alloys flowing at 1100°F.

Continuous needs require continued developments. Five new alloys have been evaluated for use with titanium alloys listed above.

Composition	Optimum Flow Temperature (°F)
Ti-47.5 Zr-5 Be	1,640
Ti-40.1 Zr-4.75 Be-5 Al	1,700
Ti-43 Zr-12 Ni-2 Be	1,470
Ti-45 Zr-8 Ni-2 Be	1,660
Ti-15 Cu-15 Ni	1,700

The last braze alloy with a flow temperature of 1700°F has shown the most consistently sound and ductile brazed joints.

Materials

One other process noteworthy of discussion is the thin film diffusion brazing process for titanium alloys. This process depends on the formation of small quantities of liquid that behave as a braze alloy. Subsequent diffusion then drastically alters the alloy composition, resulting in a metallurgically compatible titanium joint. The process is a combination of diffusion bonding and conventional brazing. Some of the advantages of this process are low cost braze material, lightweight, extremely high strength, thermal stability, and stress-corrosion resistance. Disadvantages are the close fitup requirements and control of the intermediate content needed.

Tungsten

The brazing of tungsten structures has been done in vacuum very successfully. A wide variety of brazing alloys and pure metals are potentially useful for filler materials having liquidus temperatures ranging from 1200 to 5500°F. (See Table 8.11.)

Table 8.11 Brazing Filler Metals for Tungsten.

Brazing Filler Metal	Liquidus Temp., °F	Brazing Filler Metal	Liquidus Temp., °F
Cb	4380	Co-Cr-Si-Ni	3450
Ta	5425	Co-Cr-W-Ni	2600
Ag	1760	Mo-Ru	3450
Cu	1980	Mo-B	3450
Ni	2650	Cu-Mn	1600
Pd-Mo	2860	Cb-Ni	2175
Pt-Mo	3225		
		Pd-Ag-Mo	2400
Ag-Cu-Zn-Cd-Mo	1145-1295	Pd-Al	2150
Ag-Cu-Zn-Mo	1325-1450	Pd-Ni	2200
Ag-Cu-Mo	1435	Pd-Cu	2200
Ag-Mn	1780	Pd-Ag	2400
		Pd-Fe	2400
Ni-Cr-B	1950	Au-Cu	1625
Ni-Cr-Fe-Si-C	1950	Au-Ni	1740
Ni-Cr-Mo-Mn-Si	2100	Au-Ni-Cr	1900
Ni-Ti	2350		
Ni-Cr-Mo-Fe-W	2380		
Ni-Cu	2460		
Ni-Cr-Fe	2600		
Ni-Cr-Si	2050		
Mn-Ni-Co	1870		

Fiber metal, a fairly new material concept, is finding application as a brazing filler metal. It provides capillary gaps for controlling the flow of filler metal into wide-gap joints and from interdiffusion between dissimilar alloy components of a bimetal fiber shim. The brazing filler metal is a mixture of approximately 90% tungsten particles with 10% of 57Ni-37Fe-6W particles, and brazing is performed in a vacuum furnace at 2650°F for 1 hr. At this temperature the nickel-iron-tungsten alloy becomes molten, forming a liquid metal envelope surrounding each tungsten particle and wetting the tungsten face sheets to be joined. During the 1-hr hold time, nickel and iron diffuse from the liquid into the solid tungsten particles and sheets, and tungsten diffuses the other direction to the liquid. Thus the liquid metal matrix experiences an increase in melting temperature because of its change in the composition toward pure tungsten. If complete homogeneity were to be obtained, the overall composition of the alloy (90W-6Ni-4Fe) would have very desirable service temperature capabilities. The remelt temperature has been found to be in excess of 3900°F, an increase of greater than 1250°F.

Ceramics

Ceramic materials are inherently difficult to wet with conventional brazing filler metals. Most of these filler metals merely ball up at the joint with little or no wetting occurring. When bonding does occur, it can be mechanical or chemical. The strength of a mechanical bond can be attributed to interlocking particles or penetration into surface pores and voids, whereas the chemical bond derives strength from material transfer between the filler metal and the base material.

Another basic problem in brazing ceramics results from the differences in thermal expansion between the base material and the brazing filler metal itself and, with ceramic-to-metal joints, between the two base materials to be joined. In addition, ceramics are poor conductors of heat, which means that it takes them longer to reach equilibrium temperature than it does metals. Both of these factors may result in a cracking problem when trying to make such a joint. Since ceramics generally have low tensile and shear strengths, crack propagation occurs at relatively low stresses when compared with metals. In addition, the low ductilities permit very little distribution of the stresses set up by stress raisers.

In some brazing processes the ceramic such as alumina, zirconia, or forsterite (Mg_2SiO_4) is premetallized to facilitate wetting, and filler metals such as copper, silver-copper, and gold-nickel have been used frequently.

Commercial filler metals, which have been used for brazing ceramics, are silver-copper-clad or nickel-clad titanium wires. (See Figure 8.18.) A

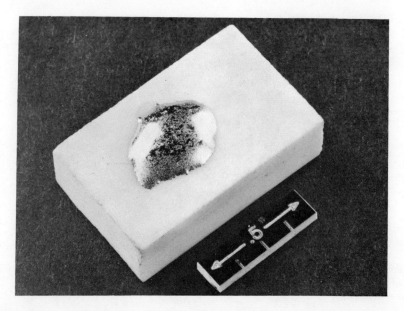

Figure 8.18 Al$_2$O$_3$ samples—ceramic wetted by the nickel-clad titanium filler metal. The filler metal was preplaced as wire and was heated in vacuum to 1050°C for 10 min, a small contact angle is evident [27].

Table 8.12 Wettability Results of Various Brazing Alloys on Al$_2$O$_3$, UO$_2$, BeO, and Graphite.

Brazing alloy, wt %	Brazing temp, °C	Wetting[a] on Al$_2$O$_3$	UO$_2$	BeO	Graphite
Commercial Alloys					
100 Cu	1100	None	None	None	None
Ni – Si – B (AMS 4778)	1040	None	None	None	None
Ni – 10 P	1000	None	None	None	None
82 Au – 18 Ni	1000	None	None	None	None
Nickel-clad titanium	1050	Good	Good	Good	Good
Experimental Alloys					
49 Ti – 49 Cu – 2 Be	1000	Good	Good	Good	Good
68 Ti – 28 Ag – 4 Be	1100	Good	Fair	Fair	Good
48 Ti – 48 Zr – 4 Be	1050	Poor	Poor	Poor	Good
46 Ti – 46 Pd – 6 Al – 2 Be	1150	Fair	Poor	Poor	Good
75 Zr – 23 Cu – 2 Be	1150	Fair	Poor	Fair	Good
95 Zr – 5 Be	1050	Fair	Good	Poor	Good
82 Zr – 6 Ni – 6 Cr – 6 Fe[b]	1150	Fair	Good	Poor	Poor

[a] Good: continuous filletting, extensive spreading; fair: intermittent filleting, little spreading; poor: no flow, wetting only at contact points.
[b] British Patent 890,971.

Figure 8.19 Demonstration compartmented Al_2O_3 assembly vacuum brazed with 49Ti-49Cu-2Be (wt%) experimental alloy; brazing conditions were 980°C for 10 min [27].

considerable amount of alloy development has been under way and the several experimental brazing alloys shown in Table 8.12 for joining refractory metals appear to be promising for brazing ceramics. Titanium and zirconium alloys exhibit especially good flow. An experimental assembly, with compartments cut into a ceramic plate of alumina, has been brazed with a 49Ti-49Cu-2Be alloy (Figure 8.19). The filler flowed well and made a sound, crack-free joint. The brazing alloy was applied in granular form. The Al_2O_3 surfaces were ground flat before brazing and were held together with molybdenum wire during the brazing cycle.

Graphite

The joining problems associated with the brazing of graphite are similar to those encountered with the ceramic materials previously mentioned. For low and intermediate temperature applications a titanium-cored silver-copper alloy is suitable for brazing graphite. One of the newest developmental brazing alloys (48Ti-48Zr-4Be-wt%), shown in Fig. 8.20, has appeared particularly promising for the joining of graphite, for its major constituents are extremely strong carbide formers. Another braze alloy, shown in Figure 8.19 (49Ti-49Cu-2Be-wt%), also has brazed graphite very successfully.

Alloys in the gold-nickel-molybdenum ternary system have been developed and found satisfactory to braze graphite-to-graphite joints. Alloys in this system containing low percentages of molybdenum can be used for graphite-to-molybdenum joints, whereas alloys with greater percentages of this carbide former can be used to braze graphite-to-graphite joints as well.

Alloys in the gold-nickel-tantalum system have also been developed that have satisfactorily flowed on graphite. Compositions suitable for brazing graphite-to-graphite joints sometime exhibited fillet cracking, but compositions containing lower percentages (less than approximately 30 wt%) of tantalum have been satisfactory for brazing graphite-to-metal joints.

The 35 Au, 35 Ni, and 30 Mo (wt%) alloy is especially attractive, for it can be used to braze both graphite-to-graphite (Figure 8.21) and graphite-to-metal joints with high integrity. It is attractive as a general-purpose alloy and can be used to braze both low and high porosity grades of graphite in which the joints have strengths at least as great as that of the graphite.

Dissimilar Metal and/or Ceramic Combinations

The problems that arise from the desire to make dissimilar metal joints depend mainly on the difference in composition among the alloys. If they are similar or are metallurgically compatible over a wide composition range, problems will not be great, assuming that good brazing practice is used.

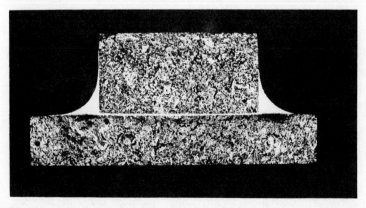

Figure 8.20 Photomicrograph of graphite-to-graphite T-joint brazed with the 48 titanium-48 zirconium-4 beryllium (wt%) alloy. Good filleting and flow of the brazing alloy in the joint can be seen [28].

Figure 8.21 Typical graphite-to-molybdenum and graphite-to-graphite joints brazed with 35 gold-35 nickel-30 molybdenum (wt%). Note the good flow attained [28].

Beryllium to Stainless Steel

Stainless steel tubes have been brazed to beryllium end caps with the experimental alloy of the composition 49Ti-49Cu-2Be (wt%). Tests have indicated that it readily wets beryllium and that the joints exhibit adequate strength for the application, provided that the brazing time is kept sufficiently short to minimize the formation of intermetallic compounds. Pure silver and silver-lithium alloys have also been successfully used to join beryllium to stainless steel.

Beryllium to Titanium

The brazing of beryllium to titanium and 6AL-4V titanium alloy has been evaluated with a 92%Ag-7.2%Cu-0.2%Li alloy which was brazed at a temperature of 1640°F and pure silver which was brazed in the temperature range of 1660 to 1720°F.

Of these two braze alloys the Ag-Cu-Li alloy produced brazes with superior tensile strength up to 1060°F.

Beryllium Oxide to Pyrolytic Graphite

The large differential in thermal expansion between platelets of BeO and pyrolytic graphite in the "X" plane results in high residual stresses in the brazed joint, and these residual stresses are sufficiently large in many instances to crack either or both of the base material components. Joint designs have been developed which prevents cracking in small (up to $\frac{1}{2}$ in. long) joints, but these designs will not prevent cracking of one or the other of the base materials in longer joints. Three brazing alloys, which successfully wet and bond to both BeO and pyrolytic graphite, have been developed: 93Ti-7Ni, 93Ti-7Fe, and 53Ti-47Cr. Ordinary brazing techniques employing vacuum atmosphere are suitable for effecting BeO-to-pyrolytic graphite bonds with the three brazing alloys.

Columbium and Other Refractory Metals

Columbium and its alloys have been successfully joined to cobalt-base alloys (HS 25) with an 82% gold-18% nickel braze alloy, Type 316 stainless steel to the Cb-1%Zr alloy using (21%Ni-21%Cr-8%Si-Rem. Co) braze alloy at 2150°F, and columbium brazed to alumina using zirconium-nickel as braze filler. Tantalum has also been brazed to molybdenum in vacuum at 1770°C using pure zirconium as the braze filler.

Titanium and Its Alloys to Steel, Refractory Metals and Others

Titanium and its alloys have been brazed to carbon steels, austenitic stainless steels, and to reactive and refractory metal alloys with the aforementioned (48Ti-48Zr-4Be alloy). Ternary braze alloys based on the titanium-vanadium-beryllium system, titanium-columbium-beryllium system, titanium-iron-vanadium system have proven satisfactory for brazing columbium to titanium. Brazing temperatures up to 2375°F have been required for these braze alloys.

Copper, ceramics, and magnetic alloys have been very successfully brazed to titanium with pure silver in the rather specialized field of electronics and high vacuum equipment. A binary titanium-beryllium filler metal and a ternary copper, titanium, beryllium filler (49Cu-49Ti-2Be) have been very satisfactory.

The latest brazing alloy for joining titanium to stainless steel is in percent by weight: 81.1 palladium, 14.3 silver, and 4.6 silicon. This alloy met several new standards for braze alloys for titanium:

- The homogeneous solid was inert to nitric acid.
- The sealed joint worked well under a high vacuum.
- The joint's tensile strength was about 75,000 psi.
- The alloy formed a metallurgical bond, with alloy penetration of 0.0015 in. into titanium and 0.003 in. into stainless.

The alloy flowed well at brazing temperatures between 1395 and 1450°F.

Stainless Steel to Tantalum and Ta-10 Tungsten

The gold-nickel alloy and pure copper have produced excellent tensile tests results in brazing type 304 stainless steel to tantalum and tantalum-10 tungsten.

Ceramics to Metals and Graphites to Metals

Ceramic to metal seals are an excellent example of brazing alumina to Ni-Fe alloys. Sealing is accomplished with one-cycle brazing, using a unique variation of the "active alloy" method which creates a chemical rather than a mechanical bond. The brazing alloy used is a 72%Ag-28%Cu eutectic alloy. One of the basic reasons for the choice of alloy is the alloy's good wettability to titanium, and a band of titanium is painted around the ceramic body at the points where joining will take place. The eutectic alloy brazing rings, which melt and flow at 780°C, take the titanium into

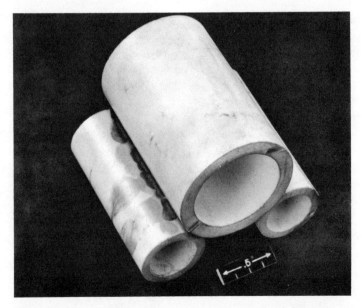

Figure 8.22 Beryllium oxide tubular assembly brazed with 49Ti-49Cu-2Be (wt%) in vacuum for 10 min at 1000°C [27].

Figure 8.23 Completed demonstration graphite-molybdenum fuel element assembly, vacuum brazed with 49Ti-49Cu-2Be (wt%) alloy at 1000°C for 5 min; excellent flow of the alloy is evident [27].

solution. The solution of braze with the titanium forms a new strong alloy that flows and wets the ceramic.

The alloy shown in Figure 8.22 as well as the 4% beryllium alloy have been very instrumental in joining graphite to metals also. (See Figure 8.23.)

When a service and strength requirements for materials change so does environment. With the advent of space applications, developments are aimed toward space power systems, and, in particular, space radiators in the concept of metallic tubes that transport a heat-transfer fluid, bonded to thermally conductive fins and armor. During actual operation in the space environment, radiator tubes require protection from meteoroid puncture by means of a satisfactory armor of minimal weight. Table 8.13 shows

Table 8.13   Thermal Stability Test Results [31].

Material Combination	Braze Alloy (wt%)	Metallographic Analysis	
		After Aging for 500 h at 1350°F (734°C)	After 500 Thermal Cycles at 350 to 1350°F (177 to 734°C)
316 SS-Graphitite-G	66Ag-26Cu-8Ti	Crack near braze	No cracking
316 SS-Expanded pyrolytic-graphite	66Ag-26Cu-8Ti	No cracking	No cracking
Cb-1Zr-Graphitite-G	66Ag-26Cu-8Ti	No cracking	No cracking
Cb-1Zr-Expanded pyrolytic graphite	66Ag-26Cu-8Ti	No cracking	No cracking
316 SS-Graphitite-G	48Zr-48Ti-4Be	Crack near braze	Crack near braze
316 SS-Expanded pyrolytic graphite	48Zr-48Ti-4Be	No cracking	No cracking
Cb-1Zr-Graphitite-G	48Zr-48Ti-4Be	No cracking	No cracking
Cb-1Zr-Expanded	48Zr-48Ti-4Be	No cracking	No cracking
316 SS-Graphitite-G	84Au-10Ni-5Fe	Crack near braze	Crack near braze
316 SS-Expanded pyrolytic graphite	85Au-10Ni-5Fe	No cracking	No cracking
Cb-1Zr-Graphitite-G	85Au-10Ni-5Fe	No cracking	No cracking

the alloy and graphite combinations and the test results of the three best braze alloys evaluated.

JOINT DESIGN AND PROPERTIES

Configuration of Brazed Joint

A brazed joint is not a homogeneous body. Rather it is a hetereogenous assembly that is composed of different materials with differing physical and chemical properties. In the simplest case it consists of the base metal parts to be joined and the added filler metal. Diffusion processes, however, can change the composition and therefore the chemical and physical properties of the boundary zone formed at the interface between base metal and filler metal. Thus in addition to the two different materials present in the simplest example given further dissimilar materials must be considered.

In determining the strength of these heterogenous joints, the simplified conceptions of the elasticity theory—valid for a homogeneous metallic body where imposed stresses are uniformly transmitted from one surface or space element to the adjacent ones—can no longer be applied. In a brazed joint formed of several materials with different characteristics of deformation-

Joint Design and Properties

resistance and deformation-speed, the stresses caused by externally applied loads are nonuniformly distributed.

In order to obtain comparative values these facts and their effects must be considered in the choice and design of specimens for determining the strength of brazed joints. The tensile strength of brazed joints on base metals with a higher strength than that of the filler metal in the as-cast condition is reported to increase when the clearance is reduced to a certain small value, depending on the area of the joint. If the clearance is reduced still further, the tensile strength of the joint again decreases, unless the test is carried out with special precautions. The decrease in strength is caused mainly by tightly fitting points due to slight surface irregularities. In these points the film of filler metal is interrupted, thus producing a reduction of the brazed area. The increase in strength with decreasing clearance is basically due to the modification of the stress conditions in the filler metal layer.

Why should small clearances be used? The smaller the clearance, the easier it is for capillarity to distribute the brazing alloy throughout the joint area. There will be less likelihood of voids or shrinkage cavities as the alloy solidifies. Small clearances and correspondingly thin brazing alloy films make sound joints and sound joints are strong joints [32]. The soundest joints are those in which 100% of the joint area is wetted and filled by the brazing alloy. They are at least as high in tensile strength as the brazing alloy itself and often higher. If brazing clearances ranging from 0.001 to 0.003 in. are designed, they are designed for the best capillary action and strongest joint.

Joint clearance here is probably the most significant factor in vacuum brazing operations. Naturally, it receives special consideration when designing a joint. Joint clearance may be defined as the distance between the surfaces of the joint at room temperature before brazing into which the brazing filler metal will be flowed. Actually joint clearance is not the same at all phases of brazing operations. It is one value before brazing, another value at brazing temperature, and still another value after brazing, especially if there has been diffusion of the filler metal into the base metal. To avoid confusion it has become general practice to specify clearance in the joint design as being a certain value at room temperature before brazing.

Optimum clearance ranges have been worked out for each type of brazing filler metal that will be suitable when brazing all similar-metal combinations with which that filler metal can be used (Figure 8.24). Generally, clearances in the lower portion of the ranges should be used in order to obtain maximum joint strength.

Recommended joint clearances are based on joints having members of similar metals and equal mass. When dissimilar and/or metals of widely

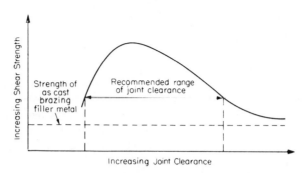

Figure 8.24 For each filler metal there is an optimum joint clearance range. Values below minimum give low strength because filler metal flow is obstructed. Clearance larger than optimum require excessive filler metal and result in loss of effective distribution of filler metal because capillary action is destroyed.

differing masses are joined by brazing, special problems arise which necessitate more specialized selection among the various filler metals and joint clearance suitable for the job at hand must be carefully determined.

In these joints the member with the higher thermal conductivity and/or greater mass tends to either increase or decrease the joint clearance depending on the shapes of the members and their relative positions in the joint. It is therefore necessary to consider the joint clearance that must exist at brazing temperature and make due allowance in the joint design to attain this usable clearance. Joint clearance at room temperature that makes possible a usable clearance at brazing temperature must be determined. At the same time a brazing filler metal must be selected that will be usable over the brazing temperature range and the changing clearances resulting. Thus a filler metal with a long brazing temperature range is best for bridging relatively great gaps that may occur before or after attaining brazing temperature. On the other hand, a filler metal having a short brazing temperature range is preferred when joints are tight fitting through the room-temperature to brazing-temperature range.

Although there are many kinds of brazed joints, the problem is not as complicated as it may seem because butt and lap joints are the two fundamental types. All others, such as the scarf joint, are modifications of them. It is identical with the butt joint at one extreme of the scarf angle and approaches the lap joint at the other extreme of scarf angle.

Selection of the type of joint to use is influenced by the configuration of the parts as well as stress requirements and other service requirements such as electrical conductivity, pressure tightness, and appearance. Also influential in selecting joint type are fabrication techniques, production

Joint Design and Properties

quantities, method of feeding filler metal, etc. Lap joints are generally preferred for brazing operations, particularly when it is important that the joints be at least as strong as the weaker member. The lap joint length should equal three times the thickness of the thinner member joined for maximum strength.

Butt joints overcome the one inherent disadvantage of lap joints in that they provide a smooth joint of minimum thickness. When smooth appearance of joint is the prime consideration or when the double thickness of the lap joint would be otherwise objectionable, butt joints should be used. However, since the butt joint is difficult to fitup for preserving the necessary joint clearances, its strength will not be as reliable as the lap joint, being particularly weak in bending.

One additional advantage the butt joint gives is that of a single thickness at the joint, which can be important in many brazing applications. However, the butt joint has a smaller bonding area than the lap joint. Consequently, it will seldom be as strong as a comparable lap joint. Butt joints offer the advantages of simple preparation and adequate strength for a great many applications.

All other things being equal, the lap joint gives greater strength than the butt joint, simply because the bonding area is greater. In many instances, its double thickness at the joint is not objectionable. And the lap joint is generally "self-supporting." Flat members seldom need special preparation; merely holding them in place with a little pressure maintains uniform joint clearance. In tubular joints the nesting of one tube inside the other automatically supports the assembly in proper alignment during brazing.

Lap joints, too, have the advantage of design flexibility. The bond area of the butt joint is limited to the cross section of one of the parts joined. With a lap joint, however, one can usually vary the lap areas as much as desired to meet any strength requirement.

A scarf joint is an attempt to achieve the smooth contours of the butt joint and the desirable properties provided by the large joint area of a lap joint. However, to secure a lap of three times the thickness of the parts joined, it would be necessary to have a very small scarf angle. Machining a scarf, particularly on thin metal, would be a difficult and costly operation, and assembling the joint members to preserve the required joint clearances would be even more difficult. Scarf joint designs are therefore seldom used.

Suppose strength and single thickness in the same joint are wanted. Both these advantages are obtained by using a butt-lap joint. It is a little more work to prepare than a straight butt or lap joint, but the extra work is often worthwhile. Maximum strength is achieved in a joint of single thickness and like an ordinary lap is usually self-supporting. When strength is required in a joint of single thickness, the butt-lap joint is usually most practical.

Based on the preceding discussion, the following mistakes should be avoided:

1. Excessive clearance that will result in filler metal waste, possible loss of capillarity, and probably joint weakness.
2. Insufficient clearance whereby loss of joint strength occurs.
3. No allowance for base metal expansion that will lead to possible metal-to-metal contact at brazing temperatures, preventing normal capillary flow. Enlargement of the joint gap may also result from thermal expansion in instances when one part is fitted inside another that has a higher coefficient of expansion.
4. No allowance for filler metal shrinkage that can create a concave joint surface or a possible notch effect that will be a potential stress raiser. Bending may also cause joint failure through tearing.

Testing a Brazed Joint

During the past ten years sufficient components that have been brazed have functioned very well under conditions of high temperatures and high stress to support the statement that brazed joints, using any of a number of "high temperature" brazing filler metals, should have an overall endurance limit equal to that of the base metal. Because of this successful history and because it is immeasurably less expensive, most testing is done using standard specimens rather than by simulating actual brazements and actual service conditions. Many test specimen designs have been made and used for strength testing brazed joints. None of these designs is perfect. If they are properly tested and properly evaluated, however, good usable results can be obtained. Brazed joint strength should be expressed as "load-carrying capacity," rather than as a basic unit strength value (such as tensile or shear strength). The load-carrying capacity of a brazed joint might be expressed as pounds per lineal inch of joint or pounds per square inch of the weakest base metal cross section.

The sleeve-type test bar has more joint restraint than the single-lap type, thus reducing the bend effect. However, the shear-stress in the joint area of both types of specimens has the same stress concentration pattern.

When ductile base metals are properly brazed, the base metal usually fails before the joint. When hardenable base metals are brazed in exactly the same way, the failure is more likely to be in the joint (Table 8.3). This is why it is necessary to express joint strength (load-carrying capacity) in terms of stress in the braze area. Again, no matter what kind of test specimen is used, it should be kept in mind that a structure is being tested, not a material, and the results should be expressed accordingly.

LIGHT METAL PROPERTIES (BERYLLIUM, TITANIUM, ALUMINUM)

Beryllium

The optimum conditions for silver brazing beryllium on the basis of tensile tests and metallographic and x-ray examination have recently been defined [33]. Investigators found that improved quality joints could be made in structural beryllium by brazing with pure silver at 900°C (1650°F) using a brazing time of 1 to 5 min and a pressure in excess of 40 psi. Starting with a very thin layer of silver, they found that closely controlling the brazing conditions increased the quality and reliability of the braze. Further improvement in the quality of the silver-brazed joint resulted from heating at 600°C (1110°F) for periods up to one week after brazing.

These experimentors [33] further found that the tensile strength of butt-brazed beryllium specimens, vacuum brazed at 900°C (1650°F) for 3 min, was significantly higher in the as-brazed condition with a silver-0.05 wt% lithium braze alloy. Other silver rich brazing alloys containing highly volatile additions such as zinc, phosphorus, and lithium have produced higher tensile strengths in brazed joints than those containing relatively nonvolatile additions such as aluminum, silicon, tin, and germanium. The presence of the volatile additive appeared to improve the wetting and brazing of the beryllium.

An 88% aluminum, 12% silicon brazing alloy provided serviceable joint strengths of 20,000 to 30,000 psi up to 350°F [34]. Beyond this temperature its strength rapidly deteriorated. Silver-base brazing alloys had essentially the same room temperature strength as the aluminum silicon alloy, but a strength above 25,000 psi could be maintained at 500°F and a strength of 10,000 psi at 1300°F. Thus for high strengths at temperatures higher than room temperature, the silver-base braze alloys should be used.

Discussed earlier was the fact that brazing is generally carried out on lap joints because they are easiest to make. In beryllium, however, these joints are undesirable because of the inherent notch present in this type of joint. Therefore joints that contain no notches are required to achieve maximum usefulness of brazed joints to beryllium. Step joints, scarf joints, or butt joints have been used when possible.

As a result of vapor evaporation and deposition of titanium [12] butt-brazed joints of beryllium have been made. Accurate temperature measurement of the brazed joint was difficult to attain; brazing, however, occurred

Table 8.14 Tensile Strength of Beryllium Butt-Brazed Joints [12].

Spec. no.	Braze filler	Ultimate tensile strength, psi	Joint gap, in.	Location of failure
1	CP[a] Al	31,730	0.003–0.005	Be-braze interface
2	CP Al	32,620	0.007–0.010	In Be remote from braze joint and grips
3	CP Al	33,620	0.0085	In Be remote from braze joint and grips
4	pure Al[b, c]	22,440	0.0088	Be-braze interface
5	pure Al[b, c]	20,100	0.0075	Braze
6	pure Al	29,570	0.005	25% in braze; 75% in Be

[a] CP—commercially pure aluminum (1100 alloy). [b] Pure Al—99.995% Al. [c] Braze joints heated to brazing temperatures 2 times (see Discussion).

approximately in the temperature range of 1700 to 1750°F. Table 8.14 is a compilation of the tensile data obtained from pure and commercially pure aluminum braze filler butt-brazed joints. That fracture occurred at high strength levels in the base material remote from the braze indicates the integrity of these joints. Some difficulty was experienced with the pure aluminum filler metal running out of the butt joints and more than one heating cycle was necessary for specimens 4 and 5 in order to obtain sound braze joints for testing. These occurrences may help to explain the lower strength obtained for these particular specimens.

Aluminum

The new aluminum alloy X7005 (1 Mg, 0.5Mn, 0.1Cr, 4.5 Zn, 0.1 Zr) is very attractive for vacuum brazing because of the very high strengths

Figure 8.25 High strength ×7005 cellular panel vacuum brazed [36].

Table 8.15

Area	Ultimate Load (lb)	Strength Stress (psi)	Type of Braze	Dimensions
0.0937	880	9,400	Vacuum	0.125 × 0.75
0.0937	930	9,800	Vacuum	0.125 × 0.75
0.0937	955	10,200	Vacuum	0.125 × 0.75
0.0937	1,030	11,000	Dip	0.125 × 0.75
0.125	1,218	9,750	Dip	0.125 × 1.0
0.125	1,147	9,180	Dip	0.125 × 1.0
0.125	1,260	10,080	Vacuum	0.125 × 1.0
0.125	1,155	9,230	Vacuum	0.125 × 1.0
0.125	1,580	12,600	Base Metal	0.125 × 1.0

possible in joint configurations that normally entrap fluxes. Figure 8.25 shows a $1\frac{1}{2} \times 6 \times 16$ in. cellular structural panel of X7005 that was evacuated and vacuum brazed at 1085°F for 2 min. Full strong fillets were obtained with 88% aluminum-12% silicon filler metal alloy. In unsymmetrically loaded beam tests the panel withstood a load of nearly 2000 lb/linear inch of width before failure.

For most brazeable aluminum alloys an overlap distance twice the thickness of the thinnest member is recommended to make the joint as strong as the base metal. This is not true for X7005, however, because of its high base-metal strength after brazing. One-eighth inch X7005 brazed at lap distances of $\frac{1}{4}$ and $\frac{1}{2}$ in. failed through the joints at 35 and 45,000 psi tensile strength, respectively. Brazed X7005 structures should be designed, when possible, so that direct tension loads are not applied to the joints. One precaution should be observed, that of holding X7005 assemblies at brazing temperature only long enough to insure complete flow of the brazing alloy because the silicon alloy filler penetrates very rapidly. Shown in Table 8.15 are the results of a recently completed program that evaluated vacuum brazing of 6061 aluminum alloy. Except for the last specimen, the results shown are 6061 lap shear specimens as-brazed.

Titanium

Brazed joints between titanium honeycomb core and titanium facings and edge members resist frictional heat and strain encountered at mach 2 plus speeds. The 95% silver-5% aluminum braze alloy has been evaluated in standard tests. In flexure or bend tests performed at room temperature, the average ultimate stress of a honeycomb test panel was 143,000 psi; at 800°F it was 93,000 psi.

To determine the compressive properties of the panel construction, end compression tests were performed. Average ultimate stress of the panel at room temperature was 140,700 psi, whereas at 800°F, it was 130,700 psi.

The room temperature shear strengths of brazed joints in titanium and titanium alloys exhibit a rather extensive range. Using silver or silver-based alloys, reported shear strengths range from 15,000 to 24,000 psi for vacuum furnace brazing. The shear strengths of dissimilar metal brazements of titanium, using the same silver-base brazing filler metals, are slightly lower than titanium brazements.

STEELS AND SUPERALLOY PROPERTIES (NICKEL, COBALT, IRON)

Peel tests of vacuum-brazed stainless steel to nickel base alloys have produced tough, ductile joints that are more ductile than any obtained by brazing in any other atmosphere.

Table 8.16 shows the tensile and stress rupture properties of René 41 brazed sheet overlap shear specimens using the Ni-Cr-Pd-Si braze alloy. These properties indicate that brazed René 41 has sufficient elevated temperature strength to meet jet engine component requirements.

The use of foils is becoming more important as the need and search for lightweight structures are continually stressed. One of these studies evaluated foils of TD-Nickel. A standard brazed T-joint specimen (0.006-in. foil) was selected for comparative testing. All brazing was done in a high-vacuum environment (1×10^{-4} torr or better). The results indicated that for service temperatures to 2400°F the TD-5 alloy and the 60Pd-40Ni alloy were the most promising braze alloys, which is in substantial agreement with previous results [37].

The new supersonic transport (SST) airplane has rigid requirements for engine components. The nickel-base Hastelloy R-235 has also been evaluated for engine applications, as shown in Figure 8.26. The "indicated" shear strength is defined as an index value of shear strength obtained by dividing the original shear area of the laps into the load in pounds required to fail the joint. The qualifying adjective "indicated" is used, for some degree of bending occurred in all specimens and the applied stress was not pure shear. The values, however, are comparable among themselves and the testing method has the virtue of simplicity [38].

The results of Table 8.8 is plotted in Figure 8.26. Three distinct curves are found when the results on the three types of joints tested are plotted. Strengths of vacuum-brazed joints are consistently higher than those of nickel plated and hydrogen-brazed joints at all temperatures, whereas

Table 8.16 Shear Strength Properties of Vacuum Brazed Unplated René 41.

Shear-Tensile

Test temp, °F	Joint area, sq in.	Ultimate load, lb.	Indicated shear strength, psi[b]	Fracture location
RT	0.0549	1730	31,750	Braze
RT	0.0575	1925	33,450	Braze
RT	0.0525	1930	36,800	Braze
RT	0.0500	1950	39,000	Braze
700	0.0426	2235	52,450	Braze
700	0.0480	2275	47,400	Braze
700	0.0415	2025	48,850	Braze
700	0.0455	1950	42,900	Braze
1250	0.0555	2270	41,000	Braze
1250	0.0504	1665	33,100	Braze
1250	0.0556	2050	36,900	Braze
1250	0.0530	2215	41,800	Braze
1450	0.0597	1880	31,500	Braze
1450	0.0573	2455	42,800	Braze
1700	0.0526	190	3,620	Braze
1700	0.0452	250	5,540	Braze

Shear-Rupture

Test temp,° F	Stress, psi	Life, hr.	P_a, C = 20	Fracture location
1250	20,000	100+	37.60+	No failure
1450	5,700	193+	42.60+	No failure
1700	2,000	32.42	46.50	Braze

[a] Heat treatment for all specimens subsequent to brazing: 1950° F—30 min—R.A.C.; 1400° F—16 hr—A.C.
[b] Indicated shear strength is obtained by dividing the ultimate load by the joint area.

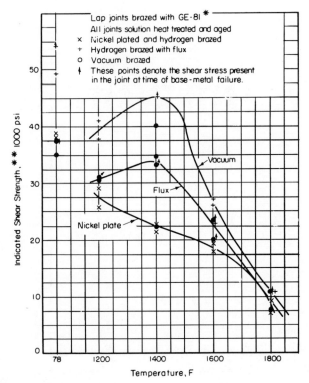

Figure 8.26 Indicated shear strength versus temperature for heat-treated brazed joints in Hastelloy R-235 [24].

strengths of fluxed hydrogen-brazed joints are intermediate. The 1400°F peak of the two higher curves reflects additional strengthening of the R-235 base material by aging during testing.

REFRACTORY METAL PROPERTIES (COLUMBIUM, MOLYBDENUM, TANTALUM, TUNGSTEN)

Of these refractory metals the alloys of columbium and tantalum have been most prominently considered for present and future uses. Therefore most testing has been performed on these alloys, especially at temperatures above 2000°F. Joints that have been mechanically tested have varied from lap to tee to honeycomb. Test results indicate an increase in lap-shear strength of Ti-33Cr braze joints from approximately 2450 psi to greater than 4450 psi at 2500°F and from zero to 1240 psi at 3000°F. Thus a diffusion treatment 300°F below the original brazing temperature actually increased joint remelt temperature by 600°F, essentially doubled lap shear strength at 2500°F, and produced excellent strength at 3000°F.

The lap joints reactive brazed with Ti-30V-4Be exhibited substantially lower strength than similar joints diffusion sink brazed with Ti-33Cr, even though remelt temperatures were similar for both systems. Lap-shear strength of Ti-30V-4Be joints was approximately one-half and one-third the strength of Ti-33Cr joints at 2500 and 3000°F, respectively. Nevertheless the Ti-30V-4Be joints exhibited attractive potential for lower stress level applications.

Lap-shear testing at 3500°F temperature of the tantalum-10 tungsten alloy using a columbium-1.3% boron braze alloy has shown the following results:

Lap Shear Stress (psi)	Base Metal Tensile Stress (psi)
4,330	12,900
3,750	12,800
4,200	12,600

Coupon testing, however, is not enough today. Actual component parts must be tested to meet the rigorous environmental demands of heat and cryogenics combined with structural loads, vibration, and other factors. Figure 8.27 shows a vacuum-brazed honeycomb panel being exposed to a rigorous heat test. Another dynamic test is depicted in Figure 8.28 and the panel is shown in the center of the test fixture.

When brazing ceramics to metals, it is important to consider joint design in order to prevent failure of the brazed assemblies. For example, whenever possible the ceramic component, because of its high compressive strength, should be placed inside the metal component. When Kovar, an alloy containing 29% nickel, 17% cobalt, and 54% iron, is used for the metal part in such assemblies it exerts a desirable compressive force on the joint as it cools. This effect results from the higher expansion of Kovar at elevated temperatures. (See Figure 8.29.) On larger parts, however, the higher expansion of the metal can introduce other problems. In these instances compensation must be made in part dimensions to ensure proper clearances for the filler metal. Eutectic brazing alloys such as 72% silver and 28% copper, for example, require 0.003 in. of clearance for satisfactory joining.

If the metal member is placed inside the ceramic, there must be sufficient clearance to permit expansion of the metal and a space allowance for filler metal. The thickness of filler metal should also allow some flexibility at the joint since the metal contracts more than the ceramic during cooling. Stresses should be controlled and compensated for to assure reliable, leak-proof joints.

Figure 8.27 Structural panel shear testing [21].

Figure 8.28 Rigorous test of D36 columbium heat shield. Temperature of part in fixture is 2300°F.

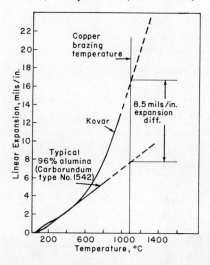

Figure 8.29 Linear expansion of Kovar and a typical 96% alumina are closely matched up to about 450°C (840°F).

When joining large diameter conductors, tubular inserts should first be joined to the metallized ceramic and the conductor then attached to the tube at one end only. This construction minimizes stresses that may result from the inherent flexibility of the tubing's thin wall. The insert reduces tensile and shear forces to allowable limits.

The same flexible construction feature is often used for outer metal components, especially when the assembly is exposed to thermal cycling. Generally, flexibility is imparted to outer metal parts by using curved sections. The forces generated in the outer area are dissipated by the curved section and stresses are not transmitted to the original brazed joint. Thin metal sections also help to relieve strains of this type.

In summary, the following advantages can be attained by designing for vacuum brazing:

1. Economical fabrication of complex and multicomponent assemblies.
2. Simple means for achieving extensive joint area or joint lengths.
3. Excellent stress distribution and heat transfer.
4. Ability to join cast materials to wrought metals.
5. Ability to join certain nonmetals to metals.
6. Ability to join widely dissimilar metal-thicknesses.
7. Ability to join dissimilar metals.
8. Ability to fabricate large assemblies in stress-free condition.
9. Ability to preserve metallurgical characteristics of metals being joined.

10. Capability for precision production tolerance.
11. Reproducibility and close process control.

APPLICATIONS AND FUTURE POTENTIAL

New and old product designs in brazing for military and commercial applications are continually being converted to vacuum brazing as vacuum furnaces become more readily available. The initial operating costs of furnaces have been substantially reduced and production rates have been increased, thereby contributing to the economical production joining of parts by vacuum brazing.

One of the largest applications for vacuum brazing is aluminum radiators for automobiles. The aluminum industry has been campaigning for years to eliminate copper and brass in radiators. A complete aluminum radiator can be brazed in one step, whereas copper and bronze radiators conventionally go through separate operations in joining sides, tanks, and bottoms to the tube and fin assembly. One auto manufacturer introduced aluminum radiators as standard equipment on some high production model cars, as shown in Figure 8.30. In using the new thin-wall tube design (which eliminates 75% of the joint structure required in previous aluminum radiator designs) the weight of an aluminum unit is $7\frac{1}{2}$ lb as compared with 18 lb for a copper radiator of comparable capacity.

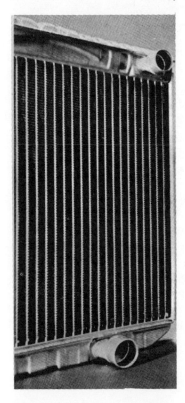

Figure 8.30 All aluminum brazed radiator for tomorrow's automobiles.

The use of aluminum for automobile radiators has also prompted the development of vacuum-brazed aluminum radiators for space applications. Future space nuclear generators powered by an isotopic fuel system that generates power and converts the heat produced by the decay of its radioisotope fuel directly into electricity through a series of thermoelectric elements next to its fuel block. Excess heat that has passed through the themoelectric

Figure 8.31 Space radiator on Saturn IV-B attached to lunar excursion model on its way to the moon.

conversion system is dissipated to the atmosphere by a closed coolant loop that runs through the generator and a radiator area. A temperature-sensitive liquid metal, NaK, flowing through the loop, picks up the excess heat from the thermoelectric conversion system and transports the heat to the radiator where it is dissipated into space. Figure 8.31 depicts an artist's conception of the future use of radiators in manned space stations of the future.

Two other areas where vacuum brazing is receiving the auto industry's attention are in the manufacturing of aluminum engine blocks and pistons. Virtually all engines use aluminum pistons. However, higher strength aluminum alloys, such as the 7005 vacuum-brazed aluminum alloy discussed previously, will be required when engine designs incorporate higher cylinder pressures.

Current vacuum-brazing programs that employ 7005 aluminum alloy are the fabrication of heat exchangers for the environmental contol system for the Apollo Lunar Excursion Module (LEM), the gas management assemblies for the Biosatellite, and earth orbiting spacecraft designed to study the effects of space flights on animal and plant life.

X-ray tube beryllium window assemblies have been vacuum brazed into monel retainers with pure silver-brazing material. All assemblies are vacuum tight and suitable for use in x-ray tubes.

Other vacuum brazing applications currently being designed, developed, tested and in use, include

(1) primary structure of honeycomb and multiwalled sandwich for antennas (aluminum, magnesium, beryllium);
(2) space hulls and interstage adapters (aluminum, magnesium, beryllium);
(3) heat exchangers where increased pressures demand high strength material and stronger fittings for heat exchangers or flow lines, landing energy absorbers, honeycomb heat shields, leading edges, and other structural-brazed components.

The role of vacuum brazing in the joining of superalloys and stainless steels for jet engines has been very important. Components such as René 41 engine nozzle assemblies, an Inconel 600 heat exchanger, and air-cooled turbine blade are shown in Figure 8.32. Others include turbine shrouds of nickel-base Hastelloy C with Inconel X honeycomb and Inconel X fans. Several other applications include components for external combustion engines and stainless steel fittings and nozzle rings.

Because of their excellent mechanical property strength at moderately elevated temperatures (1500° to 2000°F) superalloy honeycomb brazed-panels, such as René 41 and A-286 shown in Figure 8.33, have been considered for primary structural panels and leading edges and control surfaces of current and future aircraft. Figure 8.33 shows a speed brake used on the A2F "Intruder" carrier-based jet. One reason for speed brakes is that during combat or just for flying maneuverability it is often desirable to suddenly slow the speed of the aircraft. Basically, the speed brakes are panels of the fuselage that are seen swung outward laterally on hinged longerons by hydraulic motors. The speed-brake design theory has worked well enough to become a part of the airplanes basic design.

A vacuum-brazed cobalt-base alloy is currently in use in commercial jet liners as deflector vanes. The deflector vanes direct jet engine exhaust gases for thrust reverser braking and perform the same function as reverser

Inconel 600 Air-to-Air Heat Exchanger

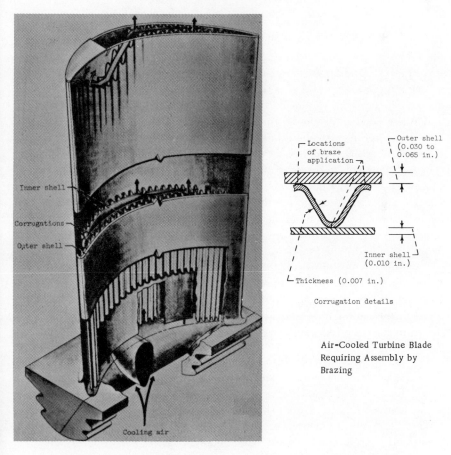

Corrugation details

Air-Cooled Turbine Blade Requiring Assembly by Brazing

Figure 8.32 Examples of nickel-base alloy products vacuum brazed.

Figure 8.33 Inside (left) and outside (right) of A2F speed brake-vacuum brazed A-286 iron-base alloy with Ni-Cr-Si filler metal.

braking. The vanes perform the same function as reverse pitch propellers on piston-engine planes, that is, they reduce landing speed and hence permit jet aircraft to land on shorter runways.

Vacuum-brazing techniques are being employed in the fabrication of curved shapes and channeled plates for heat exchangers, press platens, and other cooling purposes. In addition, vessels of vacuum clad steel are now in use in the petroleum industry.

Thus the reader can see how varied the applications have been, and there are new ones to come. Entire aircraft wing structures can be vacuum-brazed in a single operation. Other applications include stainless steel bellows and electronic power tube parts. At least one vacuum equipment manufacturer has adapted vacuum-furnace brazing to fabricate vacuum manifold components that are leak-tight to the most exacting standards as well as a hydraulic manifold system for a current ground-to-air missile.

Vacuum brazing has been the boom for refractory metals. Since various gases (nitrogen, hydrogen, pure helium, and argon) create problems of brittleness and surface contamination, the use of vacuum has substantially aided in the manufacturing development of refractory metal alloy structures. These include the following:

(1) honeycomb heat shields;
(2) structural honeycomb panels which were flown on an unmanned thermal and aerodynamic space test flight. These panels experienced temperatures of 2200°F (Figure 8.34);
(3) thermionic converters composed of various refractory metals and ion propulsion devices and molybdenum gas turbine nozzle assemblies.

From these examples, it is evident that refractory metals and vacuum brazing have become inseparable friends. Future maneuverable re-entry vehicles, uncooled rocket engines, nose cones, and thrust defectors require the application of refractory metals and vacuum brazing.

The nuclear industry has employed vacuum brazing as a means of sealing graphite-clad fuel elements and for fabricating a variety of experimental assemblies. Such a tubular fuel element assembly is shown in Figure 8.23 where metal hangers or "spiders" hold the individual capsules. The increasing use of ceramics in industrial and development applications results from their good insulating and ultra-high-temperature properties.

To keep pace with the growing scope and complexity of vacuum technology, the art of feeding electrical power or signals into high vacuum chambers is itself becoming more complex. For example, where advanced instrumentation is involved, a feed-through assembly may be required to feed very high voltage or high current without the slightest disturbance to vacuum conditions, including those in the ultrahigh range.

The performance of electrical feed-through assemblies under high vacuum conditions is directly traceable to the strength and vacuum-tightness of the ceramic-to-metal bonds.

The new fields of power development through thermionics and magnetohydrodynamics (MHD) have placed new requirements on the ingenuity of the brazing engineer. In Figure 8.35 is a ceramic-metal-seal brazed at 3000°F that has operated for 1000 hr at 2000°F. The ceramic is Al_2O_3, the metal is columbium, and the braze alloy is titanium.

Figure 8.34 D36 columbium structural honeycomb panels before their installation on Asset vehicle #6 and a flight into space [21].

Figure 8.35 Brazed seal of columbium and aluminum oxide.

Brazing of parts for electron tubes and modern high vacuum equipment has caused the development of highly refined procedures. Vacuum tubes required the maintenance of very low pressures (10^{-6} to 10^{-8} torr) for their satisfactory operation. Special high vacuum equipment can tolerate no condition which inhibits the ability to secure and maintain extremely low pressures. The demands placed on this equipment require an exceptional degree of reliability in all brazed joints as evidenced by the care taken in the assembly of a magnetron tube in Figure 8.36.

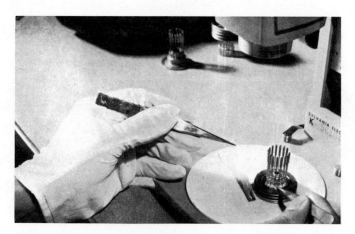

Figure 8.36 Electronic tube assembly brazed with braze alloy-68 silver-27 copper-5 palladium. Braze temperature, 1500°F.

Almost nothing makes a good, all-around furnace atmosphere. Thus, understandably, vacuum brazing is no longer a novelty, although it is not yet commonplace. However, here are some of the newer developments in progress:

- Techniques for maintaining the vacuum in the furnace while charging or removing workloads.
- Continuous vacuum brazing furnaces, with work charged and discharged at atmospheric pressure.
- The ability to quench parts after brazing from the vacuum furnace without breaking the vacuum.
- Automation of the time cycles in which brazing occurs. The furnace and control system is under automatic control from the time the operator seals the door and pushes a start button. At the completion of the brazing cycle a bell signals the operator that another load can be processed.

The future of fabrication by vacuum furnace brazing is secure because of its many advantages. However, the brazing equipment and accessories must be improved and adapted to fully realize the potentials of the brazed assemblies. There is a need for larger furnaces, more sensitive temperature control especially above 3000°F, and more rapid heating capacities. Too often improper brazing methods and techniques are employed on jobs that obviously require a specialized approach. As a result the end product is not satisfactory, and this carries the implication that the part cannot be brazed. In some cases this may be true, but with the proper design, proper selection of equipment, and proper selection of braze alloy there are still many advantages to be realized from the vacuum furnace brazing approach.

REFERENCES

[1] W. D. Hawkins, *Physical Chemistry of Surface Films,* Reinhold, New York, pp. 1–413, 1952.
[2] A. Bondi, "Spreading of Liquid Metals on Solid Surfaces; Surface Chemistry of High Energy Substances," *Chem. Rev.,* **52** (2), 417–458 (1953).
[3] R. G. Gilliland, "Wetting of Beryllium by Various Pure Metals and Alloys," *Welding J.,* 248-S–258-S (June 1964).
[4] D. R. Milner, "A Survey of the Scientific Principles Related to Wetting and Spreading," *Brit. Welding J.,* **5,** 90–105 (1958).
[5] Walter H. Kohl, "Soldering and Brazing," *Vacuum,* **14,** 175–198 (1964).
[6] W. Hack, "Which Brazing Alloys for Vacuum Systems," *Materials In Design Engineering,* 116–117 (July 1960).
[7] F. Miller, "Factors Affecting Brazed Joint Strength in High Temperature Applications," *Welding J.,* 910–915 (October 1962).
[8] F. Miller, "Importance of Purity in Manufacturing Brazing Filler Metals for High Temperature Service Applications," *Welding J.,* 821–827 (August 1961).

[9] *Braze Manual,* American Welding Society, Reinhold Publishing Co. pp. 82–83, 1963.
[10] K. G. Wikle and R. B. Magalski, "Brazing Beryllium to High-Temperature Alloy Collars," C00-310 (June 1956).
[11] "Joining of Beryllium," DMIC Memorandum 13 (March 30, 1959).
[12] S. Weiss and C. M. Adams, Jr., "The Promotion of Wetting and Brazing," *Welding J.,* 49-S–57-S (February 1967).
[13] C. W. Fox, R. G. Gilliland, and G. M. Slaughter, "Development of Alloys for Brazing Columbium," *Welding J.,* 535-S–540-S (December 1963).
[14] N. Bredzs, J. Rudy, and H. Schwartzbart, "Development of Partially Volatile Brazing Filler Alloys for High Temperature Applications and Resistance to Oxidation," WADD TR 59-404, Part II (June 1961).
[15] R. Armstrong, "Exo-Reactant Nickel Base Structural Adhesives," Bureau of Naval Weapons, Contract No. 61-0308-C.
[16] "Development of Low Temperature Brazing of Tungsten for High Temperature Service," Solar Aircraft Company, Bureau of Naval Weapons Contract No. 61-0414-C.
[17] Max Hansen, *Constitution of Binary Alloys,* McGraw-Hill, 2nd Edition, 1958.
[18] A. H. Freedman and E. B. Mikus, "High Remelt Temperature Brazing of Columbium Honeycomb Structures," *Welding J.,* 248-S–265-S (June 1966).
[19] W. R. Young and E. S. Jones, "Joining of Refractory Metals by Brazing and Diffusion Bonding," Final Report Nr. ASD-TDR-63-88, General Electric Company (January 1963).
[20] G. S. Hoppin, III, "Brazing of Molybdenum," ASM Molybdenum Fabrication Conference, Los Angeles, California (May 8, 1958).
[21] J. McCown, C. Wilks, M. Schwartz, and A. Norton, "Final Report on Development of Manufacturing Methods and Processes for Fabricating Refractory Metal Components," ASD Project No. 7-937, AF33(657)-7276, Martin Company, Baltimore, Md. (1963).
[22] K. L. Gustafson, "Development and Evaluation of Braze Alloys for Vacuum Furnace Brazing," Report NASA CR-514, Aerojet-General Corporation, Sacramento, Calif., Contract NAS 3-2555 DMIC No. 64929 (July 1966).
[23] R. M. Evans, "Joining of Nickel-Base Alloys," DMIC Report 181, 57–63 (December 1962).
[24] H. G. Popp, "Materials Property Data Compilation Part III: René 41 Products after Welding, Brazing or Fabrication, Report No. 1, AF 33(657)-8017, General Electric Company (May, 1962).
[25] R. E. Yount, R. E. Kutchera, and D. L. Keller, "Development of Joining Techniques for TD-Nickel-Chromium," AF 33(615)-3476, Interim Progress Report No. 5-8-211A, General Electric Company, Flight Propulsion Division.
[26] R. E. Yount, R. E. Kutchera, and D. L. Keller, "Development of Joining Techniques for TD-Nickel-Chromium," AF 33(615)-3476, Interim Progress Report No. 4-8-211A, General Electric Company, Flight Propulsion Division.
[27] C. W. Fox, and G. M. Slaughter, "Brazing of Ceramics," *Welding J.,* 591–597 (July 1964).
[28] R. G. Donnelly and G. M. Slaughter, "The Brazing of Graphite," *Welding J.,* 461–469 (May, 1962).
[29] K. K. Keuye, and G. R. Grow, "Brazing Beryllium Oxide to Pyrolytic Graphite," *Welding J.,* 346-S–439-S (August 1962).
[30] M. L. Torti, and R. W. Douglass, "Brazing of Tantalum and its Alloys,"

Third Technical Conference of Refractory Metals Committee, Volume 30-171-182, Part I, AIME, Los Angeles, California, Gordon and Breach Publishers (December 9–10, 1963).
[31] A. Del Grosso, "Brazed Graphite–Metal Composites for Space Radiators," Symposium on Materials for Large Space Power Systems, ANS Winter Meeting, (November 1965).
[32] J. Colbus, C. G. Keel, and G. M. Blanc, "Notes on the Strength of Brazed Joints," *Welding J.*, 413-S–419-S (September 1962).
[33] B. M. MacPherson, R. B. Magalski, and R. D. O'Rourke, "The Brazing of Beryllium," *Beryllium Research and Development Program,* Section 7 of Technical Documentary Report No. ASD-TDR-62-509, AF 33(616)-7065, Volume II (April, 1963).
[34] "Joining Beryllium," Bulletin 2145, The Beryllium Corporation, Reading, Pa.
[35] R. E. Monroe, D. C. Martin, and C. B. Voldrich, "Welding and Brazing Beryllium to Itself and to Other Metals," BMI-836 (May 28, 1953).
[36] J. H. Dudas, "Joining High Strength Aluminum Alloy X7005," *Welding J.,* 358-S–364-S (August 1965).
[37] J. W. Welty, R. J. Valdez, C. E. Smeltzer, Jr., and C. P. Davis, "Joining of Refractory Metal Foils," Report No. ASD-TDR-63-799, Part II, Solar, Division of International Harvester Company, San Diego, California, under Air Force Contract AF 33(657)–9442 (November 1964).
[38] E. G. Huschke, and G. S. Hoppin, III, "High-Temperature Vacuum Brazing of Jet-Engine Materials," *Welding J.,* 233-S–240-S (May 1958).
[39] *Applications of Welding,* Welding Handbook, Section Five, Fifth Edition, p. 95.5, 1964.

Chapter 9

DIFFUSION WELDING

Product requirements in the aircraft, missile, electronics, nuclear, aerospace, and commercial fields have given rise to many new and demanding service conditions. To meet the stringent requirements of these exacting operations, it was not only necessary to develop new materials but, equally as important, methods to fabricate them into useful engineering components. One such fabrication technique, diffusion welding, was developed to keep pace with the requirements of the advancing technology.

Fusion welding and brazing have been the workhorse techniques for such joining in the past. Each of these techniques, however, may alter the uniformity of properties in the fabricated structure and often have undesirable joint properties such as low strength and low remelt temperature. Recently, however, solid-state welding has been receiving considerable attention for many applications.

Diffusion welding per se is not a completely new joining technique. The forge welding processes have been used to join both wrought iron and low carbon steels for many years. In fact, forge welding is one of the oldest joining methods known and was the only process in common use before the nineteenth century. It is interesting to note that the famous "Damascus blade" of medieval times was made by forge welding.

Frequently, when a new process is publicized in the technical journals, many people tend to use it prematurely. This approach can be disastrous. The initial investigations are usually conducted on a laboratory scale and most likely require additional development. The reported technique may only be applicable for a specific problem, and so each new application demands additional development. Obviously, it would be wise to secure as much available information as possible on the subject and judiciously apply it to the present problem.

In addition, a number of terms are in use to describe various forms of diffusion welding: solid-state diffusion bonding (creep controlled utilizing low pressures and long bonding times), solid-phase bonding, pressure welding, gas pressure bonding, isostatic bonding, press bonding, roll welding, forge welding, explosive welding, friction welding, diffusion bonding, deformation welding (yield strength controlled utilizing high pressures and short bonding times), diffusion welding, and diffusional bonding. A more detailed differentiation of the processes mentioned is discussed under the section on mechanisms and key variables of diffusion welding.

THEORY AND PRINCIPLES OF PROCESS

Unfortunately, the term "diffusion welding" has frequently been misused and applied to processes in which melting occurs or in which the temperature approaches the solidus point that the melting of segregates can be assumed. The following general conclusions can be drawn from the literature.

Diffusion welding is a two-stage process; plastic flow produces intimate contact and disrupts oxides; diffusion and grain growth across the original interface establish the metallurgical bond [1,2]. It has been theorized that diffusion is not necessary for bonding to occur, because bonding will result by mating two clean, perfectly flat surfaces [3]. That this theory is clearly not true is shown by the lack of bonding between an alumina anvil and tungsten under conditions that cause excellent bonding of two tungsten sheets. The mutual insolubility of alumina and tungsten appears to inhibit atom transfer. It follows therefore that intimacy of contact is inadequate if diffusion-inhibiting films remain on the surfaces.

Intermediate metals with low yield stress aid in achieving intimacy of contact [1]. In addition, low yield stress intermediate metals permit deformation to be restricted to the intermediate metal so that distortion can be minimized. One of the principal effects of higher bonding temperatures is to lower the yield stress of the intermediate metal.

Surface diffusing elements accelerate diffusion bonding. Experimental work on tungsten showed that electroplates a few microinches in thickness of either nickel, palladium, rhodium, platinum, or ruthenium accelerate bonding [4]. All joints made at 900 and 950°C with a maximum heat treatment time of 4 hr failed at the joint without any fracture occurring in the base metal. At 950°C, however, recrystallization was observed after a 4-hr heat treatment. The structure produced at 1100°C is shown in

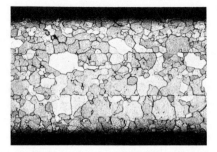

Figure 9.1 Joint from 1100°C hr with nickel on undoped tungsten. Copper ammonium sulfide etch [12].

Figure 9.1 for undoped tungsten. The 4-hr sample shows clear evidence of the joint line disappearing and penetration of the recrystallized region.

The quality of the joints and the extent of recrystallization for joints made with palladium are essentially the same as those with nickel described in the preceding paragraph. Recrystallization which accompanies bonding by this process results in a loss of room temperature ductility. Data also indicate that palladium like nickel lowers the recrystallization temperature of tungsten to 1000°C, even in the presence of a doping agent. The extent of the recrystallized zone increased with both time and temperature of heat treatment. After a 2-hour 1100°C heat treatment, the strips were completely recrystallized and the bondlines obliterated.

It is apparent from the experimental results that an "activating agent" such as nickel or palladium must be present to cause tungsten-to-tungsten bonding because no joints can be made in their absence at the temperatures employed. Evidently this process is one in the solid state since the minimum melting points in both the nickel-tungsten alloy system and in the palladium-tungsten system are those of pure nickel and palladium, respectively [5,6]. The enhancement of tungsten diffusion by the presence of nickel or palladium leads to the development of tungsten "bridges" across the joint. This diffusion process is accompanied by recrystallization at as low a temperature as 1000°C.

On the other hand, surface diffusers are usually grain boundary diffusing elements and lead to unstable grain boundaries in the metal being joined. Consequently, hot shortness, recrystallization, and grain growth are problem areas with this type of accelerator.

Volume diffusing elements with high diffusion rates accelerate diffusion welding. Small interstitial atoms can rapidly diffuse along lattice interstices [7] and along grain boundaries adjoining the interface and can irrigate (diffuse laterally) into the interstices of the lattices of the grains that border along these boundaries [8]. Substitutional atoms of different elements can volume diffuse by a vacancy jump mechanism [7] at a slower rate into adjoining base alloy materials.

If the interdiffusing elements are substitutional and soluble in each other, the conditions shown in Figure 9.2 might prevail. This figure shows that, after sufficient time at temperature, a "mixing" of the atoms has occurred with the mutual atomic attraction of the atoms across the interface producing the bond. This schematic, of course, represents only the simplest conditions that would prevail for diffusion welding to occur and does not consider other factors that may affect the diffusion processes, such as dislocations, atomic and grain misorientations, grain boundaries, voids or the possible formation of either intermetallic compounds or low-melting solid solutions, and eutectics or peritectics that could melt at the diffusion-welding temperature.

Sometimes initial diffusion of an element into a base metal may result in the formation of a low-melting solid solution, eutectic or peritectic composition. The low-melting composition may then fill the joint and constitute a "pseudobrazing alloy." The diffuser element may then again migrate into the base metal from the liquid alloy. In another type of reaction the

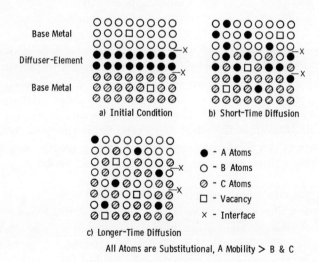

Figure 9.2 Nature of the diffusion-bonding mechanism when "primary" diffuser-element has been preplaced at the interface. Idealized condition in which atoms of the separate regions are oriented and substitutional [15].

diffuser element may migrate into the base alloy to form a high-melting solid solution with the base-metal matrix elements. This reaction would be preferred, for a joint of this nature (consisting of the base alloy elements modified by the presence of the soluble diffuser element) might be expected to possess strength and ductility properties approaching those of the base metal. With sufficient time at temperature, a joint should be produced at the interface region that approaches the base-metal composition and is altered slightly by the presence of the diffuser-element.

An ideal diffuser element, or "primary" diffuser element, for diffusion welding would possess certain properties. The primary diffuser element would be able to diffuse rapidly away from the interface into the base-metal members to facilitate bonding. In addition, the primary diffuser element would remain in solid solution in the base metal(s) so that no embrittling intermetallic compounds form at the joint interface. Also, the primary diffuser element would be easily preplaced between the materials that are to be joined before diffusion bonding.

Beryllium was found to be a promising diffuser element. The beryllium atom possesses a relatively small atomic radius that should facilitate its diffusion into base materials. A nickel-3% beryllium foil provides a carrier for beryllium as a diffusion aid to bond nickel-base alloys. General requirements of diffusion aids appear to be some solubility in the alloy being joined, formation of high-melting point solid solution with the alloy and a high diffusion rate (hence small atoms appear most favorable).

Residual cold work can accelerate diffusion bonding. This cold work may be throughout the metal or locally at the surface as a result of mechanical abrasion. In discussion of other work Kaufmann suggested that cold work in an abraded surface lowered the recrystallization temperature locally [9]. Underwood [10] in a review of superplasticity pointed out that metals have unusually high ductility at metallurgical instabilities. Abnormal ductility is found at the recrystallization temperature and is accompanied by low strength. Therefore the effect of recrystallization may be to accelerate the first stage of the bonding process by promoting intimate contact through increased plastic flow rather than by any effect on the second stage involving diffusion.

Surface preparation is less critical when the bonding is performed at such a temperature where the yield stress is low [1]; it appears to be more critical when the diffusion stage is more important in the establishment of bonds [1,2]. This general conclusion is in agreement with more recent studies by Cunningham on the mechanism of bonding [11].

The formation of diffusion barriers inhibits bonding. These barriers may take the form of interstitial compounds [1] or intermetallics [3]. The direction of diffusion may be opposite to the concentration gradients (so-called uphill diffusion), especially to form compounds such as Ti(C,N) at the interface [1]. These barriers are usually brittle.

Melting in the bond at any stage must be avoided. Because of the misuse of the term "diffusion bonding," braze processes in which increases of remelt temperature may occur on annealing have been included under this category. However, even diffusion welds may suffer from this problem, for example, as the lowering of the melting point of the bond by eutectic formation such as Ni-W or Ni-Cr-W [4].

The effect of surface roughness is difficult to evaluate. At stresses and temperatures where plastic flow is restricted increased roughness is a disadvantage [2] and smooth surfaces are preferred. At higher temperatures surface roughness is not important. The explanation may be associated with residual cold work [1] or with the higher surface energy.

Formation of a diffusion weld can be studied best by separating it into three separate stages. The first stage involves the initial contact of the interfaces. This includes whatever deformation of surface asperities (microscopic roughness) and films is needed to establish some degree of initial mechanical contact. Second, there is a time-dependent deformation of the same interfaces that further establishes intimate interfacial contact. Finally, there is a diffusion-controlled elimination of the original interface. This final step may occur as the result of grain growth across this interface, the solution or dispersion of interfacial contaminant, or by simple diffusion of atoms along or across the original interface.

The initial interfacial accommodation occurs by the plastic flow of the surface asperities and by the simultaneous rupturing or displacement of surface films. Surface films, if present, may be oxides, organic deposits, gaseous, or aqueous films. This initial deformation is caused by the compressive load applied to the interfaces to be joined. It occurs essentially instantaneously at the welding temperature of the joint or at ambient temperature if the parts are assembled before heating.

Cunningham and Spretnak [12] have shown that the initial deformation of the interfaces of two material surfaces can be related to Meyer hot hardness. Their work in OFHC (Oxygen Free High Conductivity) copper was based on observations in specially prepared, geometrically grooved surfaces with finishes from 8 to 500 rms. The tests were conducted with both isostatic and mechanically applied pressure. The unit interfacial pressures necessary to achieve intimate joints were closely related to the hot hardness at each test temperature. Hardness measures the yield strength of a material plus a work-hardening factor that prevails in the range of plastic flow undergone during the hardness test. Thus the correlation to hardness indicates that most of the deformation observed was of short duration. Not all diffusion welding experiments are conducted at pressures that exceed the materials yield strength at the welding temperature [13]. In these instances one must consider that the deformation at the interface is not instantaneous but rather time dependent. This is borne out by the fact that in low pressure experiments much longer times or much higher temperatures are needed to accomplish welds and eliminate interfacial defects. This indicates that time-dependent deformation, or creep, is a reasonable concept to include as a mechanism in stage two in this discussion. The relative significance of stages one and two is governed principally by the combined effects of time, temperature, pressure, and prior condition of the materials to be joined. When high pressure and high temperatures are used, it is likely that extensive instantaneous local plastic deformation occurs, and the necessity for creep deformation is diminished. Similarly, in cold-pressure welding applications, when creep and all diffusion-controlled processes are reduced to insignificant levels, the most important criterion is immediate and extensive plastic flow. This has been shown in numerous investigations of many types of cold-welding applications where the criterion for obtaining successful welds is simply for deformation to exceed a minimum threshold value [14]. The temperature dependence of threshold deformation in pressure welding would indicate direct comparison of this process to the work of Cunningham and Spretnak. Thus it is necessary to consider two separate forms of deformation in diffusion welding. The specific parameters of a given experimental condition govern their interrelationship.

In diffusion welding the first two stages serve to set up the third stage, the elimination of the original interface. This is essentially controlled by diffusional processes in its entirety. The influence of diffusional processes is felt in numerous ways that tend to overlap with the second stages of the welding process. The creep that dominates the second stage is largely a diffusion activated process. All proposed mechanisms of eliminating the original interface are diffusion controlled. Four possible mechanisms have been proposed to be involved in this final stage:

1. Atom transport occurs across the original interface and in doing so creates bonds between the parts [15]. This occurs by volume diffusion.
2. Recrystallization and/or grain growth occurs at the interface, resulting in the formation of a new grain structure that sweeps across the original boundary. It has also been suggested that during recrystallization the yield strength of a metal is essentially zero. Thus total accomodation of the interface by deformation can occur with little or no applied pressure [16]. This brings surface atoms into sufficiently close proximity to permit metallic bonding at the interface.
3. Surface diffusion and sintering action cause the interface to grow together rapidly [17].

4. Surface films and oxides are dissolved by the base metal and in doing so eliminate these barriers that resist the formation of normal metallic bonds. In the absence of interfering films these bonds can form spontaneously [18].

The concept of a bulk diffusion process simply transporting atoms across an interface to promote welding is rather inadequate. Ideally, atomic interchange in the bulk is not possible until interatomic distances across the interface are approximately equal to the lattice parameter. When the original interfaces reach this state of proximity, they become a grain boundary and except for contaminant atoms this state is essentially as strong as the metals to be joined that contain numerous grain boundaries. Therefore the entire interface is already in complete contact and the weld can be assumed complete.

Recrystallization and grain growth have become recognized as very significant factors in the final stages of diffusion welding [16,19]. Parks [16] even thought it appropriate to rename the process "recrystallization welding" because of the coincidence of the temperature of rapid weld formation with the recrystallization temperature in his experiments. It is apparent that recrystallization can have a very potent influence on the formation of a diffusion weld if it occurs while the interfaces are in contact under pressure. However, in some circumstances, diffusion welds have been made between materials that have been recrystallized by preweld annealing [12]. It has also been possible to make diffusion welds successfully in materials below their bulk recrystallization temperature by the application of adequately high pressure for a long enough time. The desire to obtain diffusion welds in refractory metals without recrystallization has been the subject of considerable research because recrystallization raises the ductile-brittle transition temperature significantly in most columbium, molybdenum, and tungsten alloys [20]. It has been reportedly possible to obtain strong diffusion welds in these metals without gross recrystallization. It should be noted in this aspect that it is difficult to detect recrystallization in extremely thin surface layers. As a result observations reporting recrystallation must be regarded cautiously.

These results suggest that although the occurrence of recrystallization can distinctly aid the diffusion-welding process, it is not apparently requisite to its success. It is reasonable to state that recrystallization or other grain boundary motion at the interface has a most beneficial effect and should be introduced, if possible, to aid bond development.

The results of some work at low pressure and relatively high temperature have suggested that surface diffusion may be a significant factor in the bonding mechanism. Tylecote et al. [17] calculated that the activation energy for diffusion welding copper under certain conditions was 10,000 cal/molecule. He estimated that this was the approximate activation energy for surface self-diffusion in copper. The estimate was based on the known values of volume diffusion in copper and the relative values of activation energy for volume and surface diffusion in nickel. Tylecote reasoned that if the mechanism for weld formation in their experiments was recrystallization, one would expect that the activation energy would more closely approximate that of volume diffusion. Experiments have shown that recrystallization activation energies do correspond to those of volume diffusion. In separate experiments, however, activation energies for surface diffusion in copper have yielded higher values [21,22] than Tylecote predicted. Activation energy calculations are not always indicative of mechanisms that may be complex combinations of metallurgical phenomena.

The solution of surface films is significant in alloy systems that have tenacious oxide films and that dissolve their own oxides. Oxide films retard bond formation very drastically [18]. Pressure welds made at low temperatures are particularly affected by oxide films. The importance of oxide films and other adsorbed or chemisorbed layers has been recognized generally in all diffusion welding work. In diffusion welding surface films can sometimes be dissipated by the exposure to elevated temperatures. This is particularly true if the surface film is volatile or if it can be dissolved in the base metal. However, it is customary to observe meticulous precautions during surface preparation before diffusion welding.

It can be seen from the preceding discussion that to propose a simple universal model for all types of diffusion welding is very difficult. The mechanism can involve one or any combination of a number of metallurgical and mechanical events. The specific conditions under which a diffusion weld is made, as well as the results desired from the weld, determine which of the phenomenological events are dominant. The following elementary model, however, has been proposed as a general description of the mechanism of diffusion welding. Initially two surfaces are brought together under load. If the pressure exerted on the joint is high or the temperature quite high, there is considerable plastic flow of the surface asperities until the interfaces have achieved a high degree of conformity with each other. At this point the joint has considerable strength because at certain regions of the interface metallic bonds form. If the initial pressure is lower, this same surface conformity may be achieved at longer times by creep and/or surface diffusion of atoms. During the deformation any thin surface films are seriously disrupted and some plastic work may be put into the surfaces.

Since the joint is held for some finite length of time at elevated temperature, there is a great degree of atom mobility in the joint. Recrystallization or grain boundary motion may occur to extend and further strengthen atom-to-atom bonds and cause further disruption of the surface films at the joint. Both dissolution and agglomeration of contaminants may occur and both remove the barriers to the formation of additional metallic bonds. Again, these processes occur to varying degrees depending on the temperature, time, interfacial deformation, metal characteristics, and other factors. In the last stages of joint formation some additional exchange of atoms occurs across the initial interface, thereby assisting in the process of structural and chemical homogenization of the joint area.

Another significant theory that has been proposed revolves around the essential role of diffusion. The diffusion phenomenon is of prime importance from both the theoretical and practical aspects of metallurgy. This is because of the many phase changes that take place in metal alloys involving a redistribution of the atoms present. These changes occur at rates that are dependent on the speed of the migrating atoms.

Diffusion in metal systems is usually categorized into three different processes depending on the path of the diffusing element. Each of these processes that have been discussed previously are volume diffusion, grain boundary diffusion, and surface diffusion, and each one has different diffusivity constants. The specific rates for grain boundary and surface diffusion are higher than the rate for volume diffusion. However, because of the small numbers of atoms in these regions, their contribution is small compared with the total diffusion process.

The basic equation for diffusion in metals in Fick's first law is given as

$$dm = \frac{-D(c/x)}{A\,dt}$$

where dm is number of grams of metal that cross a plane perpendicular to the direction of diffusion; D is diffusion coefficient whose values depend on the metallic system being considered (cm^2/sec) (minus sign expresses a negative concentration gradient); A is the area (cm^2) of the plane across which diffusion occurs; c/x is the concentration gradient that exists at the plane in question (c is expressed in grams per cubic centimeter). The diffusion coefficient D is not generally constant since it is a function of such dynamic variables as temperature, concentration, and crystal structure. Since at any one particular time these variables are assumed to have finite values, the coefficient D is a definite number. Several mechanisms can account for the diffusion of atoms in metals of which two are the interstitial mechanism and the vacancy mechanism.

Interstitial mechanism is concerned with the movement of atoms having small atomic radii compared with the matrix atoms. These elements move from one location to another along the interstices of the crystal lattice, hence the name interstitial elements. These moves occur within the crystal without distorting or permanently displacing the matrix atoms. The matrix or substitutional atoms use the vacancy mechanism for their mode of transportation. Because of their size, it is literally impossible for these atoms to migrate along the interstices. The only path open to them is that of vacancy sites. Although the energy required to move a matrix atom is equal to that for an interstitial element, the rate is considerably slower. This because of the fewer number of vacant locations available to the atoms.

The pronounced effect that temperature has on diffusion may be evaluated by the rule of thumb that a 20°F rise in temperature doubles the diffusion constant. It has been found that the diffusion constant changes with variations in concentration. For example, the diffusion constant of carbon in iron at 1700°F will show a threefold increase over a range of carbon from 0 to 1.4%. Crystal structure has also been found to influence the diffusion constant at a given temperature. Self-diffusion of iron occurs 100 times more rapidly in ferrite than in austenite. Directionality of a crystal also influences the diffusion constant. It has also been shown that crystal distortion due to plastic deformation usually increases with the rate of diffusion.

Definition

The widespread difference in descriptive terminology can probably be traced to a combination of factors. There has always been the desire on the part of an investigator for individuality. Thus by giving his particular program an original title, he can attain this distinctiveness. There is also the consideration of patents, particularly in a relatively new field. Still another major factor is the lack of an adequate official definition describing the bonding process. To avoid any misunderstanding the following definition is used to describe the joining technique discussed below:

Solid state or diffusion welding is a joining process wherein diffusion caused by time, temperature, or pressure is produced through the coalescence of contacting surfaces.

The two key words in the definition are "diffusion" and "coalescence." Diffusion depicts the mechanism by which actual joining occurs. Coalescence defined as "to grow or come together into one" characterizes the final condition in the joint after the bonding process is completed. The diffusion welding process is not a complicated procedure. As already stated it is fundamentally the growing together of the metals to be joined while in the solid state. The resultant bond should have essentially the same properties (mechanically and metallurgically) as the base metal. Any deviation from these properties only points to a deficiency in the bonding procedures that can be traced to the joint interface. Time, temperature, and pressure are the three essential ingredients of this process. The metals to be bonded are placed in contact with each, and this union is maintained by applying external pressure. The assembly is then heated to a specific temperature and held at this temperature for a specific time. Actual times and temperatures vary from metal to metal. While in intimate contact at temperature, the atoms in the metal acquire increased energy from the elevated temperature to which they are subjected. This energy causes the atoms to become more and more mobile and in time they migrate from one location to another. The transport of atoms across the original joint interface results in a solid-state bond after a specific period of time. The migration or diffusion takes place along several different paths and at various diffusivity rates. It is of interest to note that at temperatures close to the melting points of many metals each atom may change sites as many as 10^8 times/sec. The effect of pressure on the bonding mechanism is twofold. It is used primarily to maintain intimate contact between the parts to be joined so that diffusion can take place. A second effect is that of distorting the crystal lattice of the materials so that a greater number of vacancy sites are available for atom migration. These essential ingredients are discussed in further detail in the next section.

MECHANISMS OF PROCESS AND KEY VARIABLES OF PROCESS

Since diffusion welding is any joining process in which two or more solid phases are metallurgically joined without the creation of a liquid phase, the term "metallurgically joined" can be used to describe weld formation by the action of atomic forces rather than solely by mechanical interlocking or by a nonmetallic adhesive. Diffusion welding can be divided into two general categories: deformation welding and diffusion welding. Deformation welding includes those techniques in which gross plastic flow that promotes intimate contact and breaks up surface oxides is the principal factor in the formation of a weld and diffusion is not essential. In diffusion welding deformation occurs only on a microscale, and diffusion is the principal factor in the formation of a weld. Examples of terms used to describe joining processes in each of these classes are the following:

Deformation Welding	Diffusion Welding
Pressure welding (yield stress controlled)	Gas pressure bonding (isostatic)
Roll welding (yield stress controlled)	Vacuum diffusion bonding (creep controlled)
Forge welding	Diffusional bonding
Explosive welding	Press or die pressure bonding
Friction welding	Transient melt and eutectic Diffusion bonding

DEFORMATION WELDING

At present two theories exist that attempt to explain the mechanism of deformation welding. The "film theory" proposes that when two clean metallic surfaces are brought into intimate contact, welding occurs. Plastic deformation, in this theory, is therefore necessary to break up surface films that prevent metal-to-metal contact and weld formation. The "energy barrier" theory suggests that in addition to bringing two clean metallic surfaces into intimate contact, an energy barrier must be overcome before a weld can be created. Deformation supplies this energy. Originally, diffusion was thought to be the principal mechanism that produced the metallurgical bond in joining processes such as forge welding. However, when deformation welding at room temperature, where diffusion rates are very low, was perfected, diffusion was no longer a satisfactory explanation. At the present time no completely satisfactory mechanism has been formalized. It is generally agreed, however, that three factors are important: (a) surface contamination, (b) an "energy" barrier, and (c) elastic stresses. It is the relative importance of these three factors that is presently unresolved.

SURFACE CONTAMINATION

A real metallic surface is not clean and is rough. The influence of both surface cleanliness and roughness on the deformation-bonding process is of major importance. Concurrently with the development of cold-pressure welding, the film theory of deformation welding was developed. In essence, the theory states that if two perfectly clean surfaces were forced into intimate contact by means of deformation, a weld would be produced. Since a perfectly clean surface does not exist, in addition to bringing the surfaces into contact, the plastic flow breaks up the surface films and allows uncontaminated areas to come into contact to produce a weld. The success of wire brushing as a preweld cleaning technique for the deformation bonding of aluminum was thought to result from the formation of a hard brittle surface layer on the faying surfaces of the components. When the two layers came into contact, they adhered to each other and broke up as one layer. This resulted in a potential bonded area equal to the percentage-increase in area the composite has undergone. Therefore the maximum possible weld strength was obtained when the entire new area was welded.

The deformation at which welding is first detected is termed the threshold deformation and is a function of (a) the particular material and (b) surface cleanliness. There is a general trend toward higher threshold deformations with increasing melting point or hardness of the metal. Generally, the cleaner the surface the lower the threshold deformation for a particular metal. However, a zero threshold deformation has not been observed for even the best surface conditions obtained.

ENERGY BARRIER

The existence of a threshold deformation was one of the factors that led to the development of the energy-barrier theory of deformation welding. This theory states that an energy barrier would have to be overcome to form a weld below the recrystallization temperature even if perfectly clean metal surfaces were brought into intimate contact. The decrease in surface energy because of the elimination of the free surfaces of the two components is not sufficient to create a weld spontaneously. The energy barrier is related to the energy required to rearrange the atoms to form the weld, essentially a grain boundary. The role of deformation is to increase the energy of the system until the barrier is surrounded. As yet the theory has not been developed in sufficient detail to permit comparisons between predicted and experimental results. However, qualitatively the effects of alloying can be explained in terms of effects on the ability of atoms to move freely and are related to the degree of lattice strain and alterations in interatomic forces.

ELASTIC STRESSES

The considerations of the importance of elastic stresses in deformation welding were initiated in the field of friction and wear. Investigators in this field found that some surfaces that exhibited a friction force attributed to bonding did not show any adhesion when examined under conditions of no sliding. Using the simple geometry of a hard sphere pressed into a flat surface, calculation was done on the elastic recovery when the load is released. The recovery resulting from the elastic stresses was found to be large enough to break any weld formed under the load. The threshold deformation was then postulated to be the deformation required to exceed exactly the elastic recovery. Since the threshold deformation can be reduced greatly by improved cleaning, the effect of elastic recovery is probably rather small. It may, however, be as significant as the energy barrier previously discussed.

The discussion of deformation welding becomes more complicated when the effects of increased temperature are considered. The effects include: (a) modification of surface contaminants, (b) increased atom mobility, and (c) lower yield strength.

Increased temperatures may increase the solution of surface oxide films in the base material or even cause melting of the oxide. At elevated temperatures diffusion becomes more significant in the formation of a bond. This effect is most pronounced at an allotropic transformation or at the recrystallization temperature. As the temperature is raised the yield strength decreases until it is zero at the recrystallization temperature. This reduction in yield strength permits easier plastic flow and promotes the formation of a bond. Even though diffusion is enhanced, deformation is still required to form a bond because the time under load for most deformation welding processes is very short. There are several other factors that have their effects on deformation welding and a review of their significance and their role in welding follows.

Degree of Deformation

The bond strength of aluminum as a function of deformation when rolled at room temperature is reported by Vaidyanath and Milner [23]. No bonding was detected for deformation less than 40%. At about 65% reduction the bond strength reached the strength of a solid specimen which had been reduced the same amount. Conn reported successful roll diffusion welds in Ti-6AL-4V sandwich panels with single pass reduction of 18% [24]. The sandwich panel components were encapsulated in mild steel tooling and mill-rolled at 1750°F. Total reduction of 80% and first pass reduction up to 30% have produced uniform bonding with excellent geometric control. Several researchers have reported that the material being roll diffusion-welded determines the rolling temperature and to some extent the amount of reduction in the roll-welding process.

Surface Condition

A few investigators have attempted to measure surface cleanliness by contact resistance determinations and then to correlate these measurements with bond strengths. Vaidyanath and Milner reported the influence of surface preparation on joint strength and degree of deformation for room temperature roll welding of aluminum [23]. These investigators concluded that wire brushing after chemical cleaning was a particularly effective surface preparation because of the following:

1. It removes surface contaminants.
2. It forms hard rough layers on the surface which break up to allow virgin metal to be exposed for bonding.
3. It may harden the surface to promote recrystallization of the surface layers.

Further studies of surface influence revealed that even heavily oxidized surfaces could be bonded if they were first baked out and allowed to cool in a desiccator to prevent reabsorption of contaminants. Generally, a surface

Mechanisms of Process and Key Variables of Process

preparation that resulted in a low contact resistance permitted bonding to occur at a lower threshold deformation for any given temperature.

Conn reported that surface condition of component details and mild steel filler bars is critical in the roll diffusion welding of Ti-6AL-4V sandwich panels [24]. The titanium alloy components were cleaned as follows: condition scale, rinse, acid descale, rinse, nitric acid dip, rinse, and dry. Tests conducted with carburized steel filler bars resulted in severe cracking, reportedly from embrittlement caused by outgassing of the carburized bars. Flame spraying the titanium alloy bond interfaces with aluminum oxide prevented diffusion during the rolling operation. Because of the affinity of titanium and its alloys for atmospheric gases at bonding temperatures, deformation welding requires strict control of the atmosphere before, during, and after rolling.

Temperature

Increased temperature, as mentioned previously, results in increased atom mobility, desorption of surface contaminants, decreased yield strength, and thus improved bonding. Vaidyanath, Nicholas, and Milner reported that the threshold deformation of aluminum is reduced significantly by increasing the bonding temperature [14].

Conn reported complete bonding of Ti-6AL-4V when roll reduced at temperatures of 1650, 1750, and 1850°F [24]. The 1850°F temperature, however, resulted in a coarse acicular structure in the titanium and an undesirable surface roughness. Note the overheated condition at the edge of the vertical leg of joint shown in Figure 9.3. The 1750°F roll-welding temperature was too high for the Ti-6AL-4V alloy. Currently, no successful deformation welds of titanium have been made at room temperature.

Composition

In general, alloying additions that increase the base-metal hardness will increase the threshold deformation required for bonding. Some researchers have reported some difference in diffusion bonding characteristics related to alloying content; this is apparently because of the more rapid reabsorption of contaminants caused by some of the alloying constituents.

Prior Cold Work

The effects of this factor have been clearly defined. However, it has been reported by Tylecote that prior cold work has little effect on the deformation welding of aluminum by cold rolling [17]. Vaidyanath,

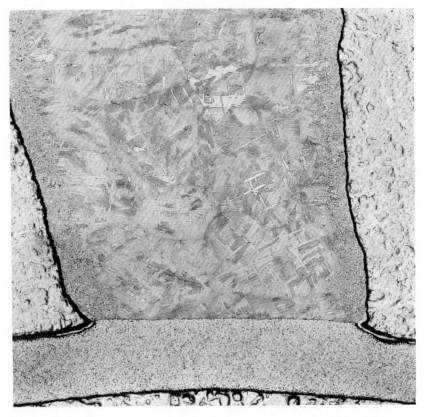

Figure 9.3 Roll welded stiffener plate (0.032 in.) –8AL-1Mo-1V Ti alloy to 6AL-4V Ti alloy. (50×)

Nicholas, and Milner reported the effect of prior cold work on the deformation welding of aluminum, which is in agreement with Tylecote [23].

Crystallographic Orientation

Any effect of crystalline structure orientation on welding is probably caused by the effect of the differing mode of deformation of the three planes on the breakdown of the oxide film.

Postheat Treatment

Subsequent to deformation welding, thermal treatments such as solution treatment and aging can be used to strengthen the welded joint. The effects of postheat treatment on the bonded strength of aluminum cold bonds from work reported by Tylecote and Wynne shows that the initial increase in strength begins to deteriorate as time and temperature is increased [25]. Finally, the above work shows that low deformation bonds exhibit a greater percentage increase in strength than high deformation bonds.

The results of postheat treatment on roll welded Ti-6AL-4V alloy was reported by Conn [24]. Postroll annealing temperatures were evaluated in the 1200 to 1800°F range before removing the steel tooling from roll welded panels. Four hours at temperatures up to 1400°F did not improve bond imperfections substantially. Four hours at 1500° resulted in major bond improvement without significant iron diffusion.

DIFFUSION WELDING

In contrast to deformation welding, diffusion welding has been classified as the process in which diffusion is the principal factor in the formation of a weld. Since diffusion rates at room temperatures are generally quite low, diffusion welding requires intimate contact of relatively clean metallic surfaces at elevated temperatures. It is therefore important to discuss what constitutes clean metallic surfaces in diffusion welding.

METAL SURFACES

A knowledge of metal surfaces is essential to understanding the principles of diffusion welding. Two dominant characteristics must be considered to describe adequately a real metal surface: (a) roughness and (b) impurities or cleanliness. A breakdown of a real surface would consist of base metal, an oxide layer, and a very thin adsorbed liquid or vapor. On a microscopic scale, surfaces are extremely rough. Surfaces prepared mechanically, such as by machining or grinding, consist of grooves or scratches. In addition to these primary or first-order scratches, secondary ridges exist on the faces of the primary scratches. Even the most highly polished surfaces will have primary peak-to-valley distances of roughly 500 Å and ground surfaces will be many times rougher than this. Therefore the initial contact area produced when two metallic surfaces are brought together are very much less than the nominal surface area. It has been estimated that the initial contact area could be 10^{-6} times the nominal area. Because of surface roughness, a load normal to the interface between two surfaces is supported only by the surface asperities that plastically deform until the actual area of contact is capable of supporting the applied force. The true area of contact is proportional to the load and, for a given material, is independent of the size or shape of the surfaces.

From this discussion it is apparent that metal surface roughness is one factor that must be dealt with in diffusion welding.

The metal atoms at the surface of a metal have incomplete shells of neighbor atoms. This leaves them with the ability to attract and hold other atoms such as oxygen from the air. All metallic surfaces that have been exposed to air are covered with an oxide film. The thickness of this film varies with the chemical properties of the metal and with the environment. The oxide film on noble metals may be only several atomic layers thick, whereas on most other metals it is much thicker. Such an oxide layer can build up to a considerable thickness before becoming visible as a film of tarnish. For example, a bright, apparently clean steel surface may have on it a 200-molecule thick layer of iron oxide. In air at atmospheric pressure, a monolayer of oxide forms on a surface in about 4×10^{-9} sec, whereas at a pressure of 10^{-6} torr, approximately 3 sec are required.

The surface molecules in the oxide film ion, similar to the underlaying metal atoms, can attract other molecules and atoms. These attractive forces, although relatively weak, lead to the adsorption of gases and moisture. The result is an adsorbed layer on top of the oxide. This layer varies in thickness, but is usually never less than 2 or 3 molecules thick.

To summarize then, an ordinary bright, metallic surface is a hill and valley matter, the hills being several hundred to many thousand atom diameters high. On this metal surface is a layer of metal oxide usually 20 to 200 molecules thick and adsorbed on this is a thin layer of moisture. Therefore we see that in diffusion welding we are not dealing with the ideal smooth, clean surface, but with rough, dirty surfaces. Obviously, then, in diffusion welding one of the first jobs is to try and get the surfaces as clean as possible. In fact, surface preparation is usually one of the most critical steps. This cleaning is done by chemical methods, such as acid baths, or by mechanical methods, such as wire brushing or grinding. Some of these methods simultaneously smooth and clean the surface, but as pointed out previously the result is far from a perfectly smooth, clean surface. Therefore each diffusion welding method must overcome these problems in order to form strong joints.

For convenience, the diffusion welding process can be considered as a two-stage process. However, in reality, the two stages can occur at the same time. The two stages are (a) microscopic plastic deformation resulting in intimate metal-metal contact and (b) diffusion and grain growth to complete the weld and ultimately eliminate the interface formed in stage (a). The definition of a diffusion weld necessarily must be arbitrary. Some authors consider that the interface must be completely eliminated before the weld is complete. However, at some point before elimination of the interface the weld develops a strength as great as that of the base metal. The definition of the diffusion weld in terms of strength has been termed the engineering definition, whereas the definition in terms of the elimination of the interface has been termed the scientific definition. In the following discussion the scientific definition is applied.

As with deformation welding, the nature of a real surface must be considered as a prime factor in diffusion welding. First, as discussed previously, a real surface is never perfectly clean. In deformation welding one of the important effects of the gross plastic flow is to break up surface films. Since only microscopic flow occurs in diffusion welding, techniques must be employed to obtain the cleanest surfaces possible before welding. Second, real surfaces are very rough on an atomic scale, so local plastic flow must be produced to create sufficient metal-to-metal contact.

PLASTIC FLOW

The factors that are important in producing enough localized plastic flow include (a) applied pressure, (b) welding temperatures, (c) surface roughness, and (d) work hardening.

Sufficient pressure must be applied to cause localized plastic flow at the points of contact along the interface. Increasing the bonding temperature lowers the yield strength and permits flow to occur at lower applied loads. Surface roughness has two opposing effects on the area in contact along the interface. Smooth surfaces brought together have a larger area in contact than rougher surfaces. As the load is applied, however, the unit pressure on the points of contact for rough surface is higher than for smooth surfaces. Therefore plastic flow occurs at a lower applied load for the rough surfaces than for the smooth surfaces. For a given load the contact area for the rough surfaces therefore increases more than that for smooth surfaces. Work hardening of the asperities in contact is only a factor in welding done below the recrystallization temperature of the material. If work hardening does occur, it reduces the amount of plastic flow for a given applied load.

The nature of the interface produced when the two metal surfaces are brought into intimate contact can be deduced from indirect evidence. Under the action of etchants, the interface etches to give the appearance of a wider, high-energy region than do grain boundaries. Diffusion along an interface has been shown to be more rapid than either volume or grain-boundary diffusion. The character of the interface must differ from point to point. The areas of initial contact would be more severely deformed than those that came into contact later in the deformation process. Voids could be formed at points that never came completely together. In summary, the interface can be pictured as a nonuniform, high-energy region along which material transport (diffusion) is rapid.

Diffusion and Grain Growth

Two mechanisms important in eliminating the interface to form a weld are diffusion and grain growth. To eliminate voids in the interface diffusion must occur. A necessary requirement is the presence of effective sinks for vacancies that must diffuse away from the interface. Moving grain boundaries and, in some instances, dislocations are effective vacancy sinks. Diffusion can be affected by both temperature and pressure. Increasing temperature increases diffusion. Pressure affects diffusion particularly when an intermetallic compound is formed. In some cases, for example, Al-Ni, high pressures restrict the width of the intermetallic zone formed during bonding. These mechanisms are discussed below.

Diffusion is a basic phenomenon found in al metals at temperatures above absolute zero. It consists of the migration of atoms or vacancies throughout the structure of the metal. The rate at which this diffusion occurs is significantly dependent on temperature: the higher the temperature the more rapid is the diffusion. When diffusion is vacancy controlled, the diffusion coefficient, D, may be expressed by

$$D = D_0 e^{-(E_a + E_m)/kT}$$

where D_0 is a constant which may be structure dependent, E_a is the energy required to create a vacancy in the metal, E_m is the energy required to move a vacancy, k is Boltzmann's constant, and T is the absolute temperature. In most metal systems diffusion is very slow at room temperature, but increases rapidly (exponentially) as temperature is raised.

To cause an atom to diffuse its energy must exceed $E_a + E_m$, the activation energy of the process. The atoms in a metal at temperature T have thermal energies occurring on a Boltzmann's distribution. As the temperature increases larger percentages of the atoms have the requisite energy to exceed the activation energy for diffusion. Time is also important in diffusion processes. In the general case, diffusion, as measured by the change in connection gradient C with distance from the interface x and time t, is expressed by Fick's second law

$$\frac{\partial c}{\partial t} = \frac{\partial}{\partial x}\left(D \frac{\partial c}{\partial x}\right)$$

For diffusion across a plane interface between adjoining columns of solute and solvent, with the columns sufficiently long so that there is no appreciable change in composition at either end and where D is also constant, the solution can be shown to be

$$c = \frac{C_0}{2}\left[1 - \frac{2}{\sqrt{\pi_0}} \int^{x/2} \sqrt{Dt}\, e^{-y^2}\, dy\right],$$

where c is the concentration after a time t at a distance x from the interface, C_0 is the initial concentration of the solute, and D is the diffusion coefficient. The second term in the bracket is the probability integral or gauss error function, and y is used as the integration variable with a purely mathematical significance.

Time and diffusion coefficient can be seen to be equally important, for the quantity Dt determines the change in concentration at any given distance from the interface. This quantity is equal to the length that the atoms will move and roughly provides an estimate of the region over which the concentration will change substantially.

Grain growth in the region of the interface results from several driving forces: strain energy, interfacial energy, and in some systems, the energy associated with allotropic transformations. The presence of voids and in some instances impurities in the interface increases the energy required for grain growth. Recrystallization furnishes additional energy and aids in elimination of the interface.

Intermediate Materials

Although not a part of the mechanism of diffusion welding, the use of intermediate materials in diffusion welding is of considerable importance. Intermediate materials can be used to

(1) promote diffusion at much lower temperatures than for self-welding;
(2) promote plastic flow and surface conformance at lower pressures;
(3) prevent the formation of intermetallic compounds;
(4) obtain clean surfaces.

Intermediate or activating agent materials are commonly used in either the form of foil, electroplates, or as vapor deposits. Foreign-atom diffusion rates are generally much greater than self-diffusion rates. Thus the intermediate layer can accelerate the diffusion stage in the bonding process. Intermediate materials with low yield strengths permit larger contact areas for a given applied pressure. In systems that form intermetallic compounds an intermediate layer is used to restrict the interdiffusion of the components and thus prevent the formation of brittle compounds. For metals like aluminum, which are difficult to clean for bonding, an electroplate of some other more readily cleaned metal, such as nickel, can be used to facilitate bonding.

Excessive diffusion may produce voids by coalescence of microvoids not detectable in the as-welded structure, or by expansion of trapped gas, or as a result of the Kirkendall effect. In addition to changes in the weld structure, extensive diffusion into the base metal may occur, causing recrystallization at a lower than normal temperature, extensive grain growth, formation of intermetallic phases, segregation of impurities at grain boundaries, or changes in any number of physical or mechanical properties related to composition and structure.

Activation Concepts

The purpose of the intermediate layer of material or activation agent is also to make bonding possible in very short times. As stated previously, welding is a two-stage process. Stage one is largely mechanical in nature and involves plastic deformation of the interfaces to achieve intimacy of contact and disrupt surface films. Stage two is the strengthening of this contact by diffusion.

The activation process for stage one should reduce the flow stress or yield stress at the interface so that deformation is limited to this region and is not general. The requirements to achieve stage one rapidly may be subdivided into the following:

1. General methods to reduce flow stress include high temperature, high pressure, and use of low flow stress intermediate metals.
2. Methods to localize flow stress are use of low flow stress intermediate metals, surface active diffuser to lower recrystallization temperature, cold work on surface to lower recrystallization temperature, and yield stress as a result of superplasticity, local dissipation of ultrasonic energy, and bonding at temperature of a metallurgical metastability (allotropic transformation, etc.) to promote superplasticity [19].

In stage two it appears that activation is less likely to be applicable because if intimate contact between clean surfaces has been established in stage one, few atom layers need to interdiffuse to establish the bond. Because the bond temperature is high, the time for these atom transfers is very small. Approaches to accelerate stage two, however, (i.e., strengthen the bond by diffusion) are the following:

- Use of intermediate metal with high diffusion rate of the components (diffusion aid)
- Use of dissimilar metals to increase concentration gradient
- Cold work (assuming cold work promotes recrystallization and hence atom movement)
- Metallurgical instability at interface (allotropic transformation, solid solution boundary, etc.)
- Use of a dissimilar metal with high surface diffusion rate
- Application of fields such as high frequency or ultrasonic [19]

Mechanisms of Process and Key Variables of Process

One other method to accelerate stage two of a diffusion welding process and reduce the time per bond would be to complete the bonding and diffusion by a furnace anneal in which large numbers of bonds could be treated simultaneously. When these approaches, however, are applied to the refractory metals, other criteria must be met. Principal among them are the following:

- Brittle compounds must not form at the interface.
- Segregation of elements from the base metal must not be caused by uphill diffusion, for example, to form interstitial compounds of titanium or zirconium.
- Diffusing elements must not raise the ductile-brittle transformation markedly.
- No low-melting point residuals can be tolerated; for example, nickel used in bonding tungsten must be diffused into solid solution to avoid melting at the Ni-W eutectic [19].

DIFFUSION WELDING PROCESS VARIABLES

A discussion of process variables must include five principal parameters: surface preparation, temperature, time, pressure, and one called special metallurgical effects.

Surface Preparation

The surfaces of parts to be diffusion welded normally are carefully prepared before application of heat or pressure. A universal philosophy which is the bylaw for the preparation of parts is maximum attainable cleanliness. Surface preparation, however, involves more than just cleanliness. It also includes generation of an acceptable finish or smoothness, removal of chemically combined films (oxides, etc.), and cleansing of gaseous, aqueous, or organic surface films.

The primary surface finish is obtained ordinarily by machining, abrading, grinding, or polishing. One property of a correctly prepared surface is flatness. A certain degree of flatness and smoothness is required in order to assure that the interfaces can achieve the necessary compliance without excessive deformation. Machine finishes, grinding, or abrasive polishing are usually adequate to obtain the surface flatness and smoothness needed. A secondary effect of the primary machining or abrading which is not always apparent is the plastic flow introduced into the surface during machining. Cold-worked surface layers have lower recrystallization temperatures than the bulk material. The importance of recrystallization during diffusion welding has already been mentioned. Furthermore, the diffusion rates in cold-worked materials are higher than for annealed material due to a much higher defect concentration [27]. A problem encountered in evaluating the effect of cold work is the difficulty in measuring the amount, depth, and reproducibility of cold work caused by machining. It has often been suggested that a cold-worked surface is an effective aid in diffusion welding [16]. Although it has been found that recrystallization is not necessary in certain cases, it has been reported that it is beneficial to prepare surfaces in such a way that it reduces the likelihood of having surface work [20]. As a result the optimum techniques for machine preparation remain somewhat controversial.

Chemical etching or pickling, commonly used as a form of preweld preparation, has two effects. The first is the removal of nonmetallic surface films, most frequently oxides. The second is the removal of part or all of the cold-worked layer that occurs during preliminary surface preparation. The benefits of oxide removal are apparent because it is difficult to cause two oxidized surfaces to adhere. Many chemical solvents are suitable for use with different metals systems. It would be prohibitive to list all pickling baths because of their number and also because of lack of general agreement on the effectiveness of a solvent for any given alloy. A useful starting point for chemical preparation might lie in references dealing with metallographic etches [28]. Manufacturers' literature may also provide useful information on pickling agents.

Degreasing is a universal part of any procedure for prediffusion weld cleaning; alcohol, trichlorethylene, acetone, detergents, and many others have been used by numerous investigators. Choice seems to rely on individual preference. Frequently, the recommended techniques are very intricate and may include multiple rinse-wash-etch cycles in several solutions.

Vacuum bake-out has also been used occasionally to obtain clean surfaces. The usefulness of vacuum bake-out depends to a large extent on the material and the nature of its surface films. Organic, aqueous, or gaseous, adsorbed layers can be easily removed by vacuum heat treatment at temperatures and pressures that cause boiling of these relatively volatile materials. Oxides are not easily dissociated by vacuum bake-out, particularly on such materials as titanium, aluminum, or alloys containing significant amounts of chromium. However, it is possible to dissolve adherent oxides in some base materials at elevated temperature. Zirconium, titanium, and others have reasonably high solubility for oxygen and will dissolve their own oxides at elevated temperature [18]. During vacuum bake-out oxygen is not available for continued formation of additional oxide. As a result surfaces may be cleaned and made ready for diffusion welding. Baking in vacuum usually requires vacuum or controlled atmosphere storage and careful handling before use to avoid the recurrence of surface adsorbed or chemisorbed layers.

Many factors enter into selecting the total surface preparation treatment. In addition to those already mentioned the specific welding conditions to be used may affect the selection. With higher welding temperatures and pressures, it becomes less important to obtain extremely clean surfaces because increased atomic mobility, surface asperity deformation, and solubility for impurity elements all contribute to the self-removal of surface contaminants during welding. As a corollary it can be stated that to lower the minimum diffusion welding temperature or pressure, it is necessary to provide better prepared cleaner and preserved surfaces.

Surface preservation is a necessary feature in a discussion on surface preparation. It is folly to exercise supreme caution in preparing surfaces before diffusion welding if they are permitted to become recontaminated during subsequent handling. One solution to this potential dilemma lies in the effective use of protective environments during diffusion welding. Vacuum protection during diffusion welding provides continued freedom from contamination as it does during surface preparation. Use of hydrogen as an atmosphere in diffusion welding helps minimize the amount of oxide formed during welding and it may reduce existing oxides. However, it will form hydrides in alloys of titanium, zirconium, hafnium, columbium, and tantalum, which may be detrimental to weld properties. Argon, helium, and possibly nitrogen can be used to protect clean surfaces at elevated temperatures. When these inert gases are used, their purity must be very high. Inert gases offer none of the advantages of chemical or physical activity that hydrogen or vacuum does. Many of the precautions and principles applicable to brazing atmospheres can be applied directly to diffusion welding.

Temperature

Temperature has received much attention as a process variable for a number of reasons.

1. Temperature is the most readily controllable and measurable process variable.
2. In any thermally activated process, incremental changes in temperatures cause the greatest changes in process kinetics compared to other parameters.
3. Virtually all the mechanisms involved in diffusion welding are temperature sensitive.
4. Physical and mechanical properties, critical temperatures, and phase transformations are important reference points in any effective use of diffusion welding.

The results of Parks, as previously mentioned, show that a very abrupt increase in strength occurs in diffusion welded joints at their recrystallization temperature. One difficulty in analyzing Parks' work is that the pressures used to obtain the welds are unknown. Finally, Parks' work showed a tendency to promote recrystallization nearer to the minimum recrystallization temperature during subsequent heating.

One of the most systematic published works on parametric evaluation of diffusion welding was prepared by Kazakov. In his work abrupt change in strength with increased temperature is not evident at any stress or time. He shows a continuous temperature-time-pressure interdependence in which increased temperature results in increased strength. Increases in pressure and time also increase the joint strength.

An interesting comparison can be drawn between the 0.45% carbon steel in Kazakov's data and the 0.20% carbon steel in Parks' work. Parks achieved welds at 820°F in 1020 steel, whereas Kazakov's conditions for good welds were between 1460 and 2020°F. This candidly illustrates how reported results can vary with small differences in composition, pressure, surface preparation, and by the effects of crystallographic changes made by allotropic transformation.

Time

Time is closely related to temperature in that most diffusion controlled reactions vary with time. The diffusion length, δ, is the average distance traveled by the average atom during a diffusion process. It can be expressed as

$$\delta = K \sqrt{dt}$$

where δ = diffusion length,
D = diffusivity,
t = time,
K = a constant.

Thus the time dependence of diffusion reactions is related to $t^{1/2}$.

Data presented by Kazakov illustrate the effect of time as a diffusion-welding variable. These data of Kazakov indicate that increasing time at temperature and pressure increases joint strength up to a point. Beyond this point no further gains are achieved. This illustrates that time is not a quantitatively simple parameter. The simple relationship, which describes the average distance traveled by an atom, does not reflect the more complex changes in structure that result in the formation of a diffusion weld. Although atom motion continues indefinitely, structural changes tend to approach equilibrium. An example of similar behavior can be found in recrystallization.

A deformed sample, when heated, first undergoes recovery, then recrystallizes. At first the formation and growth of new grains are rapid, but as time progresses the rate of grain boundary motion and physical change diminishes. The decrease is caused by the relative stabilization of microstructure through reduction of internal energy. Thus the driving force for continued structural change is also reduced. The rate of atom motion does not decrease significantly throughout this entire sequence of events.

In a practical sense time has been varied over an extremely broad range in diffusion welding from seconds to hours. Realistic factors influence the time allowed for diffusion welding such as delay periods in the apparatus necessary to provide the heat and pressure. When systems have thermal and mechanical (or hydrostatic) inertia, diffusion welding times are longer because of the impracticality of suddenly changing variables. When there are no inertial problems, welding times may be relatively short (5 min to 1 hr). This is also interrelated and certainly dependent on the temperature and other variables. For economic reasons it is desirable to reduce the time necessary for diffusion welding to increase the potential production rates.

Pressure

Pressure is an important variable in diffusion welding. It is less easy to deal with as a quantitative variable than either temperature or time. Pressure affects several aspects of the diffusion-welding process. The initial phase of bond formation is most certainly affected by the amount of deformation induced by the pressure applied. This is the most obvious single effect and probably is most frequently and thoroughly considered. Increased pressure invariably results in better joints for any given time-temperature value. The most apparent reason related to this effect is the greater interface deformation and asperity breakdown that comes from high pressures. There is also the effect of deformation on the recrystallization behavior. Increased pressure (and deformation) leads to lower recrystallization temperature. Conversely stated, the increased deformation accelerates the process of recrystallization at a given diffusion welding temperature.

In practice, diffusion welds have been reported at pressures as low as 6 psi and as high as 40,000 psi. A practical limitation, in addition to metallurgical considerations, is the apparatus that is available to apply pressure and geometry of the joint. Because the pressure needed to achieve success is related so closely to the other parameters of temperature and time, there is a great degree of latitude in the pressure needed to make good welds. Pressure has additional significance when dissimilar combinations are considered. From economic and manufacturing aspects, reductions in welding pressure are desirable. Increased pressures require costlier apparatus, greater need for control, and generally more complex part-handling procedures.

Metallurgical Factors

In addition to those parameters already discussed, there are a number of metallurgically important factors which should be considered. Two factors of particular importance in conjunction with similar metal welds are allotropic transformations and microstructural factors that result in modification of diffusion rates.

Allotropic transformations (or phase transformations) occur in a limited number of metals and alloys. Heat-treatable alloy steels are the most familiar of these, but other metals, such as titanium, zirconium, and cobalt, also undergo allotropic transformations. One reason for the importance of the transformation is that the metal is very plastic while undergoing a transformation. This tends to permit more rapid initial interface accomodation at lower pressures in much the same manner as does recrystallization. Furthermore, diffusivity may vary by orders of magnitude in the two crystal forms involved in the transformation. For example, the self-diffusion coefficient for iron in its body-centered cubic form is approximately 100 times greater than that for self-diffusion in face-centered cubic iron when projected to the same temperature. A further consideration in an alloy system undergoing phase transformation is the effect of cooling rate on the properties of these joints. The properties of welded joints are affected by the decomposition products of austenite in all heat-treatable steels.

Any other means of alteration of the diffusion rates affects diffusion welding. We have already seen that direct or indirect alterations occur as a result of temperature changes and transformations. If the rate of atomic motion can be increased by other means, the process contributing to the formation of a weld may accelerate. Diffusion rates have been found to be higher in plastically deformed metals. The merits of using cold work as a diffusion accelerator are not certain. The tendency to have recrystallization occur in worked surfaces makes it difficult to evaluate the secondary effect of enhanced diffusivity.

Another means of enhancing diffusion, which is used fairly extensively, is alloying or, more specifically, introducing elements with high diffusivity in the system. The function of the high diffusivity element is to accelerate the process of atomic motion at or across the interface. In addition to simple diffusion acceleration, the addition of these alloys may have secondary effects. The atoms that are selected as diffusion accelerating are usually those which have reasonable solubility in the metal to be joined, do not form stable compounds, and do depress the melting point locally. Melting point depression by alloying must be controlled because of the possibility of interfacial liquefaction.

When using a diffusion-activated system, it is considered desirable, either during or after the welding process, to heat treat sufficiently to disperse the high diffusivity element away from the interface. If not done, the presence of the low-melting region may produce metallurgically unstable structures. This would be particularly important for joints that would be exposed to elevated temperature service.

It has become rather common to use some form of interlayer material in many applications which have been labeled diffusion welding. In many instances the interlayer is a standard brazing alloy that merely fills the gap and subsequently undergoes a diffusion reaction with the base metal. It is the author's opinion that many of these applications are brazing processes.

Another aspect of the use of interlayers is to provide a significant difference in hardness between the base metals and the interfacial layer. A soft nickel interlayer placed between two very hard nickel alloys permits the use of lower pressures to achieve interface conformity. After the joint is achieved, the diffusion of alloy elements into the unalloyed layer results in minimal compositional and property gradients across the joint.

Activation Agents

SURFACE ACTIVITY

Rhodium, platinum, nickel, ruthenium, palladium, and rhenium have been studied as possible surface active metals in the diffusion welding of tungsten.

Diffusion-welded single lap joints have been evaluated by room temperature tensile tests and by joint remelt tests to 3200°F under a 15 psi load Strong joints resulted when nickel, palladium, platinum, and rhodium were used as surface active metals. From this work it can be concluded that surface active elements could contribute to bond strength even at bonding times of 60 sec. Therefore this concept is worthy of consideration as an aid in the development of high speed welding processes. One disadvantage of surface active elements was their effect on the recrystallization of tungsten.

DIFFUSION AIDS

Feduska and Horigan [15] reported the concept of high-diffusion rate elements as aids to diffusion bonding. Beryllium is known to diffuse rapidly in nickel because the atom is much smaller than the nickel atom. In a search for a similar element to aid in the bonding of refractory metals, boron was selected as a likely candidate because it has been stated to have appreciable solid solubility in tantalum, the small atom size was expected to favor diffusion, and no low-melting point compositions are formed with tantalum. The borides are brittle, however, so that the quantity of boron must be limited to the solid solubility limit.

The results indicate that control would be extremely difficult to limit the boron to the solid solubility limit; thus even if boron can act as a diffusion aid, it will not represent a practical process.

COLD WORK

This concept was evaluated with molydenum joints prepared with tantalum, columbium, and vanadium intermediate foils. The foils were used in both the cold-worked and recrystallized conditions. The results were not regarded to be a conclusive demonstration that the cold work in the tantalum foil promoted recrystallization and hence bonding. The lower strength values could have been the result of slight surface contamination from annealing, even though precautions were taken to prevent contamination. However, later work with columbium foil showed the reverse effect.

It was therefore concluded that the effect of recrystallization is too small to be an effective accelerator for high-speed diffusion bonding [19].

HIGH FREQUENCY FIELD

Samuel [29] has claimed that a high-frequency field (700 kc) will accelerate the diffusion of chromium into iron by up to ten times. This effect, if verified, would demand consideration in the development of high-speed bonding and would influence the choice of a heat source in the development of equipment.

Samuel claimed that high-frequency fields increase the diffusion rate. This should be modified to claim that the combined action of a high-frequency field and the special chromizing pack cause the observed effect.

ALLOTROPIC TRANSFORMATION

Titanium has an allotropic transformation at 1621°F from the hexagonal to cubic close-packed structure and was selected to bond D36 columbium alloy in a study of the effect of an allotropic transformation [19]. The recrystallization temperature of titanium is expected to be near 1400°F for heavily worked material and for anneals of a few minutes. This observation provides further support for the result previously obtained on tantalum, columbium, and vanadium under cold work.

SUBSOLIDUS ALLOY

The yield strength of an alloy decreases rapidly as the solidus is approached. Feduska and Horigan [15] found low pressures were adequate with Ni-3Be foils as the temperature approached 2400°F, and the surface preparation became less critical under these conditions. It must be possible, however, to raise the remelt temperature dramatically by subsequent diffusion to attain high-temperature strength. The lack of systems metallurgically compatible with the refractory metals has prevented complete evaluation of this concept.

In conclusion, the activation agent utilizing the use of surface active elements was shown to be an extremely powerful method to aid bonding. This concept of activation was used in the development of welding methods for tungsten. It was recognized that control of the surface active element was essential to avoid excessive recrystallization and hot-shortness.

DIFFUSION WELDING EQUIPMENT AND TOOLING

Induction Bonding and Resistance Heating

An induction bonding device [19] has recently been developed that produced excellent joints on small lap specimens. The bonding force was applied hydraulically through high-density aluminum oxide anvils. A protective atmosphere was maintained within a quartz tube, enclosing the anvils specimens, by purging with high-purity inert gas or dry hydrogen. The specimen was heated by conduction from an inductively heated tungsten disc.

The bonding pressure was applied after 330 sec of heat for as long as required (usually 60 sec). The pressure was then released and the furnace turned off.

The induction-heated bonder provides close control of temperature, pressure, and atmosphere.

GENERAL TOOLING REQUIREMENTS

The three basic requirements of tooling to accomplish diffusion welding are the following:

1. Clean material surfaces must be precisely positioned and brought together with enough force to provide intimate contact on a microscopic scale.
2. The materials to be bonded must be heated to a temperature sufficient to cause diffusion across the joint interface in a reasonable period of time.

3. A protective atmosphere must be provided to prevent oxidation and/or contamination of the materials being bonded.

The first requirement is readily met by a variety of loading devices, but when the high speed bonding requirement is superimposed, the pressure device must have other characteristics. It must be fast acting, yet provide a soft system so that pressure buildup does not occur as a result of thermal expansion when the parts are heated. A pneumatic system fits these needs best.

The second requirement offers many more alternate solutions. Heating may be by induction (either directly in the parts or indirectly to a susceptor as in the induction bonder); resistance (either by direct heating as in resistance welding or by current flow through the work from separate contacts); conduction (from heated anvils using an auxiliary heater); or by radiation (from focused radiant sources). Review of these sources of heating shows that the major objection to many of them is the difficulty of application to a structure. Yield stress controlled-deformation welding requires access to the parts for opposed-pressure anvils. If access is required, in addition, for a separate heat source, the applicability to all but the simplest structures becomes problematical. Radiant heating would require a large solid angle of access for radiation, and induction heating would require wide clearances from the induction coil to avoid unwanted heating. Conductive heating from independently heated hot anvils and direct resistance heating using electrically conductive electrodes appear to be possibilities. The major advantage of independently heated anvils is that alumina may be used to reduce electrode sticking problems. However, it has been found that sticking was not too severe with thoriated tungsten and hence this advantage did not become important. Thus independently heated hot anvils and direct resistance heating have been used.

Atmosphere is not a problem and would present little difficulty with modern methods. Accordingly, high quality atmospheres have been used to prove the diffusion-welding processes, although it was recognized that it may eventually be possible to lower the quality of the atmosphere requirements.

Two-Anvil Bonder Resistance-Heated

In addition to the method of heating, the basic differences between the induction-heated bonder and the resistance-heated bonder were the design and composition of the anvils. The induction-heated bonder used ceramic insulating-type anvils designed to sustain high compressive loads. The anvils, or electrodes, of the resistance-heated bonder were electrical conductors, possessed good compressive strength at high temperatures, and resisted stick-

Figure 9.4 Two-anvil resistance-heated bonder [19].

ing to the material being bonded. Various materials were used, ranging from graphite to tungsten alloys. The two-anvil resistance-heated bonder is shown in Figure 9.4 [19]. Table 9.1 lists some of the most suitable electrode materials that have been used. Although capable of producing a satisfactory diffusion weld, the feasibility of using a resistance spotwelder as a diffusion welding machine could not be considered suitable in its present form for this purpose. Considerable modifications are required to provide more suitable control of heating current, electrode pressure, and atmosphere purity to permit efficient operation as a diffusion welder.

Gas-Pressure Autoclave

The most advanced and commonly used diffusion-controlled welding technique is isostatic gas-pressure bonding. Because bonding pressures are well below the yield strength of the materials to be joined (a characteristic of all diffusion-controlled welding techniques), deformation is kept to a minimum and close dimensional control is possible. Furthermore, since isostatic gas-pressure bonding techniques produce uniform pressures on the part in all directions, complex shapes with joints in different planes may be fabricated in one bonding cycle. Components to be bonded are assembled into an expendable container or are edge welded to produce a pressure

Table 9.1 Electrodes Used for Resistance-Heated Bonding [19].

ELECTRODES USED FOR RESISTANCE-HEATED BONDING

Alloy	Bonding Conditions	Electrode Material
Columbium base D36, D43, and B66	2450 F 12,000 psi 20 sec	Tungsten alloys: 1 percent thoriated; 2 percent thoriated; 3 percent rhenium. All perform equally well.
Tantalum base T111	2600 F 24,000 psi 20 sec	Tungsten -3 percent rhenium distinctly superior to thoriated alloys. Approximately 30 bonds without refacing. Cost is 40 times greater than for thoriated electrodes.
Molybdenum base TZM	2450 F 12,000 psi 20 sec	Electrode problems are not critical with the bonding of either molybdenum or tungsten alloys. 1 percent thoriated tungsten was used. Approximately 50 bonds without refacing.
Nickel base TD Nickel	2100 F 12,000 psi 20 sec	Severe electrode problem. Tungsten alloy electrodes useful only below 1700 F. Can be used to 2200 F with film of graphite parting agent on face. Approximately 10 bonds without refacing.

tight, evacuated envelope. The assembled components are heated to an elevated temperature in an autoclave containing an inert gas at high pressure. The isostatic pressure is uniformly transmitted (through the container if one is used) and forces all the mating surfaces into intimate contact along any desired surface contour. The mating surfaces are held under pressure at temperature for a sufficient time to permit diffusion welding between the components.

Early experiments with gas-pressure bonding were conducted with hot-wall autoclaves. These systems are still being used today for work with aluminum and other low-melting point materials. However, the hot-wall autoclaves are generally limited to 1500°F and 5000 psi because of the loss of strength of the autoclave body; at lower temperature, 1000°F, the same type vessel might be capable of retaining 30,000 psi.

Recent trends to the use of high-strength, high-temperature materials necessitated the development of new type equipment to make the gas-pressure bonding process useful for general application. By placing a resistance-

General Tooling Requirements

heated furnace inside a larger autoclave, it is possible to obtain very high part temperatures while maintaining the autoclave wall at no more than 300°F. This arrangement permits operating pressures to be very high, the limit being dictated only by the vessel design and not by the loss of wall-strength because of temperature increases. To date, maximum operating pressures have been 50,000 psi and the highest temperatures are approximately 5500°F.

Figure 9.5 shows a typical cold-wall autoclave employed for gas-pressure bonding.

Electric Blanket Heating

This technique of diffusion welding has been used as a method of joining the outer skin to the core of honeycomb panels.

Since the actual contact or bond areas are small compared to the total

Figure 9.5 High pressure diffusion bonding autoclave [38].

Figure 9.6 Curved electric blanket for diffusion bonding [30].

sheet area, very little pressure is required. In fact, the pressure is limited to that which will not cause buckling or excessive indentation of the cover sheet between cells. Recently a diffusion-welding program utilizing blanket heating was completed [30]. Stainless steel retorts containing cobalt-base honeycomb panels were fabricated by using electric blankets at the bonding temperature of 2150°F. The internal pressure of argon was 12 psi on the face sheets. Retorts containing Cb-752 columbium alloy panels were also heated to 2200°F temperature. The differential pressure of 30 in. Hg created a core-to-face sheet interface pressure of approximately 700 psi. The curved blanket heating tool is shown in Figure 9.6.

Hot Pressure Welding (Hot-Roll Bonding)

Hot-roll bonding is frequently used to clad sheet configurations. The process is sometimes termed "pressure cladding." In this process a metal to be clad on both surfaces is first sandwiched between two sheets of the cladding material. The sandwich is then heated to an appropriate temperature and passed through rolls that force the surfaces together with a significant amount of accompanying plastic deformation. The conditions are similar to hot-pressure welding in that the pressures resulting from the rolling exceed the yield strength at the welding temperatures and the time for bonding to occur is very short. The bond is a solid phase weld, for little or no liquid is formed. The temperatures and pressures employed depend on the metals involved.

The following may be cited as an example of the procedure for roll cladding mild steel with stainless steel. A plate of the specified cladding metal, with one surface specially prepared, is placed on a steel slab. A second similarly prepared slab of steel and cladding metal is placed on top of the first one with an infusible parting compound separating the clad metal surfaces. The assembly is held together and sealed from contamination by a continuous weld around the perimeter. The composite sandwich is heated in a soaking pit to temperatures up to 2350°F, depending on the type of cladding, and rolled in a mill. The rolling pressure reduces the thickness of the composite and bonds the cladding to the steel, providing a uniform thickness ratio from the assembled slab to the rolled plate. After rolling, the four edges of the slab are sheared or cut. They are then separated into two clad plates, each of which consists of a steel backing with a layer of clad metal metallurgically bonded to one surface.

It may be necessary to employ special cleaning procedures and inert gas atmospheres or vacuum during bonding to obtain bonding at elevated temperatures. For some metal combinations, surface layers may be cold worked to promote diffusion and sometimes recrystallization across the interface. Temperature is also very important. Some metal combination will bond only in a narrow temperature range that is high enough to promote diffusion, but not high enough to allow formation of detrimental intermetallics. Metallurgical compatibility in terms of the properties formed by interdiffusion of two metals is important in itself. In pressure cladding, as in other methods, an intermediate layer of a third metal may permit cladding of two metals not otherwise compatible under given pressure and temperature conditions. A relatively small alloy addition to one or the other member of the couple may markedly affect their bonding characteristics as well as their usefulness after cladding.

Nickel has been found to be an excellent intermediate material. As an example, hot-roll bonded composites have been made of nickel clad to a molybdenum-310 stainless steel combination subsequently using a picture-frame technique that allowed all mating surfaces to be sealed and evacuated after cleaning and subsequently rolled and clad.

Roll Welding by Hot Pack

The roll-weld process allows flexibility in selecting the structural arrangement and in the fabrication procedure. The process has been applied most frequently to titanium and its alloys.

The method consists of assembling the titanium parts to be joined in a protective jacket of mild steel that is quite similar to the pressure cladding process.

The lower steel cover sheet and yoke are placed together on an assembly table after which the titanium sheets, core strips, and steel filler bars are fitted within the yoke cavity, as shown in Figure 9.7. The steel top cover is then placed over the yoke and the entire pack is clamped together and welded. The steps in processing are shown in Figure 9.8. The processing temperature will vary between 1400 and 1900°F depending on the titanium alloys being processed. The presently developed procedure is unidirection rolling (rolled parallel to the core elements) of the pack. (See Figure

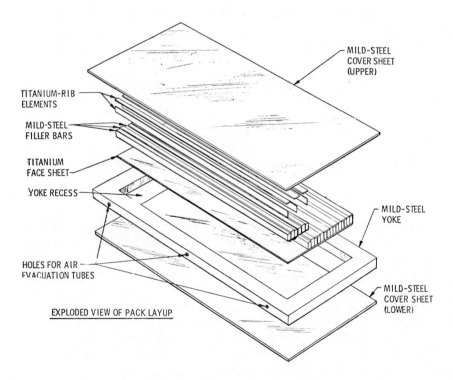

Figure 9.7 Exploded view of pack layup.

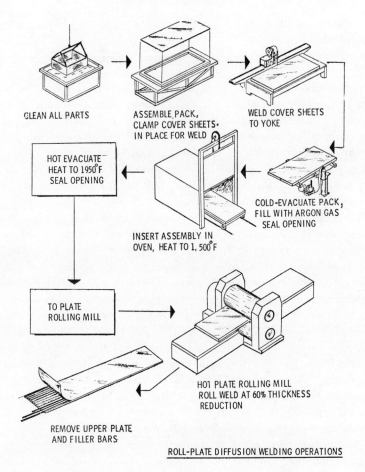

Figure 9.8 Roll-plate diffusion welding operations.

Figure 9.9 Cutaway of rolled pack with test piece of 8AL-1Mo-1V Ti alloy.

9.9). Panels as large as 3 in. high by 60 in. wide by 120 in. long have been rolled.

Vacuum Diffusion Bonding and Welding

Several techniques have been developed for creep controlled bonding parts by placing them in contact and heating under a vacuum. A mechanical load of 5 to 10 psi is often sufficient. Creep controlled bonding differs from other diffusion-welding techniques because of its low bonding pressure. The low pressure makes it suitable for bonding facings to cellular cores that are unable to withstand high pressures and for fabricating structures with large bond areas. With suitable tooling, the technique can be used for curved structures as well as flat structures.

The most commonly used creep controlled bonding technique is low pressure or vacuum diffusion bonding. Included in this class of low pressure bonding is transient melt diffusion. This class of diffusion welding relies on a thin layer of liquid phase alloy at the joint interface for promoting rapid diffusion of base metal atoms. The transient liquid phase can be provided with a low-melting point alloy or a metal that forms a low-melting eutectic with the base metal. The bonding operation can be conducted at temperatures well below the melting point of the base metal. This is significant for joining tungsten or molybdenum because the process permits bonding below the recrystallization temperature of the base metal through the use of fillers such as 50 Cr-50 Ni alloy. Subsequently, the liquid phase can be diffused away to give the joint high remelt properties. In fact, subsequent diffusion can produce metallurgical and mechanical properties in the joint that approach the base metal properties.

Heating has been done by radiant, induction, or resistance methods. Pressures may be provided by dead weight, mechanical means, or through differential vacuum. In some applications, higher bonding pressures may be used to squeeze the transient melt phase out of the joint, thereby reducing the amount in the bond.

Another low pressure bonding technique is vacuum hot pressing. Equipment for the process is simple and inexpensive and varies widely from plant to plant. Parts to be bonded are assembled and placed in an atmosphere chamber. Clean joint surfaces and accurate joint fit-up are essential. An inert gas atmosphere or a vacuum is normally required. The assembly is heated to bonding temperature and bonding pressure is applied until diffusion produces a joint. Aluminum cold plates have been fabricated this way using vacuum diffusion methods with a hydraulic press and electrically heated ceramic plates. The part temperature was held between 985 and 1020°F for 1 hr and pressures between 1000 and 500 psi were used.

Vacuum and Atmosphere Furnace

Both atmosphere and vacuum furnaces have been used to produce diffusion-welded joints. Retorts are required when a protective atmosphere is employed. High temperature vacuum furnaces have also been used extensively for this type of joining process with pressures as low as 10^{-5} torr and temperatures up to 5000°F.

High Temperature Testing Apparatus

A high temperature testing machine, which is basically a materials testing machine, has been extremely successful in producing diffusion welds using the machine as a source for heat and pressure. The heat is produced by electrical self-resistance of the material to be joined and the pressure is applied by hydraulic pistons. The temperature and pressure are automatically controlled and, once a specific set of conditions has been established, the sequence can be repeated by just a push of a button.

Quartz Lamp Heating

The use of quartz lamps to supply heat for diffusion welds in metals has ranged from aluminum to refractory metals [33]. The fixture includes complete water-cooling provisions for the gold-fired reflector system and air cooling for the lamp end seals and quartz envelopes. The design of the quartz lamp brackets and holders was tailored to the assembly being heated for bonding, and a power controller was used to control heating and temperature [34].

Press or Die Pressure Bonding

Press bonding or die pressure bonding uses die pressure to produce intimate contact of joint surfaces. Unlike isostatic gas-pressure bonding equipment requirements for press bonding are simple. Often a hydraulic press provides bonding pressure. (See Figure 9.10.) However, because the die pressure acts in only one direction, all joints in the structure must lie in parallel planes and therefore joint design is not as flexible with press bonding as with gas-pressure bonding. In addition, care is required to ensure uniform bond pressure at the interface with press bonding.

Die pressure bonding is particularly suitable for fabricating honeycomb core panels, multilayer composites, and other expandable structures with all joints lying in the same direction. The technique has been used to

Figure 9.10 Typical hydraulic press setup for diffusion bonding honeycomb core composite panels [35].

fabricate honeycomb panels from refractory materials, titanium and nickel-cobalt, and iron-base high temperature alloys. (See Figure 9.11.) Engineers in the U.S.S.R. have used press bonding to join steel and Kovar parts on a regular production basis for electronic components. They have also fabricated rotatable elbow bend junctions, end caps, and splines from steel

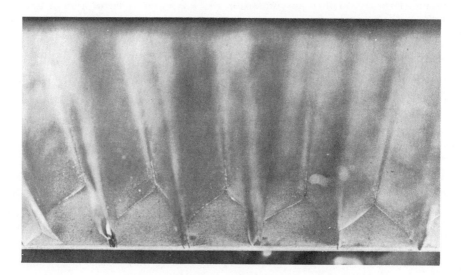

Figure 9.11 PH 15-7 Mo diffusion-welded honeycomb sandwich specimen showing facing to core bond.

using the process. A die pressure setup has been used to diffusion bond hard metal tips to cutting tools and produced tips that are highly superior to tips joined by other methods.

Tooling

A number of important considerations must be observed when selecting fixturing materials. The main criteria for selection of tooling are:

(1) ease of operation;
(2) reproducibility of bonding cycle;
(3) operational maintenance required;
(4) bond cycle time required;
(5) initial cost of tooling.

Furthermore, the fixture materials must be capable of maintaining their proper position throughout the heating cycle.

Second, suitable fixture materials are somewhat limited because of the high temperatures experienced in welding. For example, for temperatures above 2500°F, only the refractory metals themselves and certain nonmetallics have sufficient creep strength for fixturing. Tantalum and graphite have been used for fixturing tungsten and titanium, respectively. Ceramics make suitable fixturing materials provided that they are completely outgassed.

Third, since pressure is required for diffusion welding, the fixtures should be designed to take advantage of the difference in thermal expansion between the materials being joined and the fixture material itself. For example, by appropriate selection of fixture material and gap clearances, it is possible for the fixtured parts to generate at least part, if not all, of the external pressure required for welding. Thus the difference in thermal expansion between a molybdenum fixture and beryllium parts can provide all the pressure that is required to join small beryllium sections. Diffusion-welded 2219 aluminum alloy tubing to 321 stainless steel illustrated in Figure 9.12 depicts the principles. A precise method was devised to apply the correct bonding force to the tubular assembly. This method ensured application of sufficient uniform force to the mating surfaces to provide complete bonding, yet not apply so much force that excessive compressive yielding of the aluminum component would result in deformation of the bonded joint. By taking advantage of the difference between the thermal expansion coefficients of low alloy steel and stainless steel, tooling was developed that provided a uniform and reproducible bonding pressure.

Finally, precautions should be taken to avoid accidental welding of the

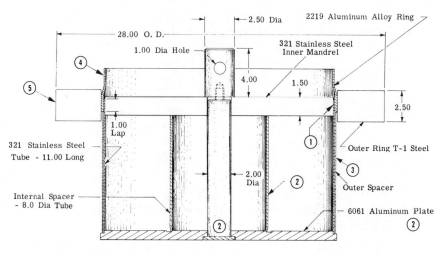

Figure 9.12 Arrangement of parts during diffusion bonding [37].

materials being joined to the fixture material by treating the fixture in the area of the joint.

What Is Ahead?

In summary, the most urgent need today is equipment. Diffusion-welding machines employing vacuum, heat, and pressure are being produced in the U.S.S.R. About fifty have already been installed in plants; twenty-seven types are available from a laboratory devoted entirely to designing this equipment. Development of portable diffusion-welding machines is "absolutely inevitable" and when introduced, they will probably be an instant success.

One possible concept for tooling and a future manufacturing method is of the most interest because of its similarity to the proven blanket brazing method. A metal gas-pressure bag is placed next to the panel assembly and then the whole tool is placed in a platen press (utilized only to react the bonding pressure in the gas bag) and finally heat and pressure are applied to bond the panel. These tools are two or three years away. Currently there are firms almost ready to announce a machine that looks and works like an ordinary spotwelder that will diffusion weld in seconds. Extremely large parts can already be joined with present pressure fixtures and large furnaces. Machines will also be developed that can handle larger parts, incorporating the necessary combinations of time, temperature, and pressure.

ADVANTAGES AND LIMITATIONS OF DIFFUSION WELDING INCLUDING COSTS

Diffusion welding is not a universal substitute for other joining techniques by any means. Pure metals and alloys of every class have been joined. Some of the advantages of diffusion welding metals include the following:

- Numerous materials are severely embrittled when joined by conventional methods from exposure to contaminating atmosphere.
- Fusion welding recrystallization damages mechanical properties.
- Defect-free joints are difficult to make by conventional welding methods.
- Design considerations dictate the method; for instance, joints are too close to other heat sensitive materials or configurations may warp when heated.
- Avoids damage to the refractory metals caused by melting or exposure to high temperatures. Welding and high temperature brazing cause various degrees of such damage.
- Bonds across the entire faying surface so that coating is not required on reentrant surfaces. Resistance-welded joints cannot be coated satisfactorily and riveted or bolted joints require coating before as well as after assembly.
- Avoids filler metals incompatible with coatings, especially refractory metals. Braze alloys are difficult to develop for compatibility with coatings, but are not required in diffusion welding or can be selected from a compatible metal (for example, tantalum foil for T-lll tantalum alloy).
- Preserves smooth aerodynamic surfaces. Fusion welding leaves surface irregularity at the bead; mechanical fasteners also leave a broken surface.
- Makes weight reduction possible by (a) avoiding large overlaps and rivets, (b) reducing the number of gages by bonding closely spaced stiffeners, and (c) avoiding need for double coating with refractory metals.
- Provides economical method to add doublers or stiffening members such as at attachment holes. Brazing is a problem because of subsequent coating and chemical milling is expensive.

Other inherent advantages applicable to other metals and alloys and that are most frequently exploited and that make it such a desirable technique when used with these difficult-to-join materials are the following:

- Diffusion welds do not disturb the metallurgical structure of the parent metal; in fact, heat treatment routines can be carried out during diffusion welding.

- The method joins dissimilar metals as easily as it combines parts made of the same metal.
- Extremely thin sections can be joined to extremely thick ones.
- Joints are as strong as base metals.
- The method can probably join every material now being handled by conventional methods.
- The lack of melting that is inherently a part of all fusion welding processes. By eliminating melting and the resultant cast structure, all the problems that beset a cast structure are eliminated. The problems of segregation, cracking, and distortion stresses are effectively eliminated. Without melting, many metallurgical difficulties characteristic of certain materials are circumvented.
- Dispersion or fiber-strengthened materials cannot be fusion welded without destroying the effective strengthening of the dispersed phase, but they can be readily diffusion welded.
- Metallurgical phenomena such as precipitation and transformational behavior are more easily controlled by the relatively stable and steady-state temperature characteristics encountered in diffusion welding.
- The complex problems in dissimilar parts geometry are often circumvented by diffusion welding. Joints between high-melting point materials and low-melting materials with physical, chemical, or metallurgical incompatibility may be joined by this process which minimizes the limitations imposed by these factors.
- Diffusion welding eliminates the potential hazard of stress corrosion cracking by molten filler metal.
- Another advantage is that significant cost savings can be realized by eliminating brazing filler metals, particularly those based on precious metals.
- In thermodynamics, heat transfer can be calculated much more accurately when the data used are based on a uniform material structure. The use of some intermediate filler metal as with brazing may warrant a redesign of a specific joint because of the thermal properties of the alloy.
- Honeycomb core made by diffusion welding has an advantage of minimum weight over adhesive bonded core. Optimum base metal properties are obtained and cell walls are straight. Foil gages and density selection are dependent on design requirements, not sandwich processing requirements because honeycomb has markedly superior crushing properties on a specific weight basis; in addition, gas-tight nodes eliminate the need for extra braze metal for sealing. (See Figure 9.11.)
- All nodes are bonded simultaneously and therefore improved core properties increase sandwich producibility.

Advantages and Limitations of Diffusion Welding Including Costs

- No intermediate low temperature alloys are involved as in brazing, and diffusion-welded parts could withstand operating temperatures equal to the parents metal's maximum temperature resistance.
- The advantages over bolted, riveted, and otherwise mechanically constructed joints are obvious. Assembly stress patterns generated by bolting or riveting are eliminated, and service stress patterns caused by service loading are evenly distributed across the diffusion-welded area rather than being confined to bolt or rivet locations. Furthermore, there is no weakening of the welded components during hole drilling, piercing, reaming, and countersinking.
- Corrosion of a joint is no longer a problem.

Along with advantages there are limitations. Several general limitations include lack of equipment, an indefinite knowledge of dimensional (size) capability, and inability to join surfaces that are inaccessible or have a peculiar configuration. Several diffusion and deformation welding processes discussed previously have definite advantages and limitations.

One of the chief advantages of pressure welding is the relatively inexpensive equipment required. Conventional hydraulic presses can supply the required welding pressures. Welding temperatures can be provided by a wide range of methods including resistance, induction and radiant heaters, or simply heating with an oxyacetylene torch. And, since many pressure welding operations can be carried out in air, additional equipment to provide protective atmospheres is not required in many applications.

Depending on the equipment available and the characteristics of the materials to be joined, bonding time can be very short, sometimes on the order of seconds. Another advantage of the process is that the finish of joint surfaces is not as critical as with other diffusion-welding techniques. However, deformation is common with pressure welding and, if subsequent machining to remove this deformation is required, joint surface preparation may reduce cost savings.

The large amount of deformation produced by pressure welding limits the type of joint configurations that can be bonded. For example, material adjacent to the joint must be designed to withstand the bonding pressure without unacceptable deformation. In addition, since bond pressure usually is exerted in only one direction, joints generally are limited to one plane (at right angles to the direction of bonding pressure).

Transient melt diffusion bonding has a very significant advantage in that low welding pressures are always used. However, there are two important limitations: prolonged diffusion treatment is required to eliminate low joint remelt temperatures and the high diffusivity of the liquid phase accelerates tendencies for void formation that can result from interdiffusion of dissimilar metals in the bond area.

The chief limitations of gas-pressure bonding stem from the high cost of equipment it requires and the limited size of workpiece that current equipment can handle. Because of high equipment costs, conventional joints by gas-pressure bonding are more expensive than those of other diffusion-welding processes. In addition, because of the slow heat up characteristics of autoclaves, bonding cycle times usually range 3 hr and more.

In press bonding the size of parts that can be joined is limited only by the size of platens available for the hydraulic press. Platens up to 72 in. by 6 in. have been used. In general, bonding cycle time can be reduced as much as 2 hr below that required for gas-pressure bonding.

The length of time (up to 24 hr) required to produce a low-pressure bonded joint is a principal drawback, and maintaining a high temperature for an extended time can be costly to the creep controlled low-pressure bonding process. In addition, preparation of joint surfaces can be expensive. In other diffusion-welding techniques, pressure produces intimate contact between joint surfaces. In low-pressure bonding, however, extremely smooth surfaces are required. Also, mechanical disruption in other diffusion-welding techiniques produces clean metal surfaces as contaminating films are broken. But, with low pressure bonding, because there is little disruption of surfaces, the contaminants must be removed before bonding.

The roll-welding (pack) process can be performed on conventional rolling mills, and the only restriction on the size of the structure that can be produced is the capacity of the rolling mill. The range of materials that can be joined by roll welding is broad. However, the limitation exists that high temperature, high strength materials may require roll pressures not possible with conventional equipment, and expensive modification of conventional rolling facilities may be required.

Now let us look at a breakdown of costs for diffusion welding, using the roll-welding technique for quantitites of a 3 ft diameter by 6 ft long cylinder constructed of a $\frac{1}{4}$ in.-thick Ti-6AL-4V sandwich. This cost breakdown is as follows:

1. Materials
   (a) Titanium sandwich components   46%
   (b) Iron pack and filler bars   15%
2. Direct fabrication labor with burden   4%
3. Rolling mill rentals   3%
4. Nitric acid and leaching cost   31%
5. Roll forming and welding   1%

This does not include the following:

1. Design and development
2. Tooling

Advantages and Limitations of Diffusion Welding Including Costs

3. Inspection and proof testing
4. Transportation cost to and from the rolling mill
5. Disposal of spent acid

These illustrate a simplified cost breakdown for a specific item analyzed. Other designs of different configuration would require individual cost analysis.

MATERIALS

A wide variety of materials have been joined by diffusion welding. Pioneer work in this area was conducted primarily in the nuclear field; in recent years, however, work has been extended to cover many structural materials. A summary of materials which have been joined follows.

Beryllium and Its Alloys

Beryllium's high strength-to-weight ratio and low neutron-absorption cross section make it especially interesting for aircraft and nuclear applications. Its usage, however, has been limited because of the difficulty of bonding or fabricating this characteristically brittle metal into desired shapes without extensive machining. Techniques have now been developed for diffusion-welding beryllium components by gas-pressure welding.

The effects of various surface preparations and welding parameters (time, temperature, and pressure) on the bonds achieved between beryllium members have been investigated. Results indicate that bonding is improved by prior cold work and that good bonding conditions are 1500 to 1650°F for times of about 4 hr at 10,000 psi.

An additional advantage gained when gas-pressure welding beryllium is associated with the safety problems involved in handling this material. In the gas-pressure welding process all bonding is accomplished with the material sealed inside a disposable metal container, thus minimizing any danger of contamination of personnel or equipment. In the preparation of surfaces, beryllium is cleaned first by a mechanical treatment, then with acetone, and sometimes with ether. It also can be prepared by grinding with wet silicon carbide paper and polishing with levigated alumina in dilute oxalic acid. Combinations of chemical polishing and etching have been found to be unsatisfactory for cleaning beryllium due to the formation of a surface film that impedes welding. Recently it was shown that strength increased rapidly with increasing temperature and reached a maximum above which strength decreased as a result of recrystallization. For polished surfaces this maximum occurred between 900 and 975°F. For ground surfaces the peak occurred at about 1050°F. Above 1050°F the resultant strength ap-

Figure 9.13 A high vacuum diffusion bond in beryllium copper was made with a silver-copper-indium filler alloy. There is no sign of the filler metal. It has diffused into the base metal. Bond strength: 108,400 psi.

peared to be independent of surface preparation. This supports the viewpoint that above 1050°F sufficient deformation of the surface occurs to achieve complete contact across the interface regardless of surface preparation [36].

The tensile strength of diffusion bonded beryllium copper parts averages 108,400 psi, which compares with an average of 40,000 psi for a brazed counterpart. To ensure continuous bonds with minimum pressure, thin layers of a silver-indium-copper alloy were preplaced and upon heating formed a continuous molten layer that promoted diffusion in areas not in intimate contact. Both hydrogen and vacuum atmospheres were employed, with the vacuum giving better results. The bonding cycle was about 30 min at 1475 to 1550°F. Parts were loaded to sustain about 6 psi during bonding. (See Figure 9.13.) The recently introduced beryllium-aluminum alloy (Lockalloy) has been diffusion welded successfully without any special preparation of the metal.

Aluminum and Its Alloys

Diffusion welding has not been extensively applied to aluminum and its alloys. Past results of experiments on unalloyed and low-alloy aluminums indicate that satisfactory joining can be achieved. Bonding parameters varied, but, in general, satisfactory welding was achieved at 700°F for relatively short times [38].

One difficulty encountered in the past in joining aluminum was the preparation of the bonding surfaces. The tenacious aluminum oxide film, which was difficult to remove, interfered with bonding and frequently prevented satisfactory bonds from being achieved.

More recent work has shown that aluminum alloys can be successfully diffusion welded at even lower temperatures (300 to 450°F).

Aluminum alloy 7075-T6 has been diffusion welded with no protective atmosphere and also in a helium atmosphere. In addition, aluminum alloys

2219-T6 and 6061-T6 have also exhibited an ability of self-bonding at 450°F. As temperature and time increased, the bond strength increased. However, the maximum shear strength obtained was only 5600 psi when diffusion welding at 870°F for 4 hr [34].

Low temperature diffusion welding of aluminum alloys using various diffusion aid materials has been successful, such as electroplate, vacuum deposit, plasma spray, loose shim material, and cladded aluminum.

VACUUM DEPOSITION

The following metals have been vapor deposited on aluminum alloys: silver, gold, nickel, aluminum, iron, magnesium, tin, and zinc. No bonding was achieved with low-temperatures diffusion welding due to poor adhesion of the vacuum deposit and the thickness of coating. Tin did develop 300 psi shear strength when bonded at 500°F for 2 hr at 16,000 psi.

ELECTROPLATING

Silver resulted in the highest joint shear strengths of 6700 psi after electroplating. Bonding parameters were 450°F for 4 hr at 16,000 psi. Silver is not desirable, however, as a diffusion aid at 450°F temperature because silver in the interface will not diffuse into the aluminum base metal and, under proper atmospheric conditions, would create a galvanic reaction. In addition, only when zinc was electroplated on aluminum alloy 6061 did any cohesion develop.

INTERLEAF MATERIAL

Interleaf material shimmed in the joint interface does appear applicable for high temperatures but not for the low-temperature bonding. The loose interleaf materials include zinc, nickel, and pure aluminum foil.

Aluminum foil has produced shear strengths of 2000 psi. An interleaf of zinc bonded at 720°F for 4 hr at 500 psi has resulted in 4700 psi shear strength.

CLAD TO CLAD

The aluminum alloy 7072 cladded to the the 7075 aluminum alloy has proved to be the best diffusion aid material for low-temperature diffusion bonding. Reproducible shear strengths of 9000 to 10,000 psi have been obtained. The optimum bonding parameter was 325°F for 1 hr at 24,000 psi interfacial pressure. The shear strength results versus bonding temperature are summarized in chart form in Figure 9.14, which also indicates four other significant variables—time, bonding pressure, cleaning procedure, and reheat treatment. These are included in the following discussions.

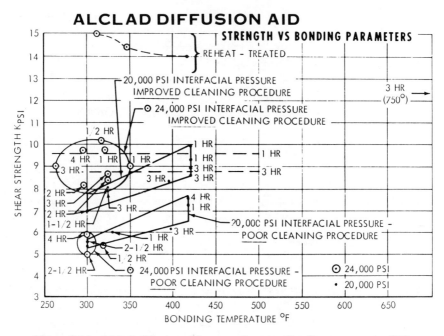

Figure 9.14 Alclad diffusion aid: strength versus bonding parameters [34].

ALCLAD CLEANING

Wire brushing was extremely beneficial, for without wire brushing, results were erratic.

BONDING PRESSURE

The bonding threshold pressure was below the compressive yield strength of the parent material up to 450°F. The maximum joint pressure at 425°F was 20,000 psi. Above this pressure the 7075-T6 base metal will start yielding. This is why 20,000 psi was used at the higher temperatures in Figure 9.14. The clad material in the interface is relatively soft and, when subjected to the high pressure, it will locally yield. This brings the interface closer than the required minimum atomic spacing for grain growth to occur.

BONDING TIME

For an alclad aluminum system time is not a critical variable. Actually, Figure 9.14 shows the shear strengths after 1 hr at bonding temperature to exceed those bonded for 3 hr. At the bonding temperature some free wandering of the upset crystal lattice elements occurs and results in diffusion at the interface.

REHEAT TREATMENT

Reheat treatment to the T-6 condition increases the joint strength by 50%.

Plasma Spray

Low temperature diffusion welds employing 100% aluminum powder plasma sprayed to the joint interface produce shear strengths of 7400 psi. The results are shown in Table 9.2 and may be summarized as follows:

1. The optimum bonding parameters are 425°F and 4 hr at 24,000 psi.
2. Particle size is important.
3. Reheat treatment results in shear strength twice that of the as-bonded strengths.
4. Plasma spray can be applied to machined-serrated joints to increase effective bond area and to increase joint strength.

Plasma Spray to Clad

A satisfactory bond has been achieved by using an aluminum (100%) plasma sprayed on one interface and the clad on the 7075-T6 aluminum on the other. The bonding parameters were the same as required for plasma

Table 9.2 Plasma Spray 100% Aluminum-Diffusion Bonding Results [34].

SHEAR STRENGTH PSI	TEMP °F	TIME HRS	JOINT PRESSURE PSI	
5400	425	3	20,000	NOT WIRE BRUSHED
4800	425	3	20,000	USED ACETIC ACID
7200	425	3	20,000	
4900	425	1	20,000	
5500	425	4	20,000	
11,300	425	4	20,000	REHEAT TREATED TO T-6
6000	425	3	24,000	NOT WIRE BRUSHED
6900	425	3	24,000	(COURSE Al MESH)
5800	425	3	24,000	(MEDIUM Al MESH)
7000	425	3	24,000	(FINE Al MESH)
7400	425	4	24,000	
11,000	425	3	24,000	LONGITUDINAL SERRATIONS

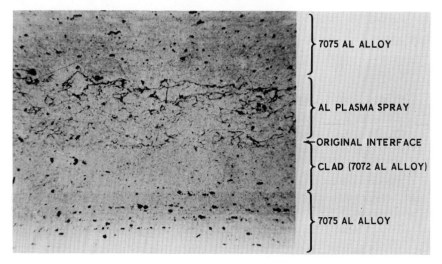

Figure 9.15 Plasma spray—100% Al to clad diffusion weld after reheat treatment to T-6 (900°F for 24 hr) [34]. (×200)

sprayed diffusion aids. The cleaning techniques used were those developed for clad and 100% aluminum spray, respectively. A typical clad-to-plasma sprayed joint in the reheat treated condition is shown in Figure 9.15.

Columbium and Its Alloys

Gas-pressure welds in unalloyed columbium have been very successful. Bonding parameters of 2100 to 2400°F and 10,000 psi for 3 hr yield high-integrity solid-phase welds. The structure and the mechanical and physical properties of the bond area are consistent with those of the parent metal.

Columbium-vanadium alloys. Results very similar to those achieved with unalloyed columbium have been obtained when columbium-vanadium alloys were gas-pressure bonded at 2400°F.

Columbium alloy Cb-752 (10W – 2.5 Zr – Cb). Electric blankets have been used in diffusion-welding columbium alloy Cb-752 honeycomb panels. A titanium foil (0.002 in.) interleaf was used between Cb-752 panel components to promote welding with a bonding cycle of 2200°F, 1×10^{-6} torr and 6 hr. Diffusion welding has also been achieved on Cb-752 with pressures of less than 1000 psi at 2000 to 2200°F in a protective atmosphere for 15 min. This work has been done using resistance-heating methods.

Materials

Columbium alloy D-36 1(0 Ti–5Zr-Cb). D36, with a comparatively low yield stress, has proved particularly amenable to diffusion welding. The D36 alloy has been bonded in the self-bonded condition and with vanadium, tantalum, and columbium intermediate foils. Bonded joints have been made both by induction and resistance. All systems produced bonding, but the self-bonded condition gave the highest joint strengths and the vanadium intermediate foil provided a good alternative system. Although surface preparation was found to be not critical when an intermediate material was used, an abraded and degreased surface was optimum in the self-bonding of the alloy.

Columbium alloy B-66 (5 Mo-5V-1Zr-Cb). The B66 alloy has been induction bonded both in the self-bonded condition and with a 0.0003-in. tantalum intermediate. Both systems have produced good bonds. Results indicate that abrasion by 3/0 emery followed by cleaning in a salt solution or a 95/5 nitric/hydrofluoric solution was the most effective surface preparation for self-bonding. Columbium, tantalum, and vanadium foils were used as intermediates in the resistance bonding of B66 alloy. The results of this work are shown in Table 9.3 which summarizes all systems, including self-bonding.

The optimum surface preparation for bonding B66 alloy with an intermediate foil was degreasing in a salt solution. This preparation produced the highest strength joints. Thus both the self-bonding and the B66-Ta systems look equally satisfactory. This was true even after specimens were coated with a silicide coating and subjected to oxidation testing at 2350°F.

Columbium alloy D-43 (10W-1Zr-.1C-Cb). High peel strengths have been obtained by induction methods in the self-bonded condition and with vanadium and tantalum intermediates. The optimum surface conditions using resistance-heating methods for both self-bonding and bonding with intermediate foils are summarized in Table 9.4.

Chromium and Its Alloys

Gas-pressure bonding of unalloyed chromium has been accomplished at 2100°F and 10,000 psi for 3 hr. The effect of bonding parameters has not been fully explored; therefore developmental work is proceeding on chromium-base alloys with about 96% chromium. The difficulty in producing chromium-base alloys, because of their inherent physical and technical properties including brittleness of diffusion-bonded joints which can become more severe in service, is viewed as a deterrent to their use; however, within five years chromium alloys will be in use.

Table 9.3 Resistance Bonding of B66 Alloy [19].

Surface Preparation	Intermediate Foil	Number of Specimens	Temperature (F)	Pressure (psi)	Peel (lb/in.) Average	Peel (lb/in.) Deviation	Parent Material Failure in Peel (%)
Etched	Tantalum (0.0003 in.)	2	2400	12,000	211[1]	–	–
		1	2500	12,000	304[1]	–	–
		2	2600	12,000	283	52	71
		9	2600	15,000	295	32	92
Etched	Columbium (0.0005 in.)	2	2400	12,000	181[1]	–	–
		2	2500	12,000	128[1]	–	–
Etched	Vanadium (0.001 in.)	2	2400	12,000	121[1]	–	–
		2	2500	12,000	370[1]	–	–
		4	2600	12,000	280	77	82
Pennsalt Degreased	Tantalum	2	2600	12,000	429	19	70
	Vanadium	2	2600	12,000	402	16	35
Abraded and Pennsalt Degreased	Tantalum	1	2600	12,000	419	–	50
	Vanadium	2	2600	12,000	306	18	25
Abraded and Pennsalt Degreased	None	8	2700	12,000	445	40	74
		9	2700	12,000	434	68	98

1. Tested in shear prior to peel test.

Parent Material: 0.012 inch B66 alloy
Joint Area: 0.050 by 0.375 inch
Total Heating Time: 18 seconds
Atmosphere: Vacuum purged argon from 2×10^{-4} Torr.

Table 9.4 Resistance Bonding of D43 Alloy [19].

Condition	Intermediate Foil	Number of Specimens	Temperature (F)	Peel (lb/in.) Average	Peel (lb/in.) Deviation	Parent Material Failure (%)
Etched	None	3	2600	27[1]	-	-
Abraded and Degreased	None	2	2600	204	30	60
Degreased Only	None	2	2700	230	10	70
As-Received and Cleaned in Acetone	None	2	2600	84	10	10
Abraded and Cleaned in Acetone	None	2	2600	170	24	40
Etched	Tantalum (0.0003 in.)	3	2400	130	-	-
		3	2600	258	-	-
Etched	Vanadium (0.001 in.)	3	2400	298	44	98
		5	2500	334	8	95
		5	2600	261	29	100
Degreased in Pennsalt 45 only	Vanadium (0.001 in.)	2	2500	322	13	90

1. Tested previously in shear

Bonding pressure - 12,000 psi
Bonding time   - 18 seconds

Copper and Its Alloys

A limited amount of work has been done utilizing copper in diffusion welding except in basic studies and as an intermediate filler material. Copper-copper and copper-nickel welds have been made by pressure welding. The high joint strength reported for copper-to-copper that was heat treated at low temperatures is a reflection of the initial joint strength. The strength of copper to nickel joints improves with postweld heat treatment up to 600°C.

Cobalt and Nickel Base Alloys

Although most applications of diffusion welding have been applied to steels, especially stainless, various nickel and cobalt-base alloys have been successfully bonded.

Nimonic 90. Recent studies on Nimonic 90, a nickel-base alloy, shows that diffusion welding was successfully done by induction and by resistance heat. In this study 0.002 in. nickel foil was used as an intermediate in the joint.

Haynes 25 or L-605. The diffusion welding of honeycomb core as illustrated in Figure 9.16 and the fabrication of panels have been developed utilizing the electric blanket process. The cobalt-base-alloy was bonded at 2150°F for 3 hr with 12 psi. Flatwise room temperature tensile tests of bonds between core and face sheet resulted in ultimate strengths from 90,000 to 117,000 psi. This compares with strengths of 125,000 to 135,000 psi for wrought material.

Inconel X. The use of an intermediate material that can increase weld strength by a diffusion mechanism rather than a simple mechanical one was developed for several alloys, including Inconel X (Inconel 600). Beryllium is excellent as an intermediate material for bonding nickel base alloys. Although beryllium forms substitutional alloys with iron, nickel, and cobalt, its relatively small atomic radius permits it to diffuse rapidly into these metals and their alloys. Joints have been made with interface strips containing 9 or 16 atomic % beryllium and formed bonds having shear strengths of 70,000 psi at room temperature by the induction method. Table 9.5 lists the bonding conditions and shear strengths of joints made in several steels and superalloys.

Thoria dispersioned: TD nickel and TD nickel-chromium. TD-Nickel, a thoria dispersion strengthened nickel-base material (2% thoria), has recently been developed for applications in the temperature range from 1800°F to 2400°F. The material has been diffusion welded by several methods. One technique utilized an evacuated steel retort similar to the electric blanket process. The bonding took place at 2025°F for 24 hr and pressure of 15 psi. A photograph of diffusion-welded sandwich panels is shown in Figure 9.17 and microsections have revealed no thoria agglomeration. Honeycomb core has also been fabricated from TD nickel.

In both induction and resistance bonding, the self-bonding of TD nickel was successful and induction bonding results are summarized in Table 9.6.

This is a new and improved TD nickel alloy which is thoria dispersion strengthened and has 20% chromium. The potential for this alloy over TD nickel is higher strength and superior oxidation resistance at 2000°F. A single-phase resistance welder has been used to produce diffusion spot welds. Diffusion bonds have produced shear strengths of 27,770 psi higher values, however, are attainable through the use of three-phase welding equipment and variations in bonding pressures and cycles.

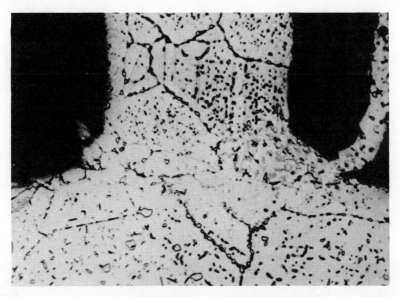

(a) Center of Structural Panel

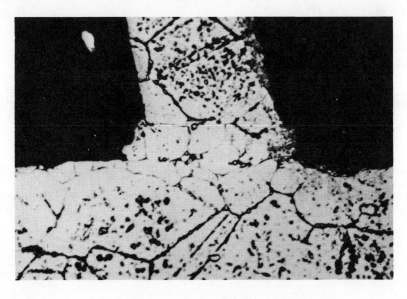

(b) Edge of Structural Panel

Figure 9.16 Photomicrographs typical diffusion bonded joints from HS-25 flat structural test panel [30]. (500×)

Table 9.5 Shear Strength of Single Lap Joint Specimens Welded with Ni-Be Intermediate Alloys [15].

Base Metal	Intermediate Alloy	Welding Temperature (°F)	Welding Time Minutes	Shear Strength (psi) Room Temperature	1500°F
AISI 410	A	2,190	2	Not welded	
AISI 410	B	2,110	5	31,500	
AISI 410	C	2,110	5	46,550	
AISI 347	A	2,175	1	Not welded	
AISI 347	B	2,110	5	33,300	
AISI 347	C	2,110	5	39,600	
AISI 410	D	2,075	4		7,400
AISI 410	D	2,010	10		8,800
AISI 410	E	1,960	10		8,500
AISI 410	E	2,100	$\frac{1}{2}$		7,750
AISI 347	D	2,080	10		15,000
AISI 347	D	2,100	10		20,000
AISI 347	E	2,080	10		18,800
AISI 347	E	2,100	10		15,300
Inconel X	B	2,110	5	35,850	
Inconel X	C	2,110	5	72,900	
Haynes 25	B	2,110	5	43,350	
Haynes 25	C	2,100	5	31,250	
Inconel X	D	2,100	10		50,950
Inconel X	D	2,075	10		32,750
Inconel X	D	2,010	10		44,400
Inconel X	E	2,030	$\frac{1}{2}$		37,250
Inconel X	E	1,995	10		30,000
Inconel X	E	2,010	10		36,500
Haynes 25	D	1,910	10		30,250
Haynes 25	D	2,055	7		25,850
Haynes 25	E	2,100	10		31,000

Intermediate Alloy Composition
A   Ni-0.28% Be
B   Ni-1.51% Be
C   Ni-3.02% Be
D   Ni-20% Cr-0.3% Mn-3.0% Be
E   Ni-5.8% Be-13.6% Cr-0.1% C-0.3% Mn-3.9% Be

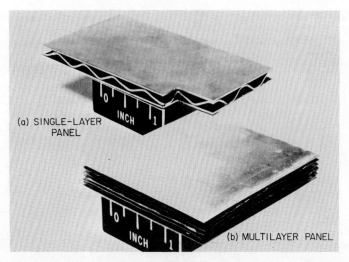

Figure 9.17 Photographs of diffusion-bonded sandwich panels [40]. *Above*—single-layer diffusion-bonded panel. *Below*—multilayer diffusion-bonded panel.

Table 9.6 Induction Bonding of TD Nickel Alloy [19].

Surface Condition	Number of Specimens	Temperature (F)	Pressure (psi)	Time (sec)	Peel (lb/in.) Average	Peel (lb/in.) Deviation	Parent Material Failure (%)
Degreased	3	1600	20,000	60	297	57	100
Degreased	4	1600	25,000	30	339	25	100
Degreased	4	1600	35,000	10	351	46	97
Degreased	4	2200	10,000	60	392	40	100
Degreased	5	2200	20,000	10	362	34	100

Parent Material: 0.012-inch TD Nickel alloy
Joint Area: 0.050 inch by 0.450 inch
Heating Time: 330 seconds
Atmosphere: Gettered argon

Molybdenum and Its Alloys

Gas-pressure bonding of molybdenum has been accomplished at 2600°F and 10,000 psi for 3 hr. Welds of high quality with complete grain growth across the original bond interface have been achieved. A loss of ductility is encountered after pressure bonding of unalloyed molybdenum; however, a 25% reduction by cold rolling restores ductility. A vacuum-hot-pressing technique has also been applied to the solid-state welding of unalloyed

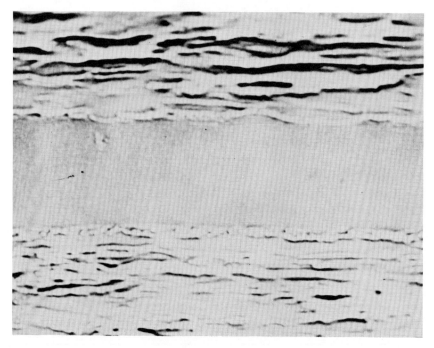

Figure 9.18 As-bonded Mo-Cb (Mo at *top*, Cb at *center*, Mo at *bottom*) couple microstructure. Specimen bonded at 1840°F with ksi pressure. Note surface conformance and solid-solution bonding [42]. (×1000)

molybdenum. Unalloyed interface metals (0.001-in. foil) have been used to obtain activated-diffusion at temperatures below the recrystallization temperature of the base metal. A surface conditioning treatment (cold grinding) of the base metal substrates has been used to enhance diffusivity and to improve bonding characteristics. Heating has also been accomplished through a resistance-heated molybdenum filament. The interface metal used for unalloyed molybdenum has been columbium.

The as-bonded microstructure of a molybdenum-columbium diffusion weld is shown in Figure 9.18. The molybdenum-columbium system, according to its phase diagram, forms a continuous series of solid-solutions and is therefore a bonding system of considerable interest. The as-bonded microstructures of molybdenum-columbium show complete solid-solution bonding of interfaces with no structural signs of inter-diffusion of substrates.

Molybdenum alloy (Mo-0.5 titanium). Gas-pressure bonding of Mo-0.5Ti has been successful using the same parameters as pure molybdenum. Using resistance-heating methods, diffusion welds between 0.5 Ti-molybdenum alloy to itself have been produced. Conditions for bonding were less than

2100°F for 15 min at 500 psi in a protective atmosphere. With vacuum hot pressing techniques, bonding parameters have been established for Mo-0.5 Ti with titanium and nickel interface metals (0.001 in.-foils). Diffusion welds involving Mo-0.5 Ti-alloy substrates have been evaluated in the as-bonded and the as-bonded plus postbond diffusion treated conditions of 30, 54, and 78 hr at 2050°F. The bonding temperature for titanium was 1680°F, whereas for nickel the temperature was 1700°F. In all instances the time-temperature cycle consisted of (a) heating to bonding temperature in 6 to 8 min; (b) holding at bonding temperature for 20 min; and (c) cooling to below 1000°F within 4 min. The as-bonded shear strength of the Mo-0.5 Ti/Ni couples was 33,800 ksi, and the as-bonded strength of the Mo-0.5 Ti/Ti couples was 20,700 ksi. Interstitial embrittlement of the Ti foil interface probably contributes to lowering the shear strength of the Mo-0.5 Ti/Ti couples.

Molybdenum alloy TZM (0.5 Ti-0.1Zr-Mo). Diffusion welding of TZM alloy has been successfully accomplished using induction bonding procedures. Columbium, tantalum, and vanadium have been the best intermediate materials. Self-bonding has been successfully carried out but only at high temperatures and pressures that have been considered impractical. Bonding was carried out at 2200°F, 10,000 psi, and 60 sec by induction. As a result of tests to establish the optimum bonding parameters for each of the systems it was found that the tantalum intermediate yielded the highest room-temperature shear strengths in the region of 650 lb/in. at optimum bonding parameters of 2600°F, 12,000 psi, and 18 sec. To select the best intermediate the coatability of each system with respect to oxidation resistance must be examined. The TZM-Cb system appears to be the most favorable with a distinctly improved oxidation life over that of TZM-Ta and TZM-V.

Steel and all Its Types

A wide variety of stainless steels, such as Types 304, 316, 347, and 410, have been successfully gas-pressure bonded. Strong metallurgical bonds have been achieved with pickled, as-rolled, belt-abraded, or milled surfaces. Bonding conditions appear to be generally the same for all types of steels. Temperatures of approximately 2100°F are usually employed with times varying from 1 to 3 hr.

Figure 9.19 shows a typical bond achieved between two pieces of Type 304 stainless steel when gas-pressure bonded at 2100°F and 10,000 psi for 3 hr. Tensile data on bonds of this type show the bond to be as strong as the parent metal. Presses conventionally employed for hot-forming work have been used to fabricate Type 321 stainless steel honeycomb panels, bonded at 2200°F at 6 psi in 3 min. Using resistance-heating methods stainless steel tubing has been joined in air in less than 3 min.

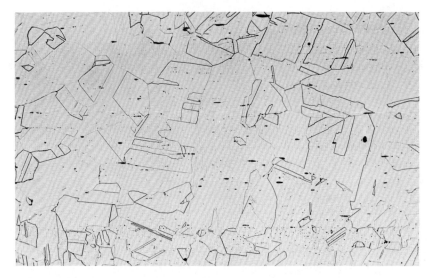

Figure 9.19 Typical self-bond obtained between stainless steel surfaces [38].

Tantalum and Its Alloys

Self-bonding with gas pressure has been successful with pure tantalum. This work produced bonding prarameters of 2350°F at 10,000 psi and satisfactory bonds as indicated in Figure 9.20. Using an intermediate material tantalum has been joined with zirconium at temperatures as low as 1600°F, well below the normal recrystallization temperatures.

Tantalum alloy T-111 (8W-2Hf-Ta). The surface preparation procedure developed for the alloy has revealed that the etching of the T-111 alloy before bonding was critical in producing good joint strengths. Abrading the metal, degreasing in a salt bath at 200°F, rinsing in water, etching in a sulfuric, nitric, and hydrofluoric acid bath for 4 sec and then water-rinsing and drying with dry nitrogen gas are the surface preparatory steps.

Self-bonded and intermediate filler joints with tantalum, columbium, and vanadium have been made using resistance-heating and the induction bonder. Results of resistance-bonding are summarized in Table 9.7.

The best bonding system selected for T-111 alloy has been 2600°F and 24,000 psi for 20 sec with a 0.0003-in. tantalum interfoil.

Tantalum alloy (90 Ta-10W). Resistance heating has produced diffusion welds in 15 min at 2500°F and 500 psi in protective atmospheres.

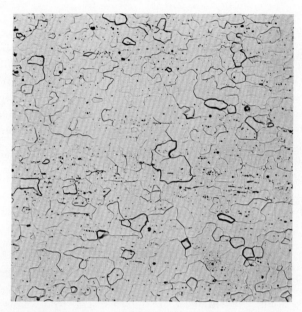

Figure 9.20 Example of typical tantalum self-bond [38].

Table 9.7 Resistance Bonding of T111 Alloy at 12000 PSI [19].

Intermediate Foil	Number of Specimens	Temperature (F)	Time (sec)	Average Peel Strength (lb/in.)	Parent Metal Failure (%)
Tantalum (0.001 in.)	5	2200	18	27	-
	3	2400	18	170	-
	2	2600	18	253	-
Tantalum (0.0003 in.)	4	2600	18	256	97
	3	2700	18	255	100
	10	2600	20-22	495	29
Vanadium (0.001 in.)	4	2500	20	815	55
	3	2600	20	863	75
	10	2700	20	731	55
Columbium (0.0005 in.)	1	2500	20	193	-
	3	2600	20	270	-
	1	2700	20	240	-
D36 (0.001 in.)	2	2500	20	820	80
	2	2600	20	877	85
	2	2700	20	771	55

Titanium and Its Alloys

By using resistance-heating techniques, pure titanium has been joined by diffusion welding at 1600°F and pressures of 200 to 600 psi in 15 min with argon or argon-7% hydrogen atmospheres as shown in Figure 9.21a. No obvious evidence of the prior interface can be observed. The uniqueness of this photograph is that there are really four different pieces of titanium diffusion-welded together. Figure 9.21b shows the interface between the four pieces a little more clearly. This same joint shown at high power (X500) reveals grain boundary growth across the interface (Figure 9.21c). Utilizing electric blanket techniques for bonding, honeycomb panels have been produced by several aerospace manufacturers.

Using gas-pressure bonding equipment, self-bonds in pure titanium have been achieved with bonding parameters of 1600°F and 10,000 psi in 4 hr.

Titanium alloy 8-1-1 (8AL-1Mo-1V). Using conventional hot-forming equipment, this alloy has been diffusion welded with 1000 psi and 1825°F temperature in 15 min. The grain structure pattern of the 8-1-1 titanium alloy is very similar to that shown in Figure 9.21a. Roll welding has been applied to the 8-1-1 titanium alloy. A typical pack lay-up is shown in Figure 9.7. Figure 9.8 indicates the procedures for rolling and the opened rolled pack. The temperature for rolling was 1800°F, the time was 30 min, and less than 10 μ total pressure existed within the pack before rolling. Roll welding and resistance heating has also been successfully used with this alloy. Temperatures for bonding, however, for this alloy are 1600 to 1650°F

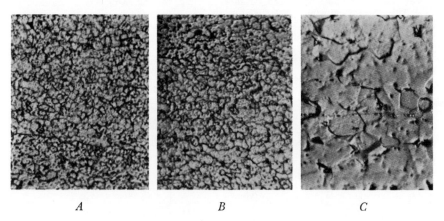

A   *B*   *C*

Figure 9.21 Solid-state bonded titanium 75A alloy. *A (left).* Four pieces bonded showing no interface center of joint (×100). *B (center).* Four pieces bonded showing interface-end of joint (×100). *C. (right).* Bonded titanium 75A (×500). All views reduced approximately 30% on reproduction [32].

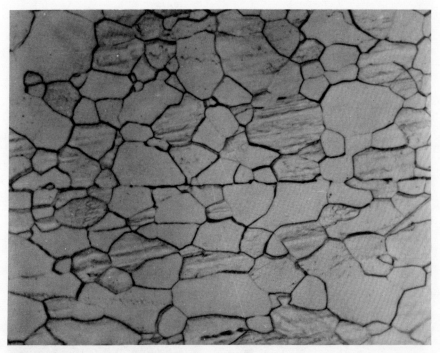

Figure 9.22 Solid-state bonded unalloyed tungsten excellent bond; no oxide present [32]. ×500 (reduced 30% in reproduction).

for roll welding and 1700°F for resistance heating. Times, pressures, and rolling schedules are similar to those used with the 8-1-1 titanium alloy.

Tungsten and Its Alloys

Up to the present limited information on the gas-pressure bonding of tungsten in the wrought condition was available. In general, emphasis has been placed on producing usable shapes from powder. It should be noted, however, that self-bonding of tungsten has been accomplished at 2800°F and 10,000 psi for 3 hr.

Diffusion welding has been achieved on recrystallizated tungsten sheet as illustrated in Figure 9.22. For all practical purposes this bond has the same properties as the base metal. Bonding conditions were less than 2800°F at 300 to 500 psi pressure for 10 min in a protective atmosphere. Figure 9.22 shows what can be accomplished when the proper techniques are followed using resistance-heating methods.

Intermediate foils and powder with wrought tungsten sheet that combined brazing and diffusion-welding procedures have successfully produced rocket nozzles. Braze filler metal was selected because the melting points of the filler metals were below the recrystallization temperature of tungsten and there was partial or complete solid solubility of one or more of the compositional elements of the filler metals that were divided into two groups. They were pure metals having partial solid solubility (elements having complete solubility were eliminated by their high melting point) and commercial alloys containing elements with both complete and partial solid solubility elements. The pure metals of titanium, nickel, and aluminum have partial solubility in tungsten. Only aluminum has a melting point below the recrystallization temperature of tungsten, and titanium and nickel form a eutectic having a melting point of approximately 1900°F. The braze filler selected was 68% manganese, 16% nickel, 16% cobalt, and 0.5% boron. The equipment used for brazing and subsequent diffusion treatments consisted of a retort box designed to contain a protective atmosphere.

The three diffusion cycles investigated are shown in Figure 9.23, these

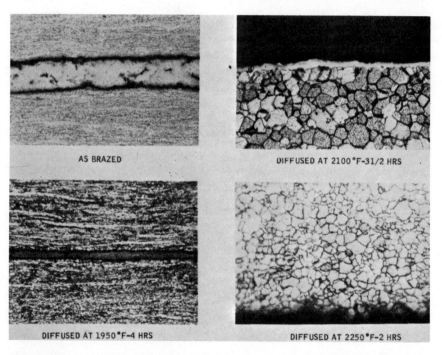

Figure 9.23 Tungsten brazed and diffusion bonded [45]. ×350 (reduced 52% on reproduction).

temperatures were selected because they covered the full brazing temperature range of the filler metal used and the atmosphere during diffusion was argon. The joint clearances varied from 0.002 to 0.005 in. Figure 9.23 illustrates the microstructure of the tungsten joints brazed and diffused with the manganese-base braze alloy. Excellent bonding is shown although small voids do occur throughout the joint. Slight alloying as evidenced by recrystallization can be observed in the 1950°F and 4 hr diffusion-treated part. Complete diffusion and alloying caused recrystallization to occur across the entire cross section of the tungsten base material for the 2100 and 2250°F diffusion cycles. The microstructures reveal equiaxed refined grains present throughout the matrix. Some residual braze filler metal may be observed in the 2100°F diffusion-cycled part; however, none remains after the 2250°F treatment. The shear strength and remelt test results are discussed in the next section under mechanical properties.

In attempting to explain the high remelt temperatures obtained in brazed and diffusion-treated joints, it is necessary to introduce other contributing factors such as mutual dissolution and volatilization. The high remelt temperatures obtained in brazed joints have been due to the high melting point of the material present in the joint during and after testing, specifically tungsten and tungsten rich alloys. Specimens subjected to high temperature tests reveal that metallurgical changes have occurred in the original brazed and subsequently diffused joint.

The mechanism for obtaining this high melting constituent in the joint and therefore a high remelt temperature is believed to have occurred through a combination of four phenomena: diffusion, dissolution, volatilization, and consolidation. These dynamic phenomena occur in three distinct stages: (a) the brazing operation, (b) the diffusion cycle, and (c) the actual elevated temperature tests of the bonded tungsten.

In order to evaluate diffusion welding techniques as a method for producing rocket motor nozzles for elevated temperature service, tungsten nozzle components have been assembled and brazed. Stacked 0.125-in. thick tungsten washers have been brazed and diffusion-cycled. The manganese-base filler metal was used and the nozzle assembly brazed at 1950°F in 5 min and diffused at 2100°F for $3\frac{1}{2}$ hr as shown in Figure 9.24. The assembly-and-brazing fixture is also shown. After final machining, it was difficult to discern the brazed joints on the inside surface of the liner, which indicated that good brazing and diffusion had occurred.

Diffusion welding of tungsten by vacuum-hot-pressing has been successful. and that for sustained high-temperature application, either a tungsten-titanium or tungsten-clumbium interleaf system, has produced minimum diffusion activity. For transient high temperature requirements the applicaton of any of these systems issuitable in the as-bonded condition.

Figure 9.24 Rocket motor nozzle diffusion bonded [45].

Materials

Zirconium and Its Alloys

Diffusion welding of zirconium has been achieved at 1300°F in vacuum at pressures of 500 psi. When joining zirconium to stronger zirconium alloys, pressures ranging around 4000 psi have been required at the same temperature.

Zirconium alloy (Zircaloy). Zirconium and its three most common alloys, Zircaloy-2, Zircaloy-3, and Zircaloy-4, have all been satisfactorily self-bonded by gas-pressure bonding in 3 to 4 hr at 1550°F and 10,000 psi. All alloys behaved in the same manner and yielded bonds similar to those shown in Figure 9.25. The best bonds have been achieved with surfaces subjected to a belt-abrasion treatment before bonding.

The transient melt diffusion bonding method has also been successfully applied to zirconium-alloy parts. A thin layer or coating of silver, copper, or iron has formed a transient-liquid layer in the interface to produce the bond. During bonding, the middle layer diffused away into the base metal and the coating formed a eutectic with the base metal.

The technique has been very effective for diffusion welding plate-type nuclear fuel elements. Before diffusion welding, the Zircaloy sheet was plated with the intermediate material, silver, iron, or copper. The best welded properties have been obtained with a 0.1 mil copper coating at

Figure 9.25 Typical Zircaloy-2 self-bond achieved with properly prepared surfaces [38].

a temperature of 1900°F for ½ to 2 hr and a positive helium pressure of 30 psi.

Dissimilar Metal and Ceramic Combinations

Using gas-pressure bonding, 100 stainless and ceramic parts have been diffusion welded at 2000°F and 10,000 psi for 3 hr. (See Figure 9.26.) Similar types of nuclear assemblies have been fabricated utilizing columbium and Zircaloy clad materials. The time, temperature, and pressure for columbium was the same as the stainless steel; the Zircaloy, however, required 1550°F temperature, 10,000 psi for 5 hr.

Various dissimilar metal combinations using resistance-heating methods have been successfully diffused welded. The 0.5% Ti-molybdenum alloy has been bonded to the columbium alloy-Cb 752. Since molybdenum and columbium are completely miscible in each other, the bonded phase is probably a solid solution of the two metals. A similar bond has been made between 0.5% Ti-molybdenum and 90% tantalum-10% tungsten alloy as well as pure molybdenum and tantalum. One of the easiest bonds formed is that between titanium and stainless steel. A narrow layer between beta titanium and stainless steel forms the titanium-nickel eutectic alloy (Ti Ni_3). Diffusion of the titanium and the nickel of the stainless steel causes the formation of this alloy at 1750 and 2050°F.

Figure 9.26 The finished assembly contains over 100 parts shown in the upper photo. All were bonded at one time at 2000°F [46].

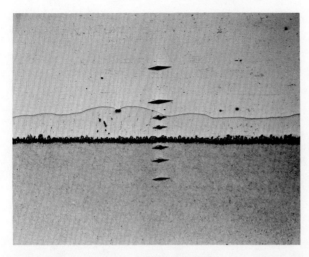

Type 304L stainless steel

Gold-nickel filler alloy

Titanium-gold-nickel intermediate alloy

6A14V titanium alloy

Figure 9.27 Transition tubes (titanium alloy 6AL-4V joined to type 304L stainless steel) are used on Gemini spacecraft. Photomicrograph shows joint [47]. (×100)

The amount of pressure to produce this Ti-Ni alloy is almost negligible. In fact, merely the weight of a piece of titanium is enough to cause its formation if the surface cleanliness and other conditions are exactly right.

Transient melt bonding has been utilized with a combination of various silver brazing alloys in bonding 6AL-4V titanium to Vascojet 1000 steel. The bonded joints can withstand pressures of 31,000 psi and meet required shear strengths of 34,800 psi. Assemblies have been successfully produced at 1700°F by using a 0.004-in. thick layer of 95% silver, 5% aluminum foil against the titanium interface, and a 0.002-in. thick layer of pure silver plus 0.2% lithium on the alloy steel interface. A gold-nickel alloy (82-18) has also been very successfully used in production to join 6AL-4V titanium to 304L stainless steel by transient melt bonding as shown in Figure 9.27.

Successful diffusion welding of 2219 aluminum alloy and 321 stainless steel has been accomplished at (a) temperature of 500 to 600°F, (b) pressures of 20 to 25 ksi, and (c) times of 2 to 4 hr. With these combinations of metals, the joint faying surfaces were electroplated with silver before bonding. The silver prevented the formation of oxide-film barriers and embrittling phases.

Utilizing these processing parameters, techniques for diffusion welding of large diameter assemblies have been successfully developed and demonstrated with the fabrication and testing of 20-in. diameter joints. Hoop stresses developed during burst testing exceeded the yield strength of the

Table 9.8 Metallic and/or Ceramic Combinations Diffusion Welded

Materials	Joined	Intermediate	Temp. (Fx 100)	Press (ksi)	Time (Min)	Atm.
Be	Be	None (—)	15–16.5	10	240	I
Al	Stl.	—	9.5	8	.5	V
Al	Cu	—	9.4	1	.25	—
Be	Be	—	15–18		60	V
Al	Al	Si	10.8	1–2	1	V
Al	Cu	—	8.8	40	4	V
Al	Zr	—	10	22–50	15	V
Al	Ni	—	9.3	22–52	4	V
Al	U	Ni Plate	9	5	—	—
7075 Al	7075 Al	Al	3.25	240	60	A
AM-355	AM-355	—	20.5	0.5	10	V
Be	Be	Ag foil	13	10	10	V
Be-Cu	Monel	Au-Cu, Au-Ag	6.5	1.5–6	180	He
Cast Iron	Cast Iron	—	14.7	0.51	10	V
Cb	Cb	—	21–24	10	180	I
Cb	Cb	Zr	15.4	10	240	A
Cb	Cb	In	20	0.13	60	V
Cb-752	Cb-752	Ti foil	14.5	5	10	V
Cb-752	Cb-752	—	20–22	.1	—	I
Cb-V	Cb-V	—	23	10	180	—
D-36(Cb)	D-36(Cb)	—	20	1.7	180	A
Cb	St. Stl.	—	20	1.4	20	—
Cb	Mo	—	25.5	0.7	10	—
Cb	St. Stl.	Iron	15–23	2.0	30	—
Cb	Ta	Zirconium	16	5	30	—
Cu	Cu	—	6	20.3	1	H
Cu	Ti	—	15.6	0.7	15	V
Cu	Al	—	9.5	1	15	V
Cu	Kovar	—	17.4	1	10	V
Cu	Cb-1Zr	—	18–19	30	30–120	V
Cu	316 St. Stl.	—	18	—	120	V
Cu	W	—	15.5	0.6	30	—
Haynes 24	Haynes 25	—	21	0.5	10	V
Inconel	Inconel	—	20–21	0.5	10	V
Mild Stl.	Mild Stl.	—	20	0.5	10	V
St. Stl.	Stl.	Ni	17	20	30	—
Cr	Cr	—	21	10	180	—
Nimonic 90	Nimonic 90	Ni	20.2	4	20	A
Mo	Mo	—	23–26	10	180	I
Mo	Mo	Ti foil	17	10	120	A
Mo	Mo	Ti foil	16	12.5	10	V
Mo-.5 Ti	Mo-.5 Ti	—	21	0.5	15	A
Mo-.5 Ti	Mo-.5 Ti	Ti Foil	16.8	10	20	V
Mo	Mo	—	29.1	1.4	20	V
Mo	St. Stl.	—	20	1.4	20	—
Mo	W	—	36.3	1.4	10	—
Mo	St. Stl.	Ni	15–23	2	30	—
René 41	René 41	Cu Foil	16	5	10	V
TZM	TZM	Pd film	20	20	1	A
TD Ni	TD Ni	—	16	24	.3	A
TD Ni	TD Ni	Ni	20	9	45	V

Table 9.8 (Continued)

Materials	Joined	Intermediate	Temp. (Fx 100)	Press (ksi)	Time (Min)	Atm.
St. Stl.	Zircaloy	—	19	Diff. Exp	3	A
St. Stl. (304, 316) (347–410)	St. Stl. (304, 316, 347,410)	—	21	10	60–180	A
St. Stl. (321)	St. Stl. (321)	—	—	—	3	Air
St. Stl. (410) (347)	St. Stl. (410) (347)	Ni+ 9–16 atomic % Be	22	0.001	5	A
Ta	Ta	—	24–26	10	180	I
Ta	Ta	Ti foil	16	10	10	V
T-111	T-111	Ta foil	22–26	10–40	1	A
Ta	Ta	Zr	16	—	—	—
Ta-10W	Ta-10W	Ta foil	26	10–20	.3	A
Ti (75A)	Ti (75A)	—	13	4–10	10	V
Ti (75A)	Ti (75A)	Al foil	10.5	10	10	V
Ti (75A) (6AL-4V) (8AL-1V-1Mo) (7AL-2Cb-1Ta)	Ti (75A) (6AL-4V) (8AL-1V-1Mo) (7AL-2Cb-1Ta)	—	17	0.5	10	A
Ti (75A)	Ti (75A)	—	15.5	10	180–240	A
U-80 Pd	U-80 Pd	—	11–12	4	10	V
W	W	Ni-Pd	18	10	90	H
W	W	Re-Ta	18.5	20	30	—
W	W	Cb	17	10	20	V
W	W	—	28	0.3–0.5	10	A
Zircaloy-2	302 St. Stl.	—	18.7	—	80	V
Zircaloy-2	302 St. Stl.	—	18.7	—	3	He
Zr	Zr	—	15.5	10	210	I
Zircaloy-2	Zircaloy-2	Copper	19	30	30–120	He
Zr	U	—	15.5	2.2	2160	I
Zr	U-10 Mo	—	12	10	360	I
Alumina	Ta	—	28	10	120	I
Ni Bonded Carbide	Ni	—	24.5	—	10	I
UC	Cb	—	21–24	10	180	I
UO_2	Zircaloy	—	15.5	10	240	I
UO_2	Zircaloy	Carbon	15.5	10	240	I
UO_2	Cb	—	21	10	180	I
UO_2	St. Stl.	—	21	10	180	I
Alum-2219 7106	321 St. Stl.	Ag	5	—	120	Air
2219 Alum	(Ti-5AL-2.5Sn)	Ag	5	—	120	Air
2219 Alum	(Ti-5AL-2.5Sn)	—	9.4	—	30	V
Inconel 600	(Ti-8AL-1Mo-1V)	—	13.2	—	10	V
Inconel 600	(Ti-8AL-1Mo-1V)	Ag	7	—	30	Air
321 St. Stl.	(Ti-8AL-1Mo-1V)	—	14.2	—	10	V
321 St. Stl.	(Ti-8AL-1Mo-1V)	Ag	7	—	30	Air
Inconel 718	321 St. Stl.	—	16.2	—	15	V
Inconel 718	321 St. Stl.	Ag	6	—	30	Air
Inconel 718	321 St. Stl.	Ni	13.2	—	30	V

Code: A = Argon, He = Helium, H = Hydrogen, I = Inert, V = Vacuum

2219-T62 aluminum alloy. The unique diffusion welding method developed utilized simple differential thermal expansion tooling that can be economically adapted for production requirements. (See Figure 9.12.)

Other dissimilar metal combinations that have been diffusion welded successfully and produced ductile joints include columbium-1Zr alloy joined to the Mo-0.5% Ti alloy, copper and columbium-1 Zr alloy, copper and Type 316 stainless steel, and the Mo-0.5% Ti alloy joined to René 41.

Eutectic diffusion bonding techniques using braze alloys have also been used to join stainless steel to Zircaloy-2 for nuclear reactor hardware. Joining has been done in a vacuum furnace at temperatures of 1870°F in 5 to 30 min and by induction heating in 30 sec to 1 min at a temperature of 1900°F. Vacuums of 3 to 5×10^{-5} torr have been used with either of these heating methods to prevent oxidation and contamination of the joint materials.

Numerous experiments have shown the feasibility of diffusion welding tungsten-graphite and pyrolytic-graphite to itself and to other ceramic materials (aluminum oxide and zirconium oxide); all work, however, has been performed on a test coupon basis.

Before discussing several miscellaneous applications and materials, Table 9.8 summarizes the combinations of materials that can be joined by diffusion welding.

MISCELLANEOUS

Honeycomb Core

Honeycomb core has been produced two ways by diffusion welding. One method is a direct metal-to-metal bond in which a stop weld material prevents welding in selected areas. In the other method, the one used most frequently, metal-to-metal bonding is done by using an intermediate material in the joint area to promote the bond, which is illustrated in Figure 9.28. In both methods pressure reduces the effects of surface roughness and provides maximum surface contact. Temperature, time, and pressure have been carefully integrated with control of atmosphere (vacuum or inert gas) to prevent the formation of surface oxides.

The honeycomb is fabricated by assembling sheets of foil into a flatpack that has regularly spaced lines of intermediate metal (or stop weld material) applied to the surface. The spacing and width of the lines determine the cell size and shape of the honeycomb, for the bond occurs only at the areas covered by the intermediate metal (or left uncovered by stop weld). In laying up the pack each succeeding sheet is placed with its line of intermediate metal midway between the line on the sheet below. The pack

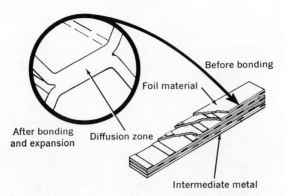

Figure 9.28 Another technique in metal-to-metal bonding is done by using an intermediate material in the joint area to promote the bond [49].

is then subjected to a protective atmosphere and heat and pressure applied. The finished pack resembles a piece of solid metal called "Hobe" (honeycomb before expansion) [49].

Cladding

In any cladding process a material possessing a certain outstanding quality is placed over the base material so that the resulting composite is better able to accomplish its task. Cladding may be applied to prevent corrosion, abrasion, oxidation, to provide special high temperature properties, etc. Various combinations are possible: metal over metal, metal over ceramic, and ceramic over metal. In each of these the bond formed may need to be an actual metallurgical bond or only one that exhibits good electrical and thermal properties.

Gas-pressure bonding has been found ideal as a cladding process. Its special advantage is that components are machined to final size and the only deformation is that required to bring the mating surfaces into intimate contact. For this reason, very brittle materials, such as sintered oxides and hydrides, can be clad without danger of fragmenting. Likewise, metallic materials of widely differing properties can be pressure bonded with little difficulty; bonds formed in this instance can be varied from purely mechanical to purely metallurgical in nature by controlling the conditions of the operation.

Numerous cladding operations of both metallic and ceramic components have been successful. Typical examples of some of these systems are shown in Figures 9.29, 9.30, and 9.31. In Figure 9.29 the ceramic material,

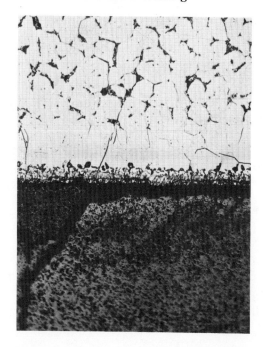

Figure 9.29 Zircaloy-clad uranium dioxide with no barrier material between the core and cladding [38].

uranium dioxide, is permitted to react with the metallic cladding to form a metallurgical bond.

Figure 9.30 represents a specimen of interest to those concerned with protection of refractory metals from high temperature oxidation. An impervious coating of alumina is soundly bonded to a tantalum substrate, completely protecting the tantalum from environmental conditions. Cermets can also be used for bonding ceramics to metals. By varying the concentrations of the constituents, a continuous transition can be made from pure ceramic to pure metal. Resulting bonds would have strength equal to that of ceramic material.

Zirconium has been used to clad a uranium-bearing alloy for nuclear-fuel-element applications, and metallurgical bonding is required for structural support as well as good thermal conductivity. Bonds between columbium and zirconium and molybdenum and titanium such as shown in Figure 9.31 are significant not only for demonstrating metallic cladding but also for showing a convenient technique for joining refractory metals below their recrystallization temperature.

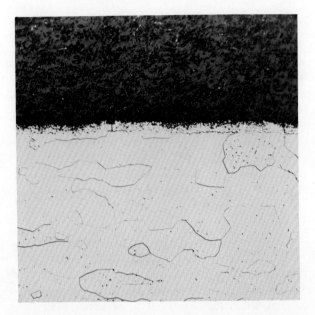

Figure 9.30 Example of bonding alumina to tantalum [38].

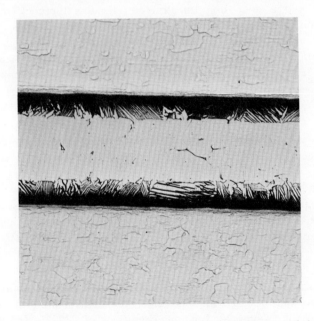

Figure 9.31 Example of bonding between pieces of molybdenum and titanium [38].

Several examples of specific configurations utilizing cladding that have been successfully fabricated by gas-pressure bonding are the following:

1. Rod and tubular-shaped assemblies have generally consisted of a core material that is clad with a second material. A bond between the core and cladding may be produced or prevented, depending on the requirements.
2. A unique technique has been developed for preparing rod-type fuel elements without using preformed tubing. In this application columbium sheet has clad sintered pellets of uranium dioxide. The columbium is formed around the uranium dioxide pellets with the required columbium end plugs and then inserted in a Type 304 stainless steel tube, sealed, and pressure bonded.

Densification

The gas-pressure bonding process has been utilized for the densification of ceramic, cermet, and dispersion materials. The densities of these materials compacted by this technique are equal to or better than those that have been achieved by other methods. This densification has generally been achieved at much lower temperatures by gas-pressure bonding than by sintering; therefore the gas-pressure bonded ceramic materials have a finer grain size and a more uniform structure. Because there is no extensive deformation and the pressure is applied equally from all sides in the gas-pressure bonding process, cermet and dispersion materials do not exhibit fragmentation or stringering of the dispersed particles. The feasibility of densifying and cladding these materials in one-step pressure bonding operations has been developed.

CERAMICS—URANIUM DIOXIDE (UO_2)

High-density uranium dioxide with final densities up to 99.5% of theoretical have been realized at conditions of 3 hr at 2100 to 2700°F and 10,000 psi.

On the basis of the fine grain size achieved, helium permeability, and thermal-expansion and thermal-conductivity tests, the gas-pressure-bonded material has proven to be as good or better than pressed and sintered oxide of equivalent density.

ALUMINUM, BERYLLIUM, AND MAGNESIUM OXIDES

These materials behave very much the same during gas-pressure bonding. Densification has resulted in fine-grained bodies with uniform densities of 98 to 99% of theoretical. Bonding conditions of 2350°F and 10,000 psi for 3 hr has been required for aluminum and beryllium oxides, whereas for magnesium oxide the temperature was lowered to 2100°F and produced dense bodies.

URANIUM NITRIDE (UN) AND URANIUM CARBIDE (UC)

Uranium nitride and high-purity uranium carbide have been compacted from powders with uniform densities of 95 to 96% of theoretical at conditions of 2700°F and 10,000 psi. Also, uranium carbide powders with slight impurities such as oxygen have been densified by this process of 99% of theoretical at 2460°F and 10,000 psi.

CERMETS

The gas-pressure bonding technique has proven ideal for the preparation of cermets. Because of low deformation, stringering and fracture of the dispersed phase do not occur. In addition, since densification was achieved below normal sintering temperatures, grain size was controlled and reactions between the matrix and the dispersed phase have been minimized. The temperatures for bonding vary from 2500 to 2600°F, 10,000 psi, and 3 hr.

Cermets have been prepared from most of the ceramic materials mentioned and with numerous matrix metals. Examples of typical matrix metals include stainless steel, chromium, columbium, molybdenum, tungsten, and rhenium. In all instances densities approaching theoretical have been achieved.

Densification and Cladding

Most of the work on the densification of metals and alloys by gas-pressure bonding has been conducted with refractory metals and alloys. These are the most difficult to process by conventional powder-metallurgy techniques. It has been possible to compact these materials by gas-pressure bonding without introducing contaminating materials. Beryllium, chromium, columbium, tungsten, tantalum, molybdenum, and rhenium have been densified from as-packed or cold-pressed powders and clad simultaneously.

JOINT DESIGN; TEST RESULTS

It has been a difficult problem to devise tests for determining mechanical properties and suitable joint designs for diffusion welding. For example, the configuration of the weld interface in clad plate makes it difficult to determine the joint strength experimentally. For this reason joints are usually evaluated by indirect or qualitative methods such as bending, twisting, or pulling the clad sheet in tension to determine whether the bond will separate. In one qualitative test a sharpened cold chisel is forced between the sheets in an attempt to cause separation. If peeling occurs or the failure follows the original interface, the joint is considered unsatisfactory or incompletely bonded.

Another test used is a plug-and-socket test specimen configuration. In this test the plug is tapered and then pressed into the socket at room temperature. Elastic stressing and friction in the joint maintain a portion of this stress at the welding temperature after the pressing load is removed. The completed joint is tested in shear by simply pulling the plug from the socket.

Whereas this above test is applied to similar materials, a different joint design has been used with dissimilar materials. Zircaloy and stainless steel, for example, have utilized a lap joint or scarf joint. The design and relative placement of each material in the lap joint are important in forming the joint and in the performance of the joint. First, close tolerances are required to insure a uniform contact around the joint. For joints in approximately $\frac{1}{4}$-in. diameter tubing a maximum clearance of 0.001 in. is desirable. When a cylindrical joint interface is used, as shown in Figure 9.32a, it is necessary to place the material with the higher coefficient of expansion, in this case stainless steel, on the inside so that the differential expansion on heating produces a net compressive pressure on the interface. If the opposite ar-

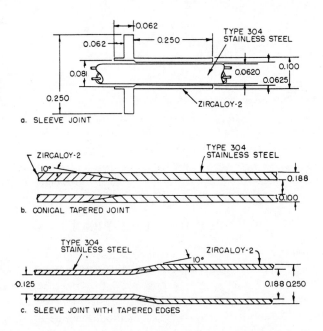

Figure 9.32 Joint designs used in various eutectic brazing applications [36].

rangement is used with a cylindrical joint, the material of higher coefficient of expansion expands more, opening the clearance between parts. Under these circumstances and with zero applied pressure, no joint is formed.

When the stainless steel is placed on the inside to produce a compressive force on heating, the differential contraction creates a residual tensile stress on the joint when it is cooled. Some of this stress is relieved during cooling by the yielding of the Zircaloy component. To use a cylindrical joint it is important that the free surfaces at the overlap taper almost to a knife edge to minimize residual stresses, increase flexibility, and remove stress concentrations.

A residual compressive stress on the joint interface is desirable because it reduces the likelihood of cracking during and after cooling. This is achieved by employing a conical tapered joint as shown in Figure 9.32b and places the stainless steel on the outside. Tapered joints must match each other within 0.001 in. to provide a good straight bond. A small or moderate axial compressive force applied to the joint maintains the necessary surface contact for diffusion and eutectic formation to occur.

Joints produced with this configuration have been tested in thermal cycling, hydraulic pressure, and pressure cycling. Properly joined samples usually fail first by splitting the center Zircaloy section. The crack formed then propagates to the bonded joints and causes their failure.

By utilizing the simple and unique application of differential thermal expansion, stainless steel and aluminum have been diffusion welded. Literature shows that 2219 aluminum alloy has a thermal contraction that is approximately 30% greater than stainless steel. Therefore, since the aluminum alloy has a greater contraction in the temperature range (77°F to −423°F), the aluminum alloy should be on the outer surface to give the most reliable joint. The joint configuration that was selected for diffusion welding is similar to a lap shear joint used in brazing and is designed to have an overlap of approximately 1.0 in. This type of joint prevents the entrapment of corrosive cleaning solutions and foreign particles in the joint faying surface and allows for variation of joint interface area, depending on the shear strength required. Room temperature single and double lap shear strengths greater than approximately 15,000 psi have consistently been obtained for diffusion-welded joints. At test temperatures of −320 and −423°F the joint strengths increased approximately 20 and 30%, respectively.

Applying the techniques established on prototype parts, 20-in. diameter joint assemblies have been diffusion welded at 500 and 600°F and successfully passed cyclic pressure tests at RT and −320°F and exhibited burst test pressures up to 670 psig at −320°F. This burst pressure is equivalent to a hoop stress of 53,600 psi, which exceeds the yield strength of 2219-T62 aluminum alloy at —320°F. These results demonstrate that steel-to-aluminum joints of high structural integrity can be produced by diffusion welding [51].

The utilization of roll diffusion welding which was discussed earlier has been mainly applied to titanium alloys to date. With this technique, however, certain structural geometries can be designed. The minimum weight of this geometry depends on the type of loading involved for the specific structural component, which illustrates that a variety of structural shapes are usually required for a minimum weight design. For example, a corrugation and strip and key design (Figure 9.33) are two different construction methods for roll diffusion welding a truss core sandwich design. As can be seen in Fig. 9.33 edge members can be easily incorporated into the design. Diffusion bonds of 100% efficiency have been obtained with properties equal to the basic material when processed by roll welding and new mechanical properties determined for the structural characteristics of the special thin foil sandwich shapes [52].

Work recently concluded on pure tungsten has shown that remelt temperatures of brazed and subsequently diffusion cycled alloys produce unique data which can be utilized in future designs. The highest temperature attained without remelting of a brazed and diffused joint is approximately 6000°F. The results of shear strength at room and elevated temperature tests show 55 to 70.8 ksi ultimate tensile strengths of tungsten joints and failures at 2930°F temperature with the location of fracture at room and elevated temperatures always in the parent metal.

The designs of the various joint configurations as well as the development of new metals reflect some of the confusion today in measurement, for many properties have never been measured and test procedures are not generally available.

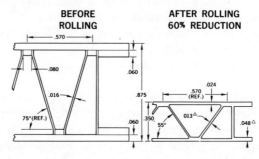

Figure 9.33 Strip and key design procedure [43].

Rather than listing numerous other charts, tables, and curves denoting properties and strengths, the author recommends References 51, 52, 33, 37, and 41.

With the initiation and development of data and criteria for structural applications the engineer can now start to institute diffusion welding into his designs that can be manufactured into efficient and useful hardware.

How are efficient joint configurations developed for diffusion welding? Let us take the refractory metals, for example, and develop some joint requirements for several structural systems. The limitations are the following:

- Operate between 2000 and 4000°F temperature.
- Use material gages from 0.006 to 0.012 in.

With these limitations which include the combination of temperature and material thickness parameters mentioned above all structures other than external surfaces on glide re-entry vehicles are eliminated. The various approaches to the design of a high temperature metallic structure for a glide re-entry vehicle can be categorized as hot monocoque and radiation shields plus insulated or insulated and cooled substructure.

Hot Monocoque Structure

Hot monocoque structures are characterized by systems that employ the hot external surface as the load-carrying surface. Because the external surface must take all loads, this surface must be rigid. Relief of thermal stresses by floating panels and corrugations becomes difficult.

Joints that are suitable for diffusion welding are the following:

- Joint of corrugation to face sheet
- Joint of doubler to corrugation
- Joint of panels to spar cap
- Joint of corrugated spar web to spar cap

Radiation Shield Plus Insulated and/or Cooled Structure

Unlike the hot monocoque structure in which the external structure must react all loads while operating hot, this form of construction separates the two effects. Light floating panels of refractory metals act as radiation shields, rejecting large quantities of heat back to the atmosphere. Since the panels are floating, they take only air pressure loads that are light. The main load-carrying structure is located below the shields and often separated from them by layers of insulation. In this fashion both the hot,

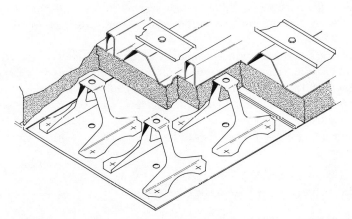

Figure 9.34 Typical heat shield panels for glider re-entry vehicle [19].

lightly loaded radiation shields and the cold, primary structure can be efficiently designed. This type of composite structure has been shown to be extremely efficient for many typical glide re-entry vehicle configurations.

Current heat shield designs fall into three categories: sandwich construction, corrugation-stiffened flat sheet, and multiple channel, Figure 9.34.

A similarity exists between the various heat shield concepts. Joints applicable to diffusion welding processes are the following:

- Edge closure of honeycomb heat shields
- Joint of corrugations to face sheets
- Joint of support clips to heat shield panel

A survey of typical structures suggests the selection of joint systems which are presented in Figure 9.35 for heat shields.

Heat shields of Type I are generally heavier than the others shown in Figure 9.35. By using diffusion welding, weight savings can result from smaller laps on the corrugation and the elimination of rivets which is the current method of joining. Reliability can be increased because of the simplicity of coating the diffusion-welded assembly.

The Type II shield can be made more efficiently with diffusion-welded joints because of the higher panel strength obtainable with continuous seam joints between the sheet and corrugation. Subsequent coating sequences are simplified by this procedure.

The Type III shield is currently made with bolted connections between the channels and the brace. A diffusion-welded joint again eliminates the weight of bolts and simplifies coating processes.

The Type IV shield cannot be made any lighter through the use of diffusion welding. Currently, panel edges and tabs are brazed while the honeycomb panels are being brazed. If this process proves to be unsatisfactory, the clips and edges can be joined by diffusion welding.

Structural joints are capable of joining honeycomb sandwich panels and

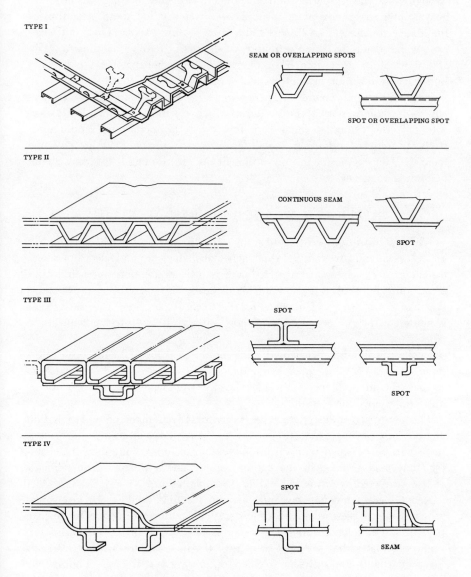

Figure 9.35 Typical joints for heat shields [19].

must be continuous and able to withstand high normal and planar shear loads. Their design is particularly significant because the theoretical efficiency of sandwich construction for a hot structure depends largely on the ability to join these panels with light, strong joints. Current methods of joining panels are welding and bolting. Welding may damage the wrought material properties and introduce distortions; bolting can be quite heavy. The main advantage of diffusion welding for these structures is the ability to fabricate stiffeners tailored to joining panels. These stiffeners cannot be made by any technique other than welding at present. The ability to build up a joint of properly placed members allows a latitude in design optimization not now available in a honeycomb structure.

In conclusion, an engineer can now realize the availability of the numerous joint designs and applications in which diffusion welding is feasible. Although the examples used were for refractory metals, the aluminums, steels, titaniums, and other metal families can be examined in the same manner. The engineer can now utilize in his future designs this new media of joining materials.

APPLICATIONS AND FUTURE POTENTIAL

Where is diffusion welding heading? If the results obtained to date are any criteria, industries other than aerospace, nuclear, and electronics will be taking a closer look at them. There is little doubt that some industries could put the process to use advantageously. The extent of the benefits that can be derived will depend in large measure on how well the process is understood and accepted and how effectively industry pursues their objectives.

In the missile and rocket field some motor parts have already been diffusion welded. With much of the industry's emphasis on high temperature service, the industry need not look far for suitable materials that can be diffusion welded successfully.

The aerospace industry, too, has need for any number of materials that are difficult or impossible to join by conventional methods. Now components can be designed with diffusion welding in mind. Tungsten, for example, may very well be in the running now that its metallurgical ailments can be corrected by diffusion welding. (See Figure 9.24.)

Reactor cores are naturals for diffusion welding. And in the electronics industry, the process might be applied to high temperature parts in electron tubes made of tungsten and molybdenum. The Apollo spacecraft has aluminum cold plates that are used to control the heat buildup in the electronic equipment and are diffusion welded on a production basis. Figure 9.36 displays the cold plate details and completed assembly of aluminum alloys

Figure 9.36 Ingredients of Apollo cold plate process used at North American Rockwell. In the center is a machined core with the face sheets to be bonded to each side. They are flanked by glide sheets used during the process. The open retort is at upper right, finished part to the left [47].

6061-T6 and 5052. Diffusion welding has provided both strength and freedom from expulsion or capillary filling of the narrow passage.

The aircraft field will also benefit from diffusion welding. Specifically, honeycomb sandwich structures have a good future in the aircraft industry. (See Figure 9.37.) With the inevitable development of supersonic aircraft, diffusion-welded sandwich panels will no doubt be examined for increasing applications. Diffusion-welded sandwich structures are already under investigation [53]. Any number of structural applications that are currently being joined by conventional methods could be adapted to diffusion welding.

The nuclear industry has used diffusion welding for many years. Fuel elements that are difficult to join by conventional means have been produced by this method. Because of the very close dimensional tolerances demanded by a reactor core, structural sections of the reactor have been diffusion welded.

A very common shape for modern nuclear-reactor fuel elements is the flat plate, which consists of a receptacle surrounding the core material and two cover plates placed on the top and bottom of the receptacle. The cladding material often has properties drastically different from those of the core, thereby making roll welding impractical. Gas-pressure bonding is well suited for this application. Because of the minimum of deformation incurred during gas-pressure bonding, even the most brittle ceramic mate-

Figure 9.37 Diffusion bonded 6 AL-4V titanium honeycomb panel. Temperature and time: 1600°F for 8 hr. Pressure: $\frac{3}{8}$ psi on core; 2 psi on laminate.

rials can be clad without fracture. Figure 9.26 demonstrates one of the flat plate assemblies that has been produced by gas-pressure bonding. Diffusion welding has also been experimentally used to assemble tube bundles for reactors. In the reactor design Zircaloy-2 tubes were joined to tube spacers (ferrules) of the same material.

Under electric blanket methods, ten titanium chambers containing 17,000 diffusion-welded joints have been fabricated for a zero gradient synchrotron with 54-ft long chambers containing 20 miles of joints.

The electronics industry, particularly those firms engaged in vacuum tube technology, have utilized diffusion-welding techniques for high temperature components. Mechanical means have been used to hold some of these components together. This approach has required, in some instances, the machining of threads and fabricating a refractory metal nut to hold a washer or other high temperature metal part in place. Diffusion welding of the two original pieces can eliminate much of the cost and the added weight of the extra mechanical joint and thus improve the design. The bonding of molybdenum rod to tungsten sheet is a typical assembly finding applications in the vacuum tube industry.

Diffusion welding is now being used to join sintered steel-bonded carbide to itself and to conventional steels. The wear-resistant carbide, Ferro-Tic C, is a powder metallurgy product consisting of fine particles of titanium carbide embedded in a tool steel matrix.

Metallurgical inspection cannot reveal the joint line currently being made between pieces of the machinable carbide joined by diffusion-welding techniques. As a result of this homogenity the joint is as machinable as the main body of carbide.

The welding process makes it possible to make machinable carbide structures that more closely approximate the needed configuration of a tool or die, thus reducing machining time and material waste. Furthermore, highly wear resistant tools have been made by facing a carbon steel part with the steel-bonded carbide. The densification of powders is an important new application of diffusion welding that currently is being applied to the production of large diameter electrodes for subsequent arc melting or of billets for subsequent forging and rolling. Diameters of 4 and 5 in. are practical with existing equipment. These electrodes or billets are of sufficient density and strength that they are readily handled without fear of flaking or cracking.

Another potential application that appears promising is the preparation of full-density, complex configurations from powder. This would eliminate the expense and waste incurred in the forming and machining of wrought material. Preforms can be produced by standard powder-metallurgy techniques or they may be plasma-sprayed into the desired shape. Gas-pressure bonding is then employed to densify the material to near full density, thereby improving the mechanical and physical properties.

The various methods of diffusion welding, described earlier, have also been investigated for the following applications:

    Elements of lifting re-entry vehicles
    Pivot mechanisms
    Cryogenic tankage and tubing
    Microwave components
    Space antennas and probes
    Composite filament structures
    Torpedoes
    Deep submergence manned vehicles
    Airborne and personnel armor
    Boundary layer control components
    Air-cooled turbine blades
    Helicopter blades
    Jet engine ducts, tail cones, and thrust reversers
    Rocket motor nozzle skirts
    Heat exchangers
    Masts, spars, and booms for high-performance sailing yachts
    Missile fins and components

462 Diffusion Welding

Fuel tank sections and 33 ft "Y" ring segments for space booster vehicles

Coatings

Future lifting re-entry and hypersonic cruise vehicles will consist of a rib-, truss-, and honeycomb-supported wing structure designs. Structures such as those mentioned have been produced from beryllium, steel, copper alloys, stainless steels (300 series, PH 15-7 Mo, AM 350); superalloys (nickel-base, cobalt-base and iron-base), and refractory metals (tantalum, molybdenum, and columbium). (See Figure 9.38.)

Control surface pivot mechanisms for future lifting entry vehicles will require joining ceramic or graphite-to-tungsten or tantalum-bearing retainer structures. Diffusion welding by creep-controlled methods appears to be a promising technique. In addition, leading edge control surfaces can also be fabricated by these techniques.

Engineers have diffusion-welded double-walled aluminum orthogonally ribbed configurations that simulated a torpedo hull construction. (See Figure 9.39.) In addition, they have fabricated aluminum heat exchangers, heat rejection systems, and heat-exchanger cold plates as shown in Figure 9.40. The assemblies used a 7075-T6 aluminum core sprayed with 100% aluminum and the face sheets were 0.020 in. alclad 7075-T6 aluminum.

Military requirements for modern heat exchangers demand lightweight

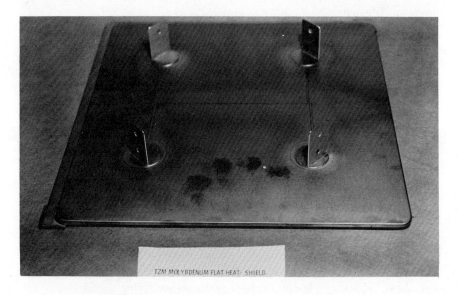

Figure 9.38 TZM molybdenum flat heat-shield.

Figure 9.39 Three inch diameter channel section rings. Diffusion-bonded assembly and disassembled rings [34].

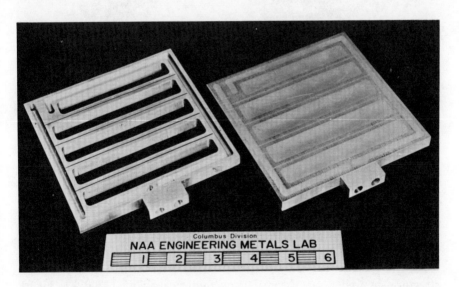

Figure 9.40 Cold plate "heat exchanger." *Left,* core; *right,* bonded assembly [34].

463

Diffusion Welding

Figure 9.41 Fracture surfaces of diffusion heat-treated corrugated sheet samples after tensile tests at room, 1000°F, and 1200°F temperatures.

structures as well as an elevated temperature capability (1200°F). Figure 9.41 shows the configuration and the results of testing the diffusion weldment.

The jet engine and propulsion industry is taking advantage of diffusion welding. By using eutectic melting techniques, molybdenum has been welded

Applications and Future Potential 465

to tungsten with a nickel interleaf and electric propulsion engines have operated at joint temperatures of 2000°C with no failures.

One of the first applications of diffusion welding in jet engines was a pump distributor plate that had two halves that were perfectly joined. The distributor plates were made of Type 410 stainless steel. One contained distribution channels for hydraulic fluid (it resembles a printed circuit). The other half also contained holes and ducts and sealed the port channels after they were machined. The end result was a metallurgical bond so nearly perfect that microscopic examination often could not detect where one piece ended and the other began. Figure 9.42 is an example of a TD-Nickel jet engine blade which has been diffusion welded.

The diffusion-welding process has joined the halves of a hollow quartz sphere without changing the material's bulk properties and has allowed designers to take advantage of the dielectric properties of quartz to reduce cost and increase the reliability of an electrically suspended gyro. First step in producing the joint was to apply a three-layer metal film to the joint surfaces of the hollow quartz hemispheres. The first layer was chromium, chosen because it adhered well to quartz. An intermediate layer of copper acted as a filler material to insure a hermetic seal. Finally, a thin alloying layer of silver was applied as a diffusion aid to reduce welding time and temperature.

Tensile tests of samples failed not in the joint but in the quartz at loads greater than 2000 psi. The interface was not affected by exposure to 500°F and survived five cycles of thermal shock between −78 and 100°F

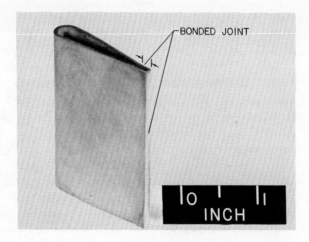

Figure 9.42 Diffusion-bonded jet-engine blade.

Figure 9.43 Fluid logic element 15 layers 0.0015 in. foil-Inconel X-diffusion welded.

without fracture. In addition, no leaks were detected at 1×10^{-8} torr using a helium probe [54].

The joining of detail parts (0.001 in. thick) of fluidic devices for use in the hot areas of aircraft engines is a potentially new application for fluidics as well as diffusion welding. Most of these devices today are glued or bolted together. As the temperature requirements on fluidic devices increase, however, a one piece fluidic component will be needed and diffusion welding will fill that need as shown in Figure 9.43.

In the design of space antennas dissimilar metals are required for the purpose of controlling heat flow from the hot side to the cold side of the structure. In this application of diffusion welding metal strips having high conductivity (copper or silver) are required to be joined to aluminum or nickel alloys. The joints must provide high conduction and operate in a temperature range of 180 to 300°F. Space probes such as Voyager have been designed to provide an internal temperature that will remain within a closely controlled operating range. Dissimilar metal joints are required at contact points between the outer shell and inner structure. Materials presently used, depending on performance requirements, which will increase or decrease heat flow to the inner structure are diffusion welded.

The increasing requirements for structural efficiency associated with advanced aerospace vehicle systems place overwhelming demands on conventional materials. The concept of filament-reinforced metal matrix composites provides a realistic solution for more efficient use of available materials. By embedding low-density, high-strength, and high-modulus filaments in a ductile and tough metal matrix, a new class of materials with physical and mechanical properties somewhere between the two composite components is created.

Applications and Future Potential

The usefulness of metal matrix composites and justification for their development are based on their relative structural advantage over conventional available materials. Some of the more important factors to be considered are tensile, compressive and shear strength, elastic modulus, fatigue strengths, and creep strength. The low density of the composites makes the specific strength and modulus of the composites of even greater significance. Other considerations of equal importance are relative cost, methods of joining, transverse strength, biaxial strength, corrosion resistance, and fracture toughness.

The intense interest in filamentary composite development programs stems from recent availability of filamentary material such as boron (B), silicon carbide (SiC) and beryllium (Be) in continuous length and production quantities with attractive physical and mechanical properties. Incorporation of these filaments in an aluminum or a titanium alloy matrix makes it possible to combine the outstanding properties of the filaments with the capability of the metallic matrix for carrying shear loads. The ultimate structural potential of the composite, its cost effectivity, and its chemical and mechanical compatibility govern the choice of a composite combination for specific application.

In fabrication, filaments are sandwiched between commercially available wrought alloy sheets; heat and pressure are applied to cause the matrix metal to flow and fill between the filaments. The fabricated composite has a wrought high-strength metal matrix integrally bonded to form an efficient structural material. The process appears ready for scaling up for fabricating composite panels in production size and quantities. To date the production of fibrous composites joined by diffusion welding include the following four systems, all using continuous filaments.

Matrix	Filament	Abbreviated Designations
X7002 Aluminum	Beryllium	Al-Be
X7002 Aluminum	Boron	Al-B
Ti-6AL-4V	Boron	Ti-B
Ti-6AL-4V	SiC	Ti-SiC

The Al-Be and Al-B systems are ready for scale up to sizes consistent with aircraft component fabrication. Of particular interest is the observation that with unidirectionally oriented filaments, reinforcement occurs with respect to room-temperature tensile and compressive strength, elastic modulus, elevated-temperature strength, and fatigue properties. In addition, tests of creep rupture strength have been made on the Al-Be composites and improvements over the matrix alloy noted for temperatures as high as 600°F.

The Ti-B composites are not as strong as the matrix, apparently because of degradation of the boron filaments during processing. The SiC filaments, by contrast, show no evidence of degradation.

An evaluation of process economics indicates that the diffusion-welding process is economically feasible for tonnage production of all four composite systems; the processing cost itself is between $3 and $6 per pound. The cost of the filaments is the principal variable factor [55].

In future cryogenic tank designs diffusion welding of dissimilar metals would prove advantageous. Primarily, transition joints that are required in tubing lines carry cryogens. These joints are needed primarily to provide design flexibility in choosing the best materials (based on total system weight and heat loss) for the pressure vessel, plumbing lines, and outer vacuum shell. In providing electrical pass throughs, plumbing inlet and outlet lines and vacuum gages, the following alloy combinations are needed and are all capable of being welded together:

1. 2219 Aluminum alloy to 321 stainless steel.
2. 2219 Aluminum alloy to titanium alloys.
3. Stainless steel to titanium alloys.
4. Inconel 718 to stainless steel or 2219 aluminum alloy.

In these applications the dissimilar metal joints must be pressure and vacuum tight and operate at a temperature of 150 to −423°F.

Currently, the most familiar titanium alloys (6-4 and 8-1-1) are being roll diffusion welded into prototype hardware for selected components of the Saturn rocket. These include a cylindrical thrust structure for the booster, a base-gore segment for the 33-ft diameter fuel tank bulkhead, a huge crossbeam for the thrust structure, and other parts. If successful, this development might lead to as much as a 38% weight savings over similar configurations that now are made of aluminum alloys. The cylindrical tank-skin stringer structure will reach dimensions of 20 and 30 ft in length and 8 ft wide, as shown in Figure 9.44.

Finished shapes have been made in shorter time than forgings by laminating sheets by diffusion welding and eventually may require only 0 to 10% machining. Figure 9.45 shows an XB-70 airplane titanium trunnion 12 in. long made by diffusion welding.

Diffusion welding appears ideal for applying impervious coatings for high temperature oxidation resistance on refractory metals and alloys. This approach is particularly attractive when one considers that frequently the structural members might well be joined together in the same operation in which the coating is being applied. Similar applications in the chemical

Applications and Future Potential

Figure 9.44 Diffusion bonded titanium structure for next generation boosters was made by North American Rockwell.

industry, such as the manufacture of duplex tubing and corrosion-resistant liners, also appear promising.

Diffusion welding offers great promise in overcoming many problems including weight, performance, and so forth that are foreseen by engineers in advanced aerospace systems. Metallic composites of the future would lose much of their potential if diffusion welding were not available for joining.

As we look into the future ten to twenty years from now, diffusion welding units may well be standard equipment on future space flights. It is not inconceivable that space platforms or laboratories will be constructed with diffusion welded joints; the vacuum of space is an ideal environment for

Figure 9.45 Diffusion-welded titanium fitting.

promoting the diffusion of metals. Welders are a definite part of the government's future space plans for joining. Therefore, if you assume that our space wizards will invent a practical, portable tool to make these joints, diffusion welding's potential rivals the imagination.

REFERENCES

[1] R. J. Davies and N. Stephenson, "Diffusion Bonding and Pressure Brazing of Nimonic 90 Ni-Cr-Co Alloy." *Brit. Welding J.*, **9**, 139–148 (1962).
[2] E. S. Hodge, "Gas Pressure Bonding of Refractory Metals." *Metals Engineering Quarterly (ASM)*, **1**, (4), 3–20 (November 1961).
[3] R. M. Evans, *Notes on the Diffusion Bonding of Metals.* DMIC Memorandum 53, Battelle Memorial Institute, Columbus, Ohio, 5 pp. 23 ref. (April 20, 1960).
[4] J. R. Brophy, H. Heideklang, P. Kovach, and J. Wulff, *Joining of Refractory Metals–Tungsten.* MIT Final Report WAL 460.54/1-2, Massachusetts Institute of Technology, Boston, Mass., 52 pp. 34 ref. (March 30, 1962).
[5] F. H. Ehlinger and W. P. Sykes, *Trans. ASM,* **28,** 619 (1940).
[6] M. A. Tylkina, B. P. Polykova, and E. M. Savitski, "Palladium Tungsten Phase Diagram," *Jnl. Inorg. Chem.,* **6,** 14 (1961).
[7] H. B. Huntington, "Mechanisms of Diffusion," Atom Movements, American Society for Metals, Cleveland, pp. 69–73, 76 (1951).
[8] A. H. Cottrell, *Theoretical Structural Metallurgy,* St. Martins Press, pp. 188–190, 1955.
[9] S. J. Paprocki, E. S. Hodge, and C. B. Boyer, *Gas Pressure Bonding of Fuel Assemblies,* USAEC Report TID-7559, Part 1, Atomic Energy Commission, Washington, D.C. pp. 213–232 (August 1959).
[10] E. E. Underwood, "A Review of Superplasticity and Related Phenomena." Journal of Metals, pp. 915–919 (December 1962).
[11] G. W. Cunningham, *The Mechanism of Pressure Bonding.* Presented at AIME Session of Western Metal Congress and Exposition (March 21, 1963).
[12] G. W. Cunningham, and J. W. Spretnak, "Mechanism of Gas-Pressure Bonding," BMI-1512.
[13] N. F. Kazakov, and Novikov, "Diffusion Bonding Magnetic Alloys," *Vestnik Mashino Strvenic,* 43 (1963).
[14] L. R. Vaidyanath, M. G. Nicholas, D. R. Milner, *Brit. Welding J.,* 6(1), 13 (January 1959).
[15] W. Feduska and W. L. Horigan, *Welding J. Res. Supp.,* 28–35 (January 1962).
[16] J. M. Parks, *Welding J. Res. Supp.,* 209–222 (May 1953).
[17] R. G. Tylecote and E. J. Wynne, *Brit. Welding J.,* **10**(8), 385–394 (August 1963).
[18] C. F. Wilford and R. F. Tylecote, *Brit. Welding J.,* **7**(12), 708–712 (December 1960).
[19] A. G. Metcalfe and T. B. Lindemer, "Diffusion Bonding of Refractory Metals," Final Report, Solar Aircraft, AF 33 (657) 8789 (December 1964).
[20] W. R. Young and E. S. Jones, "Joining of Refractory Metals by Brazing and Diffusion Bonding," ASD-TDR-63-88 (January 1963).
[21] F. J. Bradshaw, R. H. Brandon, and C. Wheeler, *Acta Metallurgia,* **12,** 1057 (September 1964).

References

[22] J. Y. Choi, and P. G. Shewmon, *Trans. AIME,* **224,** 589 (1962).
[23] L. R. Vaidyanath and R. D. Milner, "Significance of Surface Preparation in Cold-Pressure Welding," *Brit. Welding J.,* **7**(1), 1–6 (1960) and C. L. Cline, "An Analytical and Experimental Study of Diffusion Bonding," *Welding J. Supp.,* November 1966.
[24] C. E. Conn, "Liquid Rocket System Conjugate Structure and Tank Program," Progress Report No. 3, AFRPL-TR-66-98 (April 1966).
[25] R. F. Tylecote and E. J. Wynne, "Effect of Heat Treatment on Cold-Pressure Welds," *Brit. Weld J.,* **10**(8), 385–394 (1953).
[26] M. J. Furey, "Surface Roughness Effects on Metallic Contact and Friction," *ASLE Trans.,* **6,** 49–59 (1963).
[27] R. E. Smallman, *Modern Physical Metallurgy,* Butterworth, 1963.
[28] S. Storchheim, J. L. Zambrau, and H. H. Hausner, *Trans. AIME, J. of Metals,* 269–274 (February 1954).
[29] R. L. Samuel and J. D. Samuel, *The Effect of High Frequency Heating on the Diffusion Properties of One Metal into Another,* AD 239-866, 44 pages (April 10, 1960).
[30] J. P. King and A. Toy, "Diffusion Bonded Honeycomb Sandwich Panels," Air Force Contract AF 33(657)-8788, Final Engineering Report AFML-TR-64-329, (October 1964).
[31] *Applications of Welding,* Welding Handbook, Section Five, Fifth Edition, p. 95.3, 1964.
[32] M. J. Albom, "Solid State Bonding," *Welding J.,* 491 (June 1964).
[33] D. Hugill and B. Gaiennié, "Solid State Diffusion Bonding Tantalum Alloy Honeycomb Panels," Air Force Contract AF 33(615)-2777, Norair Co. IR-8-214II (September 1, 1965 to November 30, 1965).
[34] I. M. Barta, *Welding J. Res. Supp.,* 241 (June 1964).
[35] G. F. Blank, "A Practical Guide to Diff. Bonding," *Matls. Engr.,* p. 76–79 (October 1966).
[36] J. W. Gerken and W. Owczarski, TRW Report ER 6563 (June 23, 1965).
[37] C. Crane, D. Lovell and W. Baginski, "Research Study for Development of Techniques for Joining of Dissimilar Metals," Boeing Co., NAS 8-11307, Control #DCNI-4-50-01068-01 (1F).
[38] S. J. Paprocki, E. S. Hodge and P. Gripshover, Jr., "Gas Pressure Bonding," DMIC #159, pp. 1–30 (September 25, 1961).
[39] K. J. B. McEwan and D. R. Milner, *Brit. Welding J.,* **9**(7), 406–420 (July 1962).
[40] C. R. Manning, Jr., R. Royster and C. Braski, "Investigate New TD-Ni Alloy," NASA TN-D-1944 LRC.
[41] R. F. Yount, "TD-Ni-Cr Joining," Interim Progress Reports 1-8-211A, 2-8-211A, 3-8-211A, General Electric, AF 33(615)-3476.
[42] W. T. D'Annessa, *Welding J. Res. Supp.,* 232 (May 1964).
[43] J. E. Leach, Jr., "Production Diffusion Bonding of Titanium Structure Using a Formable Expendable Vacuum Chamber," SAE National Mtg., Douglas Paper No. 3429 (April 12–16, 1965).
[44] *Steel Magazine,* p. 109, April 19, 1965.
[45] M. J. Albom, "Diffusion Bonding Tungsten," Paper presented at the AWS 43rd Annual Meeting held in Cleveland, Ohio, April 9–13, 1962, *Welding J.,* November 1962.
[46] *Iron Age,* p. 83, July 18, 1963.

[47] *Steel Magazine,* p. 60, November 8, 1965.
[48] W. A. Owczarski, *Welding J. Res. Supp.,* 78–83, (February 1962).
[49] E. C. Vicars, "Diffusion Bonding: What's it All About," *Welding & Design Fabrication,* pp. 42–44 (December 1963).
[50] R. J. Davies and N. Stephenson, *Brit. Welding J.,* 9(3), 139–147 (March 1962).
[51] C. Crane, D. Lovell, W. Baginski and R. Torgerson, "Study of Dissimilar Metal Joining by Solid State Welding," NAS 8-20156, Boeing Co., DCN1-5-54-01169 (1F) (October 1, 1965).
[52] J. J. King, "Diffusion Bonding of Titanium and Beryllium, Air Force Contract AF 33(615)-3047, North American Aviation, IR-8-318 (IV).
[53] Douglas Aircraft Company, AF 33(615)-1399, IR8-212 & DMIC Review, December 10, 1965.
[54] *Mats. Eng.,* p. 29 (December 1966).
[55] A. Toy, D. G. Atteridge and D. I. Sinizer, "Development and Evaluation of the Diffusion Bonding Process as a Method to Produce Fibrous Reinforced Metal Matrix Composite Materials," AFML TR 66-350 (November 1966).

INDEX

AC and DC current, fusion spot, 227
Accelerating voltage, electron beam, 9, 16
Activity, diffusion weld, agents, 399
  concepts, 392
Aerospace structures, diffusion weld, 455
Aircraft, electron beam, 88
Aluminum and its alloys, diffusion weld, 420
  electron beam, 48
  exothermic, 274
  fusion spot, 240
  plasma arc, 159
  vacuum braze, 317
Aluminum oxide (alumina), electron beam, 53, 70
Anode, electron beam, 7
Applications, blanket braze, 186
  cold cathode, 109
  diffusion weld, 458–470
  electron beam, 83–94
  exothermic, 275–277
  fusion spot, 246–249
  laser, 217–221
  nonvacuum electron beam, 137
  partial vacuum electron beam, 118
  plasma arc, 163–167
  vacuum braze, 360–367
Arc, column, plasma arc, 145
  initiation and stabilization, fusion spot, 228
  length, fusion spot, 232
  modes, plasma arc, 144
  shaping, plasma arc, 150
  welds, electron beam, 48

Atmosphere control, nonvacuum electron beam, 126

Base metal characteristics, vacuum braze, 291
Beam, current, electron beam, 16
  focusing, cold cathode, 104
  formation, cold cathode, 102
  transfer, nonvacuum electron beam, 125
Beryllium and its alloys, diffusion weld, 419
  electron beam, 53
  vacuum braze, 316
Blanket heating, diffusion weld, 405
Bonding, exo-adhesive, 265
  exo-flux, 264
Brazing, electric blanket, 168
  electron beam, 94
  exothermic, 259
  radiant heat, 250
  vacuum, 278

Cathode, cold cathode, 103
  electron beam, 7
  gun, 12
Ceramics, diffusion weld, 442
  electron beam, 70
  vacuum braze, 338
Chambers, electron beam, 28
Chromium and its alloys, diffusion weld, 425
Cladding, diffusion weld, 447
Cold cathode, electron beam, 100–110
  applications, 109
  equipment, 105
  materials, 107
  mechanisms, 101

473

beam focusing, 104
beam formation, 102
cathode, 103
electron origin, 105
grid control, 103
Columbium and its alloys, diffusion weld, 424
electron beam, 56
fusion spot, 241
vacuum braze, 318
Computers, 93
Controls, plasma arc, 152
Copper and its alloys, diffusion weld, 427
electron beam, 55
plasma arc, 159

Deflection, electron beam, 27
Deformation welding, 382
Densification, diffusion weld, 450
Design, electron beam, 12
Diffusion welding, 370–470
activation, agents, 399
concepts, 392
advantages, 415–419
applications, 458–470
cladding, 447–450
densification, 450
honeycomb core, 446
blanket heating, 405
definition, 380
deformation welding, 382
elastic stress, 383
energy barrier, 383
surface condition, 384
surface contamination, 382
gas pressure, 403
grain growth, 390
hot pack, 408
induction bonder, 401
intermediate materials, 391
joint design, 451–455
aerospace structures, 455–458
limitations, 415–419
materials, 419
aluminum and its alloys, 420
beryllium and its alloys, 419
ceramics, 442
chromium and its alloys, 425
columbium and its alloys, 424
copper and its alloys, 427

dispersion hardened alloys, 428
dissimilar metals, 442
molybdenum and its alloys, 431
steels, 433
superalloys, 427
tantalum and its alloys, 434
titanium and its alloys, 436
tungsten and its alloys, 437
zirconium and its alloys, 441
metallurgical factors, 398
metal surfaces, 387
plastic flow, 389
pressure, 334
resistance bonder, 402
roll bonding, 406
surface preparation, 393
temperature, 395
theory, 371–380
time, 396
tooling, 413
vacuum bonding, 410
Dispersion hardened alloys, diffusion weld, 428
vacuum braze, 331
Dissimilar metal and/or ceramic combinations, vacuum braze, 341–346
Dissimilar metals, diffusion weld, 442
electron beam, 72
fusion spot, 244

Economics, blanket braze, 178
electron beam, 45
repair welding, 46
Efficiency, laser, 204
Electric blanket brazing, 168–188
applications, 186
economics, 178
equipment, 173–177
joint design, 182–186
materials, 180
principles, 169
theory, 169
tooling, 173
Electrodes, fusion spot, 230
Electron, optics, electron beam, 7
origin, cold cathode, 105
scatter, nonvacuum electron beam, 123
theory, electron beam, 6
Electron beam welding, 1–141
accelerating voltage, 16
advantages, 41

Index

applications, 83–94
  aircraft, 88
  brazing, 94
  computers, 93
  electronics, 89
  gears and wheels, 84
  nuclear, 83
    ion propulsion, 84
    pressurized water reactor, 83
  pressure vessels, 89
  repair welding, 90
  saw-blades, 93
  underseas, 84
cold cathode, 100–110
economics, arc welds versus electron beam, 48
repair welding, 46
equipment, 25
  chambers, 28
  hand-held welder, 30
  high voltage, 25
    column, 25
    deflection, 27
    focal length, 27
  low voltage, 31
    control devices, 32
filler metal additions, 33
heat-affected zone, 22
high voltage, 11
joint design, 76
  filler metal addition, 80
  gap, 80
  joint efficiency, 78
  lap joint, 79
  longitudinal stress, 79
  mechanical properties, electronics, 82
    refractory metals, 82
    steel and superalloys, 81
  stress, 77
  T-joint, 78
limitations, 44
  X-rays, 45
low voltage, 11
materials, 48
  aluminum and its alloys, 48
  beryllium and its alloys, 53
  ceramics, 53, 70
  columbium and its alloys, 56
  copper and its alloys, 55
  dispersion hardened alloys, 60
  dissimilar metals, 72
  magnesium and its alloys, 56
  molybdenum and its alloys, 56
  space-gun welded, 76
  steels, 60
  superalloys, 57
  tantalum and its alloys, 64
  titanium and its alloys, 64
  tungsten and its alloys, 68
  zirconium and its alloys, 69
nonvacuum, 121–141
partial vacuum, 111–120
penetration, 10, 19
plasma, 100–110
principles, 3, 8
  accelerating voltage, 9
  kinetic energy, 8
  penetration, 10
safety, 34
  X-rays, 34
theory, 3, 6
  anode, 7
  cathode, 7
  electron, 6
  electron optics, 7
  gun, 5
    annular, 6
    beam current, 16
    cathode, 12
    columnar, 6
    design, 12
    fixed, 18
    movable, 18
    self-accelerated, 6
    space-charge-limited, 15
    work accelerated, 6
  Pierce, 5
  Steigerwald, 4
  Stohr, 4
  thermal energy, 6
  Wyman, 5
thermal gradients, 22
tooling, 35
  honeycomb, 36
  microwelding, 38
  pressure vessels, 37
  titanium tracks, 37
  vacuum tubes, 36
vacuum environment, 18
weld distortion, 21

476 Index

Electronics, electron beam, 89
Equipment, blanket braze, 173
  cold cathode, 105
  diffusion weld, 401
  electron beam, 25
  exothermic bonding, 268
  exothermic brazing, 267
  fusion spot, 237
  laser, 206
  nonvacuum, 131
  partial vacuum, 113
  plasma arc, 151, 154
  radiant heat, 254
  vacuum braze, 300
Exothermic joining, 259–277
  applications, 275–277
  bonding, adhesive, 265
    exo-flux, 264
    exo-flux equipment, 268
  brazing, equipment, 267
    ignition temperature, 262
    mechanisms, 259
    reaction duration, 263
  materials, 272
    aluminum, 274
    magnesium, 274
    refractory metals, 272
    steels, 273
    superalloys, 273

Filler metal, additions, electron beam, 33, 80
  characteristics, vacuum braze, 293
Fixed gun, electron beam, 18
Flow, vacuum braze, 286
Focal length, electron beam, 27
Focusing, plasma arc, 148
Fusion spot welding, gas metal and tungsten arc, 223–249
  applications, 246–249
  arc length, 232
  equipment, 237
  gas shielding, 233
  joint design, 245
  materials, 240
    aluminum and its alloys, 240
    columbium and its alloys, 241
    dissimilar metals, 244
    magnesium and its alloys, 241
    steel, 243

    titanium and its alloys, 243
  mechanisms, 227
    AC current, 227
    arc initiation, 228
    arc stabilization, 228
    DC current, 227
    electrodes, 230
    heat transfer, 229
    shape, 231
  nozzle design, 234
  principles, 224
  theory, 224

Gap, electron beam, 80
Gas pressure, diffusion weld, 403
Gas shielding, fusion spot, 233
Gases, plasma arc, 150
Gas metal spot welding, 223–249
Grain growth, diffusion weld, 390
Graphite, vacuum braze, 340
Grid control, cold cathode, 103
Gun, electron beam, 5
  annular, 6
  columnar, 6
  self-accelerated, 6
  work accelerated, 6

Hand-held welder, electron beam, 30
Heat-affected zone, electron beam, 22
Heat transfer, fusion spot, 229
High voltage, electron, 11, 25
Honeycomb diffusion weld, 446
  electron beam, 36
Hot pack, diffusion weld, 408

Ignition temperature, exothermic, 262
Induction bonder, diffusion weld, 401
Intermediate materials, diffusion weld, 391

Joint design, blanket braze, 182
  diffusion weld, 451
  electron beam, 76
  fusion spot, 245
  laser, 210
  plasma arc, 160
  vacuum braze, 299, 346
Joint efficiency, electron beam, 78
Joint properties, vacuum braze, 351

Keyhole, plasma arc, 147

Index

Kinetic energy, electron beam, 8

Lap joint, electron beam, 79
Laser welding, 189–221
  applications, 217–221
  equipment, 206
  joint design, 210
    ribbon-to-ribbon, 213
    wire-to-ribbon, 213
    wire-to-wire, 210
  materials, 210–217
  mechanisms, 196–206
    efficiency, 204
    materials, 200
    pump energy, 203
    weld effect, 204
  principles, 194
  theory, 191
  tooling, 207
Limitations, diffusion weld, 415
  electron beam, 44
  X-rays, 45
Longitudinal stress, electron beam, 79
Low voltage, electron beam, 11, 31
  control devices, 32

Magnesium and its alloys, electron beam, 56
  exothermic, 274
  fusion spot, 241
  vacuum braze, 322
Materials, blanket braze, 180
  cold cathode, 107
  diffusion weld, 419
  electron beam, 48
  exothermic, 272
  fusion spot, 240
  laser, 210
  nonvacuum electron beam, 133
  partial vacuum electron beam, 117
  plasma arc, 159
  radiant heat, 256
  vacuum braze, 315
Mechanical properties, electron beam, 81–82
  electronics, 82
  refractory metals, 82
  steel and superalloys, 81
  nonvacuum electron beam, 133
  partial vacuum electron beam, 117
Mechanisms, cold cathode, 101
  exothermic, 259
  fusion spot, 227
  laser, 196
  nonvacuum electron beam, 122
  partial vacuum electron beam, 112
  plasma arc, 147
Metallurgical factors, diffusion weld, 398
Metal surfaces, diffusion weld, 387
Microwelding, electron beam, 38
Molybdenum and its alloys, diffusion weld, 431
  electron beam, 56
  vacuum braze, 322
Movable gun, electron beam, 18

Needle arc, plasma arc, 154
Nickel and its alloys, plasma arc, 159
Nontransferred arc, plasma arc, 146
Nonvacuum electron beam, 121–141
  applications, 137
  equipment, 131
  materials, 133
  mechanical properties, 133
  mechanisms, 122
    atmosphere control, 126
    beam transfer, 125
    electron scatter, 123
    weld, penetration, 128
    speed, 129
Nozzle design, fusion spot, 234
Nuclear, electron beam, 83

Partial vacuum electron beam, 111–120
  applications, 118
  equipment, 113
  materials, 117
  mechanical properties, 117
  mechanisms, 112
  pressure, 113
  tooling, 113
Penetration, electron beam, 10, 19
Pierce, electron beam, 5
Plasma arc welding, 143–167
  applications, 163–167
  arc column, 145
    nontransferred, 146
    transferred, 146
  arc modes, 144
  controls, 152
  equipment, 151, 154

Index

needle arc, 154
joint design, 160
materials, 159
  aluminum, 159
  copper, 159
  nickel and its alloys, 159
  steels, 160
  titanium and its alloys, 160
mechanisms, 147–150
  arc shaping, 150
  focusing, 148
  gases, 150
  keyhole, 147
power supply, 153
torches, 151
Plasma electron beam, 100–110
Plastic flow, diffusion weld, 389
Power supply, plasma arc, 153
Pressure, diffusion weld, 334
  partial vacuum electron beam, 113
Pressure vessels, electron beam, 37, 89
Principles, blanket braze, 169
  electron beam, 3, 8
  fusion spot, 224
  laser, 194
Pump energy, laser, 203

Quartz lamp brazing, 250–258

Radiant heat brazing, 250–258
  equipment, 254
  materials, 256
Refractory metals, exothermic, 272
Repair welding, electron beam, 46, 90
Resistance bonder, diffusion weld, 402
Roll bonding, diffusion weld, 406

Safety, electron beam, 34
  X-rays, 34
Shape, fusion spot, 231
Space-charge-limited, 15
Space-gun welded materials, electron beam, 76
Steels, diffusion weld, 433
  electron beam, 60
    low alloy ultra high strength, 62
      carburizing and nitriding grade, 62
      4340, 62
      H-11, 64
      maraging, 64

precipitation hardened, 61
stainless, 60
exothermic, 273
fusion spot, 243
plasma arc, 160
vacuum braze, 324
Steigerwald, electron beam, 4
Stohr, electron beam, 4
Stress, electron beam, 117
Superalloys, diffusion weld, 427
  electron beam, 57
  exothermic, 273
  vacuum braze, 327
Surface preparation, diffusion weld, 393
  vacuum braze, 296

Tantalum and its alloys, diffusion weld, 434
  electron beam, 64
  vacuum braze, 333
TD-nickel, electron beam, 60
Telefocus, electron beam, 13
Temperature, diffusion weld, 395
  vacuum braze, 300
Theory, blanket braze, 169
  diffusion weld, 371
  electron beam, 1, 6
  fusion spot, 224
  laser, 191
  vacuum braze, 281
Thermal energy, electron beam, 6
Thermal gradients, electron beam, 22
Time, diffusion weld, 396
  vacuum braze, 300
Titanium and its alloys, diffusion weld, 436
  electron beam, 64
  fusion spot, 243
  plasma arc, 160
  vacuum braze, 335
T-joint, electron beam, 78
Tooling, blanket braze, 173
  diffusion weld, 413
  electron beam, 35
  laser, 207
  partial vacuum electron beam, 113
  vacuum braze, 308
Torches, plasma arc, 150
Transferred arc, plasma arc, 146
Tungsten and its alloys, diffusion weld, 437
  electron beam, 68
  vacuum braze, 337

Tungsten-arc spot welding, 223–249

Underseas, electron beam, 84

Vacuum bonding, diffusion weld, 410
Vacuum brazing, 278–367
 applications, 360–367
 base metal characteristics, 291
 equipment, 300
  elements, 305
  radiant-heat shields, 307
 filler metal characteristics, 293
 joint design, 299, 346
 joint properties, 351
  aluminum, 352
  beryllium, 351
  refractory metals and alloys, 356
  steels and superalloys, 354
  titanium, 353
 materials, 315
  aluminum and its alloys, 317
  beryllium and its alloys, 316
  ceramics, 338
  columbium and its alloys, 318
  dispersion hardened alloys, 331
  dissimilar metal and/or ceramic combinations, 341–346
  graphite, 340
  magnesium and its alloys, 322
  molybdenum and its alloys, 322
  steels, 324
  superalloys, 327
  tantalum and its alloys, 333
  titanium and its alloys, 335
 tungsten and its alloys, 337
 surface preparation, 296
 temperature, 300
 theory, brazing, 281
  flow, 286
  vacuum, 279
 time, 300
 tooling, 308
 variables, 291
Vacuum environment, electron beam, 18
Vacuum theory, vacuum braze, 279
Vacuum tubes, electron beam, 36
Variables, vacuum braze, 291
Von Pirani, electron beam, 3

Weld distortion, electron beam, 21
Weld effect, laser, 204
Welding, cold cathode, 100
 diffusion, 370
 electron beam, 1
 fusion spot, 223
 laser, 189
 nonvacuum electron beam, 121
 partial vacuum electron beam, 111
 plasma arc, 143
Weld penetration, nonvacuum electron beam, 128
Weld speed, nonvacuum electron beam, 129
Wyman, electron beam, 5

X-rays, electron beam, 34, 45

Zirconium and its alloys, diffusion weld, 441
 electron beam, 69